Franz Zotter · Matthias Frank

Ambisonics

A Practical 3D Audio Theory for Recording, Studio Production, Sound Reinforcement, and Virtual Reality

Franz Zotter
Institute of Electronic Music and Acoustics
University of Music and Performing Arts
Graz, Austria

Matthias Frank
Institute of Electronic Music and Acoustics
University of Music and Performing Arts
Graz, Austria

https://doi.org/10.1007/978-3-030-17207-7

Preface

The intention of this textbook is to provide a concise explanation of fundamentals and background of the surround sound recording and playback technology Ambisonics.

Despite the Ambisonic technology has been practiced in the academic world for quite some time, it is happening now that the recent ITU,[1] MPEG-H,[2] and ETSI[3] standards firmly fix it into the production and media broadcasting world.

What is more, Internet giants Google/YouTube recently recommended to use tools that have been well adopted from what the academic world is currently using.[4,5]

Last but most importantly, the boost given to the Ambisonic technology by recent advancements has been in usability: Ways to obtain safe Ambisonic decoders,[6,7] the availability of higher-order Ambisonic main microphone arrays (Eigenmike,[8] Zylia[9]) and their filter-design theory, and above all: the usability increased by plugins integrating higher-order Ambisonic production in digital audio workstations or mixers.[7,10,11,12,13,14,15] And this progress was a great motivation to write a book about the basics.

---

[1] https://www.itu.int/rec/R-REC-BS.2076/en.

[2] https://www.iso.org/standard/69561.html.

[3] https://www.techstreet.com/standards/etsi-ts-103-491?product_id=1987449.

[4] https://support.google.com/jump/answer/6399746?hl=en.

[5] https://developers.google.com/vr/concepts/spatial-audio.

[6] https://bitbucket.org/ambidecodertoolbox/adt.git.

[7] https://plugins.iem.at/.

[8] https://mhacoustics.com/products.

[9] https://www.zylia.co.

[10] http://www.matthiaskronlachner.com/?p=2015.

[11] http://www.blueripplesound.com/product-listings/pro-audio.

[12] https://b-com.com/en/bcom-spatial-audio-toolbox-render-plugins.

[13] https://harpex.net/.

[14] http://forumnet.ircam.fr/product/panoramix-en/.

[15] http://research.spa.aalto.fi/projects/sparta_vsts/.

The book is dedicated to provide a deeper understanding of Ambisonic technologies, especially for but not limited to readers who are scientists, audio-system engineers, and audio recording engineers. As, from time to time, the underlying maths would get too long for practical readability, the book comes with a comprehensive appendix with the beautiful mathematical details.

For a common understanding, the introductory section spans a perspective on Ambisonics from its origins in coincident recordings from the 1930s, to the Ambisonic concepts from the 1970s, and to classical ways of applying Ambisonics in first-order coincident sound scene recording and reproduction that have been practiced from the 1980s on.

In its main contents, this book intends to provide all psychoacoustical, signal processing, acoustical, and mathematical knowledge needed to understand the inner workings of modern processing utilities, special equipment for recording, manipulation, and reproduction in the higher-order Ambisonic format. As advanced outcomes, the aim of the book is to explain higher-order Ambisonic decoding, 3D audio effects, and higher-order Ambisonic recording with microphones or main microphone arrays. Those techniques are shown to be suitable to supply audience areas ranging from studio-sized to hundreds of listeners, or headphone-based playback, regardless whether it is live, interactive, or studio-produced 3D audio material.

The book comes with various practical examples based on free software tools and open scientific data for reproducible research.

Our Ambisonic events experience: In the past years, we have contributed to organizing Symposia on Ambisonics (Ambisonics Symposium 2009 in Graz, 2010 in Paris, 2011 in Lexington, 2012 in York, 2014 in Berlin), demonstrated and brought the technology to various winter/summer schools and conferences (EAA Winter School Merano 2013, EAA Symposium Berlin 2014, workshops and Ambisonic music repertory demonstration at Darmstädter Ferienkurse für Neue Musik in 2014, ICAD workshop in Graz 2015, ICSA workshop 2015 in Graz with PURE Ambisonics night, summer school at ICSA 2017 in Graz, a course at Kraków film music festival 2015, mAmbA demo facility DAGA in Aachen 2016, Al Di Meola's live 3D audio concert hosted in Graz in June 2016, and AES Convention Milano 2018.

In 2017 (ICSA Graz) and 2018 (TMT Cologne), we initiated and organized *Europe's First* and *Second Student 3D Audio Production Competition* together with Markus Zaunschirm and Daniel Rudrich.

Graz, Austria
February 2019

Franz Zotter
Matthias Frank

Acknowledgements

To our lab and colleagues: Traditionally at the Institute of Electronic Music and Acoustics (IEM), there had been a lot of activity in developing and applying Ambisonics by Robert Höldrich, Alois Sontacchi, Markus Noisternig, Thomas Musil, Johannes Zmölnig, Winfried Ritsch, even before our active time of research. Most of the developments were done in pure-data, e.g., with `[iem_ambi]`, `[iem_bin_ambi]`, `[iem_matrix]`, `CUBEmixer`. Dear colleagues deserve to be mentioned who contributed a lot of skill to improve the usability of Ambisonics: Hannes Pomberger and his mathematical talent, Matthias Kronlachner who developed the `ambix` and `mcfx` VST plugin suites in 2014, and Daniel Rudrich, who developed the `IEM Plugin Suite`, that also involves technology that was elaborated together with our colleagues Markus Zaunschirm, Christian Schörkhuber, Sebastian Grill.

We thank you all for your support; it's the best environment to work in!

First readers: We thank Archontis Politis (Aalto Univ., Espoo and Tampere Univ., Finland), Nicolas Epain (b<>com, France), and Matthias Kronlachner (Harman, Germany/US), to be our first critical readers, supplying us with valuable comments.

Open Access Funding: We are grateful about funding from our local government of Styria (Steiermark) Section 8, Office for Science and Research (Wissenschaft und Forschung) that covers roughly half the open access publishing costs. We gratefully thank our University (University of Music and Performing Arts, "Kunstuni", Graz) for the other half, *transition to open access* library, and vice rectorate for research.

kunst uni graz

Das Land Steiermark
Wissenschaft und Forschung

Outline

First-order Ambisonics is nowadays strongly revived by internet technology supported by Google/YouTube, Facebook 360°, 360° audio and video recording and rendering, as well as VR in games. This renaissance lies in its benefits of (i) its compact main microphone arrays capturing the entire surrounding sound scene in only four audio channels (e.g., Zoom H3-VR, Oktava A-Format Microphone, Røde NT-SF1, Sennheiser AMBEO VR Mic.), and (ii) it easily permits rotation of the sound scene, allowing to render surround audio scenes, e.g., on head-tracked headphones, head-mounted AR/VR sets, or mobile devices, as described in Chap. 1.

Auditory events and vector-base panning: Chapter 2 of this book is dedicated to conveying a comprehensive understanding of the localization impressions in multi-loudspeaker playback and its models, followed by Chap. 3 that outlines the essentials of practical vector panning models and their extensions by downmix from imaginary loudspeakers, which are both fundamental to contemporary Ambisonics.

Harmonic functions, Ambisonic encoding and decoding: Based on the ideals of accurate localization with panning-invariant loudness and perceived width, Chap. 4 provides a profound mathematical derivation of higher-order Ambisonic panning functions in 2D and 3D in terms of angular harmonics. These idealized functions can be maximized in their directional focus (max-$\boldsymbol{r}_\mathrm{E}$) and they are strictly limited in their directional resolution. This resolution limit entails perfectly well-defined constraints on loudspeaker layouts that make us reach ideal measures for accurate localization as well as panning-invariant loudness and width. And what is highly relevant for practical decoding: All-Round Ambisonic decoding to loudspeakers and TAC/MagLS decoders for headphones are explained in Chap. 4.

The Ambisonic signal processing chain and effects are described in Chap. 5. It illustrates the signal flow from source encoding through Ambisonic bus to decoding and where input-specific or general insert and auxiliary Ambisonic effects are located. In particular, the chapter describes the working principles behind frequency-independent manipulation effects that are either mirroring/rotating/re-mapping, warping, or directionally weighting, or such effects that are frequency-dependent. Frequency-dependent effects can introduce widening, depth or diffuseness, convolution reverb, or feedback-delay-network (FDN)-based diffuse reverberation. Directional resolution enhancements are outlined in terms of SDM/SIRR pre-processing of recorded reverberation and in terms of available tools such as HARPEX, DirAC, and COMPASS for recorded signals.

Compact higher-order Ambisonic microphones rely on the solutions of the Helmholtz equation, and their processing uses a frequency-independent decomposition of the spherical array signals into spherical harmonics and the frequency-dependent radial-focusing filtering associated with each spherical harmonic order, which yield the Ambisonic signals. The critical part is to handle the properties of radial-focusing filters in the processing of higher-order Ambisonic microphone arrays (e.g., the Eigenmike). To keep the noise level and the sidelobes in the recordings low and a balanced frequency response, a careful way for radial filter design is outlined in Chap. 6.

Compact higher-order loudspeaker arrays oppose the otherwise inwards-oriented Ambisonic surround playback, as described in Chap. 7. This outlooking last chapter discusses IKO and loudspeaker cubes as compact spherical loudspeaker arrays with Ambisonically controlled radiation patterns. In natural environments with acoustic reflections, such directivity-controlled arrays have their own sound-projecting and distance-changing effects, and they can be used to simulate sources of specific directivity patterns.

Contents

Chapter 1
XY, MS, and First-Order Ambisonics

Directionally sensitive microphones may be of the light moving strip type. [...] the strips may face directions at 45° *on each side of the centre line to the sound source.*

Alan Dower Blumlein [1], Patent, 1931

Abstract This chapter describes first-order Ambisonic technologies starting from classical coincident audio recording and playback principles from the 1930s until the invention of first-order Ambisonics in the 1970s. Coincident recording is based on arrangements of directional microphones at the smallest-possible spacings in between. Hereby incident sound approximately arrives with equal delay at all microphones. Intensity-based coincident stereophonic recording such as XY and MS typically yields stable directional playback on a stereophonic loudspeaker pair. While the stereo width is adjustable by MS processing, the directional mapping of first-order Ambisonics is a bit more rigid: the omnidirectional and figure-of-eight recording pickup patterns are reproduced unaltered by equivalent patterns in playback. In perfect appreciation of the benefits of coincident first-order Ambisonic recording technologies in VR and field recording, the chapter gives practical examples for encoding, headphone- and loudspeaker-based decoding. It concludes with a desire for a higher-order Ambisonics format to get a larger sweet area and accommodate first-order resolution-enhancement algorithms, the embedding of alternative, channel-based recordings, etc.

Intensity-based coincident stereophonic recording such as XY uses two figure-of-eight microphones, after Blumlein's original work [1] from the 1930s, with an angular spacing of 90°, see [2–4]). Another representative, MS, uses an omnidirectional and a lateral figure-of-eight microphone [2]. Both typically yield a stable directional playback in stereo, but signals often get too correlated, yielding a lack in depth and diffuseness of the recording space when played back [5, 6] and compared to delay-based AB stereophony or equivalence-based alternatives.

Gerzon's work in the 1970s [7] gave us what we call first-order Ambisonic recording and playback technology today. Ambisonics preserves the directional mapping

F. Zotter and M. Frank, *Ambisonics*, Springer Topics in Signal Processing 19,
https://doi.org/10.1007/978-3-030-17207-7_1

by recording and reproducing with spatially undistorted omnidirectional and figure-of-eight patterns on circularly (2D) or spherically (3D) surrounding loudspeaker layouts.

1.1 Blumlein Pair: XY Recording and Playback

The XY technique dates back to Blumlein's patent from the 1930s [1] and his patents thereafter [4]. Nowadays outdated, manufacturers started producing ribbon microphones that offered means to record with figure-of-eight pickup patterns.

Blumlein Pair using 90°-angled figure-of-eight microphones (XY). Blumlein's classic coincident microphone pair [3, Fig. 3] uses two figure-of-eight microphones pointing to ±45°, see Fig. 1.1. Its directional pickup pattern is described by $\cos\phi$ when ϕ is the angle enclosed by microphone aiming and sound source. Using a mathematically positive coordinate definition for X (front-right) and Y (front-left), the polar angle $\varphi = 0$ aiming at the front, the figure-of-eight X uses the angle $\phi = \varphi + 45°$ and Y the angle $\phi = \varphi - 45°$, so that the pickup pattern of the microphone pair is:

$$g_{\mathrm{XY}}(\varphi) = \begin{bmatrix} \cos(\varphi + 45°) \\ \cos(\varphi - 45°) \end{bmatrix}. \tag{1.1}$$

Assuming a signal s coming from the angle φ, the signals recorded are $[X,\ Y]^{\mathrm{T}} \boldsymbol{g}(\varphi)\, s$. Sound sources from the left 45°, the front 0° and the right −45° will be received by the pair of gains:

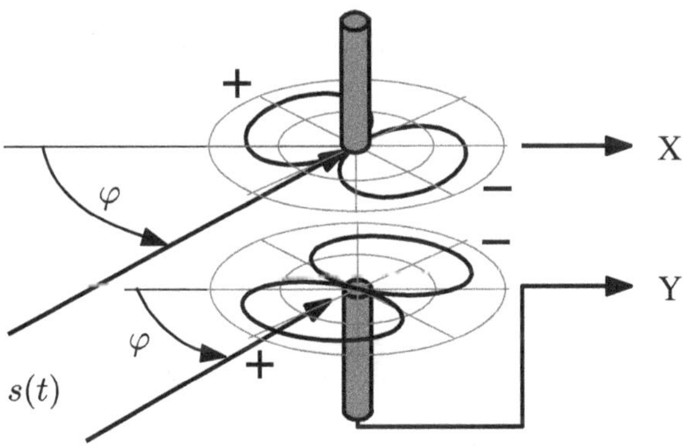

(a) Blumlein XY pair

(b) Picture of the recording setup

Fig. 1.1 Blumlein pair consisting of 90°-angled figure-of-eight microphones

$$\text{right}: \ \boldsymbol{g}_{\mathrm{XY}}(-45^\circ) = \begin{bmatrix}1\\0\end{bmatrix}, \quad \text{center}: \ \boldsymbol{g}_{\mathrm{XY}}(0^\circ) = \begin{bmatrix}\frac{1}{\sqrt{2}}\\\frac{1}{\sqrt{2}}\end{bmatrix}, \quad \text{left}: \ \boldsymbol{g}_{\mathrm{XY}}(45^\circ) = \begin{bmatrix}0\\1\end{bmatrix}.$$

Obviously, a source moving from the right -45° to the left 45° pans the signal from the channel X to the channel Y. This property provides a strongly perceivable lateralization of lateral sources when feeding the left and right channel of a stereophonic loudspeaker pair by Y and X, respectively.

However, ideally there should not be any dominant sounds arriving from the sides, as for the source angles between $-135^\circ \leq \varphi \leq -45^\circ$ and $45^\circ \leq \varphi \leq 135^\circ$ the Blumlein pair produces out-of-phase signals between X and Y. The back directions are mapped with consistent sign again, however, left-right reversed. It is only possible to avoid this by decreasing the angle between the microphone pair, which, however, would make the stereo image narrower.

Therefore, coincident XY recording pairs nowadays most often use cardioid directivities $\frac{1}{2} + \frac{1}{2}\cos\varphi$, instead. They receive all directions without sign change and easily permit stereo width adjustments by varying the angle between the microphones.

1.2 MS Recording and Playback

Blumlein's patent [1] considers sum and difference signals between a pair of channels/microphones, yielding M-S stereophony. In M-S [8], the sum signal represents the mid (omnidirectional, sometimes cardioid-directional to front) and the difference the side signal (figure-of-eight). MS recordings can also be taken with cardioid microphones and permit manipulation of the stereo width of the recording.

MS recording by omnidirectional and figure-of-eight microphone (native MS). Mid-side recording can be done by using a pair of coincident microphones with an omnidirectional (mid, W) and a side-ways oriented figure-of-eight (side, Y) directivity, Fig. 1.2. The pair of pickup patterns is described by the vector:

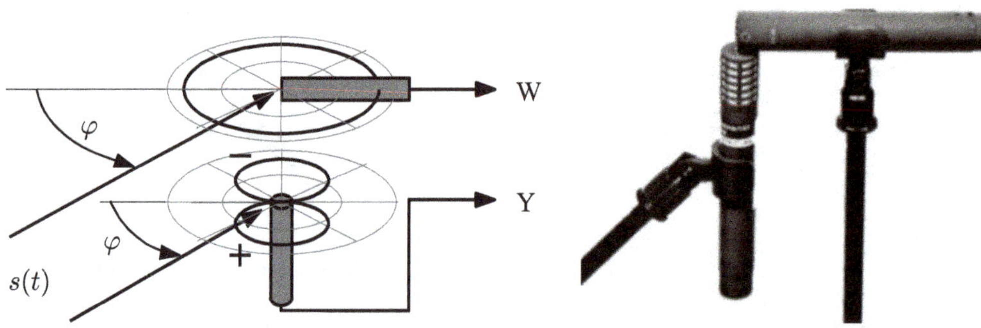

(a) Native MS recording (b) Picture of the recording setup

Fig. 1.2 Native mid-side recording with the coincident arrangement of an omnidirectional microphone heading front and a figure-of-eight microphone heading left

$$g_{WY}(\varphi) = \begin{bmatrix} 1 \\ \sin(\varphi) \end{bmatrix} \tag{1.2}$$

that depends on the angle φ of the sound source. Equation (1.2) maps a single sound s from φ to the mid W and side Y signals by the gains $[W,\ Y]^T = \boldsymbol{g}(\varphi)\, s$

$$\text{left: } g_{WY}(90°) = \begin{bmatrix} 1 \\ 1 \end{bmatrix}, \quad \text{right: } g_{WY}(-90°) = \begin{bmatrix} 1 \\ -1 \end{bmatrix} \quad \text{center: } g_{WY}(0°) = \begin{bmatrix} 1 \\ 0 \end{bmatrix}.$$

MS recording with a pair of 180°-angled cardioids. Two coincident cardioid microphones (cardioid directivity $\frac{1}{2} + \frac{1}{2}\cos\varphi$) pointing to the polar angles 90° (left) and −90° (right) are also applicable to mid-side recording, Fig. 1.3. Their pickup patterns

$$g_{C\pm 90°}(\varphi) = \frac{1}{2}\begin{bmatrix} 1+\cos(\varphi-90°) \\ 1+\cos(\varphi+90°) \end{bmatrix} = \frac{1}{2}\begin{bmatrix} 1+\sin(\varphi) \\ 1-\sin(\varphi) \end{bmatrix} \tag{1.3}$$

are encoded into the MS pickup patterns (W,Y) by a matrix

$$g_{WY}(\varphi) = \begin{bmatrix} 1 & 1 \\ 1 & -1 \end{bmatrix} g_{C\pm 90°}(\varphi). \tag{1.4}$$

The matrix eliminates the cardioids' figure-of-eight characteristics by their sum signal, and their omnidirectional characteristics by the difference. We obtain the MS signal pair (W,Y) from the cardioid microphone signals as

$$\begin{bmatrix} W \\ Y \end{bmatrix} = \begin{bmatrix} 1 & 1 \\ 1 & -1 \end{bmatrix} \begin{bmatrix} C_{90°} \\ C_{-90°} \end{bmatrix}. \tag{1.5}$$

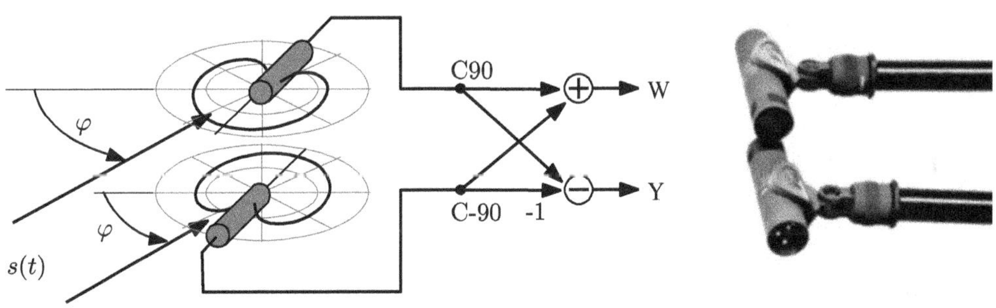

(a) 180°angled cardioid microphones (b) Picture of the recording setup

Fig. 1.3 Mid-side recording by 180°-angled cardioids

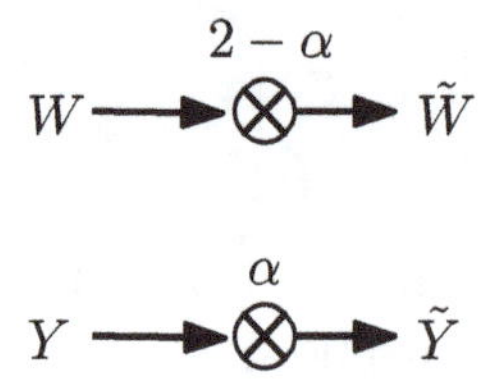

(a) Changing the MS stereo width

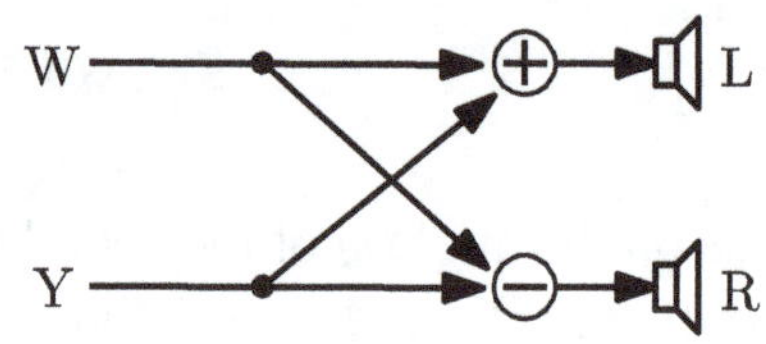

(b) MS decoding to loudspeakers

Fig. 1.4 Change of the stereo width by modifying the balance between W and Y signals of MS (left). Decoding of the M/S signal pair (W, Y) to a stereo loudspeaker pair (right)

Decoding of MS signals to a stereo loudspeaker pair. Decoding of the mid-side signal pair to left and right loudspeaker is done by feeding both signals to both loudspeakers, however out-of-phase for the side signal, Fig. 1.4b:

$$\begin{bmatrix} L \\ R \end{bmatrix} = \frac{1}{2} \begin{bmatrix} 1 & 1 \\ 1 & -1 \end{bmatrix} \begin{bmatrix} W \\ Y \end{bmatrix}. \tag{1.6}$$

An interesting aspect about the 180°-angled cardioid microphone MS is that after inserting the XY-to-MS encoder Eq. (1.5) into the decoder Eq. (1.6), a brief calculation shows that matrices invert each other. In this case, the cardioid signals are directly fed to the loudspeakers $[L,\ R] = [C_{90^\circ},\ C_{-90^\circ}]$.

Stereo width. Modifying the mid versus side signal balance before stereo playback, using a blending parameter α, allows to change the width of the stereo image from $\alpha = 0$ (narrow) to $\alpha = 1$ (full), Fig. 1.4a, see also [9]:

$$\begin{bmatrix} L \\ R \end{bmatrix} = \frac{1}{2} \begin{bmatrix} 1 & 1 \\ 1 & -1 \end{bmatrix} \begin{bmatrix} 2-\alpha & 0 \\ 0 & \alpha \end{bmatrix} \begin{bmatrix} W \\ Y \end{bmatrix}. \tag{1.7}$$

In stereophonic MS playback, the playback loudspeaker directions at ±30° are not identical to the peaks of the recording pickup pattern of the side channel (Y) at ±90°. Ambisonics assumes a more strict correspondence between directional patterns of recording and patterns mapped on the playback system.

1.3 First-Order Ambisonics (FOA)

After Cooper and Shiga [10] worked on expressing panning strategies for arbitrary surround loudspeaker setups in terms of a directional Fourier series, the notion and technology of Ambisonics was developed by Felgett [11], Gerzon [7], and Craven [12]. In particular, they were also considering a suitable recording technology.

Essentially based on similar considerations as MS, one can define first-order Ambisonic recording. For 2D recordings, a Double-MS microphone arrangement is suitable and only requires one more microphone than MS recording: a front-back oriented figure-of-eight microphone. The scheme is extended to 3D first-order Ambisonics by a third figure-of-eight microphone of up-down aiming. Oftentimes, first-order Ambisonics still is the basis of nowadays' virtual reality applications and 360° audio streams on the internet. In addition to potential loudspeaker playback, it permits interactive playback on head-tracked headphones to render the acoustic sound scene static to the listener.

First-order Ambisonic recording has the advantage that it can be done with only a few high-quality microphones. However, the sole distribution of first-order Ambisonic recordings to playback loudspeakers is typically not convincing without going to higher orders and directional enhancements (Sect. 5.8).

1.3.1 2D First-Order Ambisonic Recording and Playback

The first-order Ambisonic format in 2D consists of one signal corresponding to an omnidirectional pickup pattern (called W), and two signals corresponding to the figure-of-eight pickup patterns aligned with the Cartesian axes (X and Y).

Native 2D Ambisonic recording (Double-MS). To record the Ambisonic channels W, X, Y, one can use a Double-MS arrangement as shown in Fig. 1.5.

2D Ambisonic recording with four 90°-angled cardioids**.** Extending the MS scheme for recording with cardioid microphones, Fig. 1.3, cardioid microphones could be used to obtain the front-back and left-right figure-of-eight pickup patterns by corresponding pair-wise differences, and one omnidirectional pattern as their sum, Fig. 1.6. However, the use of 4 microphones for only 3 output signals is inefficient.

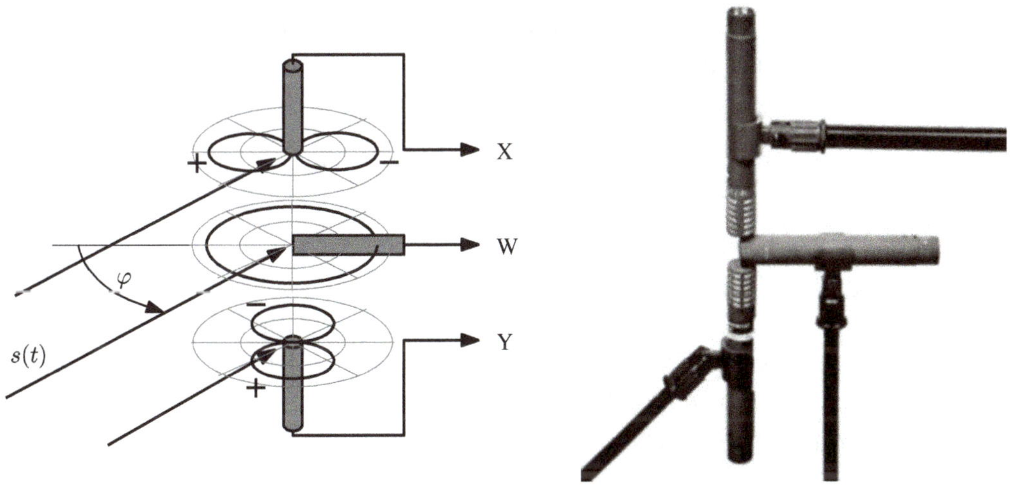

(a) Native 2D FOA recording (b) Picture of recording setup

Fig. 1.5 Native 2D first-order Ambisonic recording with an omnidirectional and a figure-of-eight microphone heading front, and a figure-of-eight microphone heading left; photo shown on the right

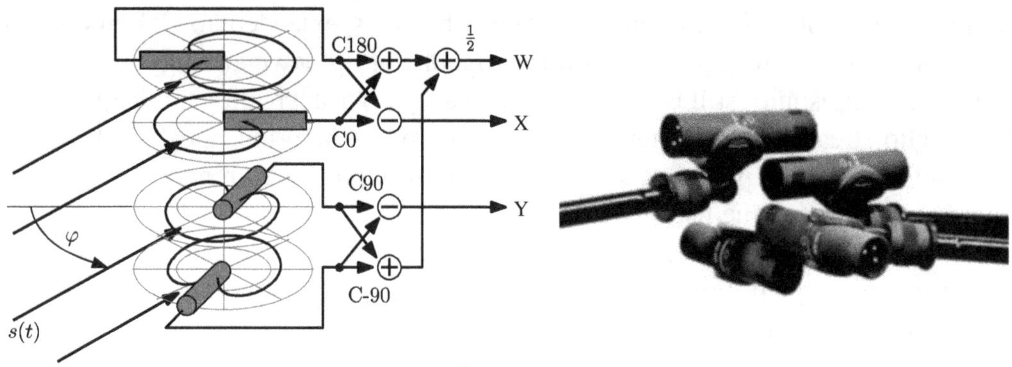

(a) 2D FOA with 4 cardioid microphones (b) Picture of recording setup

Fig. 1.6 2D first-order Ambisonic recording with four 90°-angled cardioid microphones, sums and differences between them (front±back, left±right)

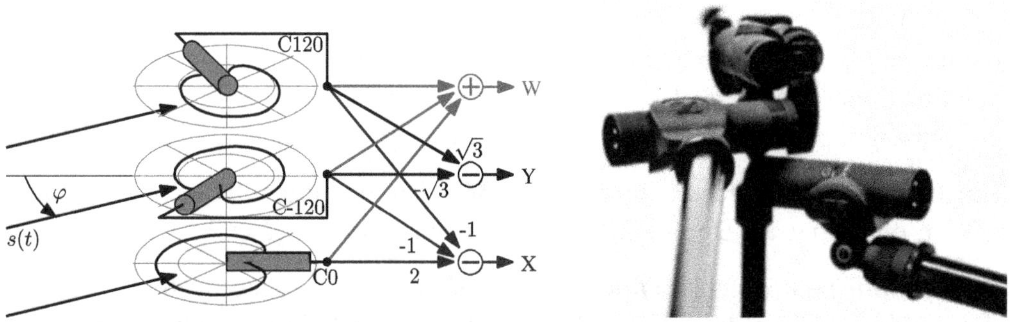

(a) 2D FOA with 3 cardioid microphones (b) Picture of recording setup

Fig. 1.7 2D first-order Ambisonics with three 120°-angled cardioid microphones

2D Ambisonic recording with three 120°-angled cardioids. Assuming 3 coincident cardioid microphones aiming at the angles 0°, ±120° in the horizontal plane, cf. Fig. 1.7, we obtain as the pickup pattern for the incoming sound

$$\mathbf{g}(\boldsymbol{\theta}) = \frac{1}{2} + \frac{1}{2}\begin{bmatrix}\cos(\varphi)\\ \cos(\varphi+120^\circ)\\ \cos(\varphi-120^\circ)\end{bmatrix}.$$

Combining all the three microphone signals yields an omnidirectional pickup pattern as $\sum_{k=0}^{N-1}\cos(\varphi+\frac{2\pi}{N}k)=0$. Moreover introducing the differences between the front and two back microphone signals and between the left and right microphone signals yields an encoding matrix to obtain the omnidirectional W and the two X and Y figure-of-eight characteristcs

$$\frac{2}{3}\begin{bmatrix}1 & 1 & 1\\ 2 & -1 & -1\\ 0 & \sqrt{3} & -\sqrt{3}\end{bmatrix}\mathbf{g}(\varphi) = \begin{bmatrix}1\\ \cos(\varphi)\\ \sin(\varphi)\end{bmatrix}. \tag{1.8}$$

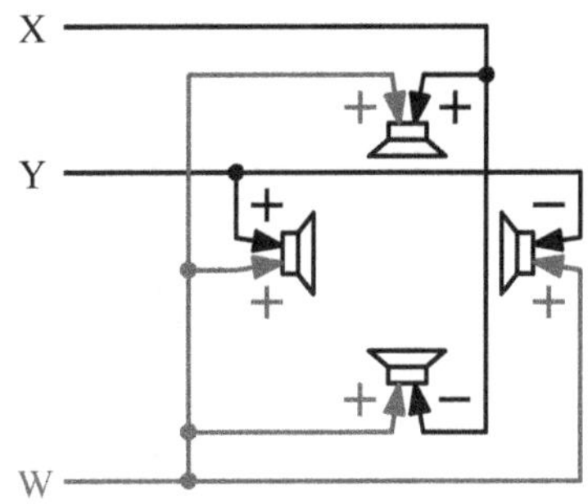

Fig. 1.8 2D first-order Ambisonic decoding to 4 loudspeakers

2D Ambisonic decoding to loudspeakers. The W, X, and Y channel of 2D first-order Ambisonics (Double-MS) can easily be played on an arrangement of four loudspeakers, front, back, left, right. While the omnidirectional signal contribution is played by all of the loudspeakers, the figure-of-eight contributions are played out-of-phase by the corresponding front-back or left-right pair of loudspeakers, Fig. 1.8.

$$\begin{bmatrix} F \\ L \\ B \\ R \end{bmatrix} = \begin{bmatrix} 1 & 1 & 0 \\ 1 & 0 & 1 \\ 1 & -1 & 0 \\ 1 & 0 & -1 \end{bmatrix} \begin{bmatrix} W \\ X \\ Y \end{bmatrix}. \tag{1.9}$$

The decoding weights obviously discretizes the directional pickup characteristics of the Ambisonic channels at the directions of the loudspeaker layout. Consequently, if the loudspeaker layout is more arbitrary and described by the set of its angles $\{\varphi_l\}$, the *sampling decoder* can be given as

$$\begin{bmatrix} S_{\varphi_1} \\ \vdots \\ S_{\varphi_L} \end{bmatrix} = \frac{1}{2} \begin{bmatrix} 1 & \cos(\varphi_1) & \sin(\varphi_1) \\ \vdots & \vdots & \vdots \\ 1 & \cos(\varphi_L) & \sin(\varphi_L) \end{bmatrix} \begin{bmatrix} W \\ X \\ Y \end{bmatrix}. \tag{1.10}$$

To achieve a panning-invariant and balanced mapping by this decoder, loudspeakers should be evenly arranged. Moreover, it can be favorable to sharpen the spatial image by attenuating W by $\frac{1}{\sqrt{3}}$ to map a sound by a supercardioid playback pattern.

Playback to head-tracked headphones and interactive rotation. In headphone playback, the headphone signals are generated by convolution with the head-related impulses responses of all four loudspeakers contributing to the left and the right ear signals

$$\begin{bmatrix} L_{\mathrm{ear}} \\ R_{\mathrm{ear}} \end{bmatrix} = \begin{bmatrix} h_{\mathrm{L}}^{0^\circ}(t)* & h_{\mathrm{L}}^{90^\circ}(t)* & h_{\mathrm{L}}^{180^\circ}(t)* & h_{\mathrm{L}}^{-90^\circ}(t)* \\ h_{\mathrm{R}}^{0^\circ}(t)* & h_{\mathrm{R}}^{90^\circ}(t)* & h_{\mathrm{R}}^{180^\circ}(t)* & h_{\mathrm{R}}^{-90^\circ}(t)* \end{bmatrix} \begin{bmatrix} F \\ L \\ B \\ R \end{bmatrix}. \tag{1.11}$$

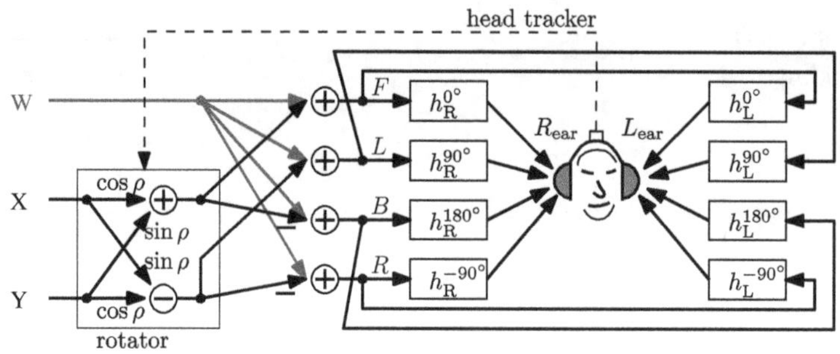

Fig. 1.9 2D first-order Ambisonic decoding to head-tracked headphones

To rotate the Ambisonic input scene of the decoder, it is sufficient to obtain a new set of figure-of-eight signals by mixing the X, Y channels with the following matrix depending on the rotation angle ρ, keeping W unaltered

$$\begin{bmatrix} W \\ \tilde{X} \\ \tilde{Y} \end{bmatrix} = \begin{bmatrix} 1 & 0 & 0 \\ 0 & \cos\rho & -\sin\rho \\ 0 & \sin\rho & \cos\rho \end{bmatrix} \begin{bmatrix} W \\ X \\ Y \end{bmatrix}. \tag{1.12}$$

This effect is important for head-tracked headphone playback to render the VR/360° audio scene static around the listener. A complete playback system is shown in Fig. 1.9. The big advantage of such a system is that rotational updates can be done at high control rates and the HRIRs of the convolver are constant.

1.3.2 3D First-Order Ambisonic Recording and Playback

The first-order Ambisonic format in 3D consists of a signal W corresponding to an omnidirectional pickup pattern, and three signals (X, Y, and Z) corresponding to figure-of-eight pickup patterns aligned with the Cartesian coordinate axes.

In three dimensions, we cannot work with figure-of-eight patterns described by $\sin\varphi$ or $\cos\varphi$ of the azimuth angle only, anymore. It is more convenient to describe the arbitrarily oriented figure-of-eight characteristics $\cos(\phi)$ using the inner product between a variable direction vector (direction of arriving sound) and a fixed direction vector (microphone direction). Direction vectors are of unit length $\|\boldsymbol{\theta}\| = 1$ and their inner product corresponds to $\boldsymbol{\theta}_1^{\mathrm{T}}\boldsymbol{\theta} = \cos(\phi)$, where ϕ is the angle enclosed by the direction of arrival $\boldsymbol{\theta}$ and the microphone direction $\boldsymbol{\theta}_1$. Consequently, a cardioid pickup pattern aiming at $\boldsymbol{\theta}_1$ is described by $\frac{1}{2} + \frac{1}{2}\boldsymbol{\theta}_1^{\mathrm{T}}\boldsymbol{\theta}$.

Native 3D Ambisonic recording (Triple-MS). To record the Ambisonic channels W, X, Y, Z, one can use a Triple-MS scheme as shown in Fig. 1.10. With the transposed unit direction vectors representing the aiming of the figure-of-eight channels $\boldsymbol{\theta}_{\mathrm{X}}^{\mathrm{T}} = [1,\ 0,\ 0]$, $\boldsymbol{\theta}_{\mathrm{Y}}^{\mathrm{T}} = [0,\ 1,\ 0]$, $\boldsymbol{\theta}_{\mathrm{Z}}^{\mathrm{T}} = [0,\ 0,\ 1]$, to produce the direction dipoles $\boldsymbol{\theta}_{\mathrm{X}}^{\mathrm{T}}\boldsymbol{\theta}$,

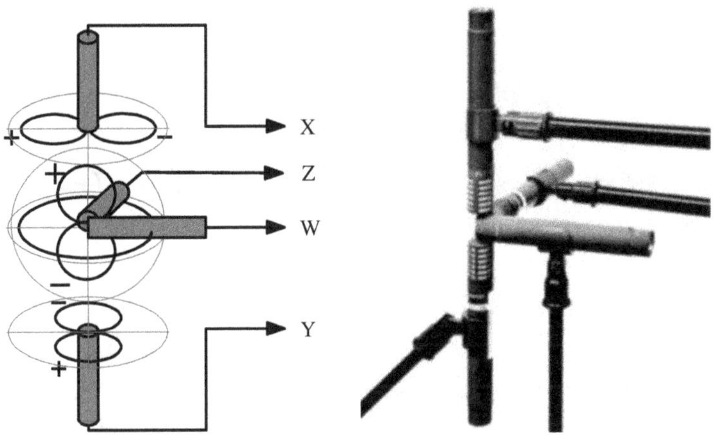

(a) Native 3D FOA recording (b) Picture of the recording setup

Fig. 1.10 Native 3D first-order Ambisonic recording with an omnidirectional and three figure-of-eight microphones aligned with the Cartesian axes X, Y, Z

$\boldsymbol{\theta}_{\mathrm{Y}}^{\mathrm{T}}\boldsymbol{\theta}$, and $\boldsymbol{\theta}_{\mathrm{Z}}^{\mathrm{T}}\boldsymbol{\theta}$, we can mathematically describe the pickup patterns of native 3D first-order Ambisonics as

$$g_{\mathrm{WXYZ}}(\boldsymbol{\theta}) = \begin{bmatrix} 1 \\ \begin{pmatrix} \boldsymbol{\theta}_{\mathrm{X}}^{\mathrm{T}} \\ \boldsymbol{\theta}_{\mathrm{Y}}^{\mathrm{T}} \\ \boldsymbol{\theta}_{\mathrm{Z}}^{\mathrm{T}} \end{pmatrix} \boldsymbol{\theta} \end{bmatrix} = \begin{bmatrix} 1 \\ \begin{pmatrix} 1 & 0 & 0 \\ 0 & 1 & 0 \\ 0 & 0 & 1 \end{pmatrix} \boldsymbol{\theta} \end{bmatrix} = \begin{bmatrix} 1 \\ \boldsymbol{\theta} \end{bmatrix}. \tag{1.13}$$

3D Ambisonic recording with a tetrahedral arrangement of cardioids. The principle that worked for three cardioid microphones on the horizon also works for a coincident tetrahedron microphone array of cardioids with the aiming directions FLU-FRD-BLD-BRU, see Fig. 1.11, and [12],

$$g(\boldsymbol{\theta}) = \frac{1}{2} + \frac{1}{2} \begin{bmatrix} \boldsymbol{\theta}_{\mathrm{FLU}}^{\mathrm{T}} \\ \boldsymbol{\theta}_{\mathrm{FRD}}^{\mathrm{T}} \\ \boldsymbol{\theta}_{\mathrm{BLD}}^{\mathrm{T}} \\ \boldsymbol{\theta}_{\mathrm{BRU}}^{\mathrm{T}} \end{bmatrix} \boldsymbol{\theta} = \frac{1}{2} + \frac{1}{2} \frac{1}{\sqrt{3}} \begin{bmatrix} 1 & 1 & 1 \\ 1 & -1 & -1 \\ -1 & 1 & -1 \\ -1 & -1 & 1 \end{bmatrix} \boldsymbol{\theta}. \tag{1.14}$$

Encoding is achieved there by the matrix that adds all microphone signals in the first line (W omnidirectional), subtracts back from front microphone signals in the second line (X figure-of-eight), subtracts right from left microphone signals in the third line (Y figure-of-eight), and subtracts down from up microphone signals in the last line (Z figure-of-eight), see also Fig. 1.11,

$$g_{\mathrm{WXYZ}}(\boldsymbol{\theta}) = \frac{1}{2} \begin{bmatrix} 1 \quad 1 \quad 1 \quad 1 \\ \sqrt{3} \begin{pmatrix} 1 & 1 & -1 & -1 \\ 1 & -1 & 1 & -1 \\ 1 & -1 & -1 & 1 \end{pmatrix} \end{bmatrix} g(\boldsymbol{\theta}). \tag{1.15}$$

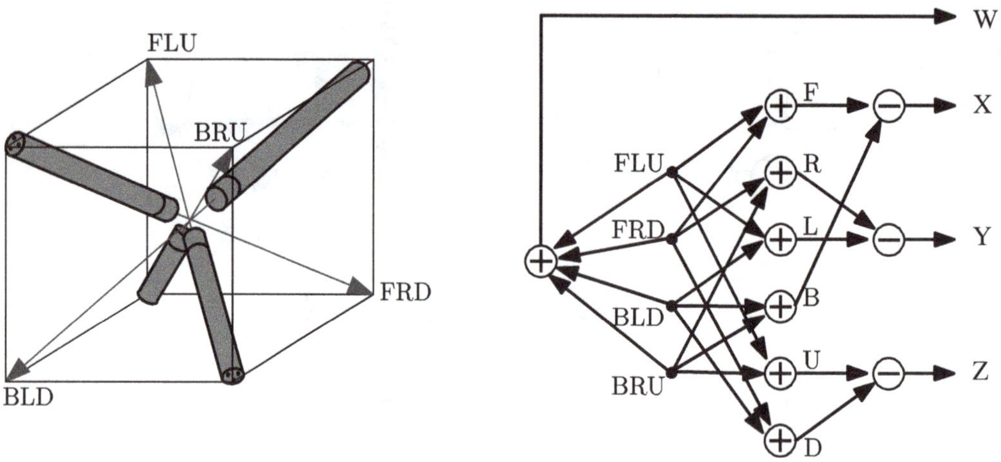

(a) Tetrahedral array with four cardioids (b) Encoder of microphone signals

Fig. 1.11 Tetrahedral arrangement of cardioid microphones for 3D first-order Ambisonics and their encoding; microphone capsules point inwards to minimize their spacing

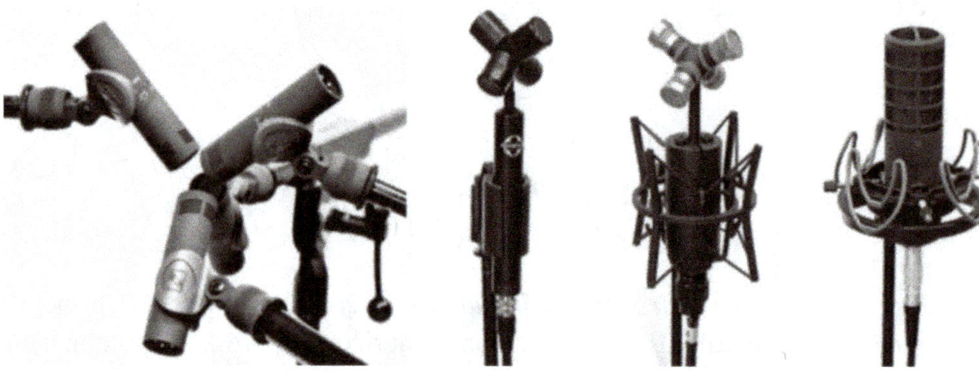

(a) Tetrahedral array of 4 cardioids (b) SPS200 (c) MK4012 (d) ST450

Fig. 1.12 Practical tetrahedral recording setups with cardioid microphones; the Soundfield SPS200, Oktava MK4012, and Soundfield ST450 offer a fixed 4-capsule geometry. Equally important: Zoom's H3-VR, Røde's NT-SF1, Sennheiser's AMBEO VR Mic

As Fig. 1.12 shows, practical microphone layouts should be as closely spaced as possible. Nevertheless for high frequencies, the microphones cannot be considered coincident anymore, and besides a directional error, there will be a loss of presence in the diffuse field. Typically a shelving filter is used to slightly boost high frequencies. Roughly, a high-shelf filter with a 3 dB boost is sufficient to correct timbral defects at frequencies above which the microphone spacing exceeds half a wavelength, e.g., 5 kHz for a 3.4 cm spacing of the microphones. More advanced strategies are found, e.g., in [7, 13–15].

3D Ambisonic decoding to loudspeakers. As before in the 2D case, a *sampling decoder* can be defined that represents the continuous directivity patterns associated with the channels W, X, Y, Z to map the signals to the discrete directions of the loudspeakers. Given the set of loudspeaker directions $\{\boldsymbol{\theta}_l\}$ and the unit-vectors to X, Y, Z, the loudspeaker signals of the sampling decoder become

$$\begin{bmatrix} S_1 \\ \vdots \\ S_L \end{bmatrix} = \frac{1}{2} \begin{bmatrix} 1 & \boldsymbol{\theta}_1^T\boldsymbol{\theta}_X & \boldsymbol{\theta}_1^T\boldsymbol{\theta}_Y & \boldsymbol{\theta}_1^T\boldsymbol{\theta}_Z \\ \vdots & & & \\ 1 & \boldsymbol{\theta}_L^T\boldsymbol{\theta}_X & \boldsymbol{\theta}_L^T\boldsymbol{\theta}_Y & \boldsymbol{\theta}_L^T\boldsymbol{\theta}_Z \end{bmatrix} \begin{bmatrix} W \\ X \\ Y \\ Z \end{bmatrix} = \underbrace{\frac{1}{2} \begin{bmatrix} 1 & \boldsymbol{\theta}_1^T \\ \vdots & \vdots \\ 1 & \boldsymbol{\theta}_L^T \end{bmatrix}}_{D} \begin{bmatrix} W \\ X \\ Y \\ Z \end{bmatrix}. \tag{1.16}$$

Equivalent panning function/virtual microphone. The sampling decoder together with the native Ambisonic directivity patterns $\boldsymbol{g}_{\mathrm{WXYZ}}^{\mathrm{T}}(\boldsymbol{\theta}) = [1,\ \boldsymbol{\theta}^{\mathrm{T}}]$ yields the mapping of a signal s from the direction $\boldsymbol{\theta}$ to the loudspeakers to be

$$\begin{bmatrix} S_1 \\ \vdots \\ S_L \end{bmatrix} = \frac{1}{2} \begin{bmatrix} 1 & \boldsymbol{\theta}_1^T \\ \vdots & \vdots \\ 1 & \boldsymbol{\theta}_L^T \end{bmatrix} \begin{bmatrix} 1 \\ \boldsymbol{\theta} \end{bmatrix} s = \frac{1}{2} \begin{bmatrix} 1 + \boldsymbol{\theta}_1^T\boldsymbol{\theta} \\ \vdots \\ 1 + \boldsymbol{\theta}_L^T\boldsymbol{\theta} \end{bmatrix} s, \tag{1.17}$$

This result means that the gain of a source from $\boldsymbol{\theta}$ at each loudspeaker $\boldsymbol{\theta}_l$ corresponds to evaluating a cardioid pattern aligned with $\boldsymbol{\theta}$. Consequently, the Ambisonic mapping corresponds to a signal distribution to the loudspeakers using weights obtained by discretization of an Ambisonics-equivalent first-order panning function.

Equivalently, Ambisonic playback using a sampling decoder is comparable to recording each loudspeaker signal with a virtual first-order cardioid microphone aligned with the loudspeaker's direction $\boldsymbol{\theta}_l$.

It is decisive for a panning-independent loudness mapping and balanced performance that the directions of the loudspeaker layout are well chosen. Also, it can be preferred to reduce the level of the omnidirectional channel W by $\frac{1}{\sqrt{3}}$ to map a sound by the narrower supercardioid playback pattern instead of a cardioid pattern, which is rather broad.

Decoder design problems were early addressed by Gerzon [16], Malham [17], and Daniel [18]. A current solution for higher-order decoding is given in Sect. 4.9.6 on All-round Ambisonic decoding.

3D Ambisonic decoding to headphones. 3D Ambisonic decoding to headphones uses the same approach as for 2D above, except that additional rotational degrees are implemented to compensate for any change in head orientation. Rotation concerns the three directional components X, Y, Z

$$\begin{bmatrix} \tilde{X} \\ \tilde{Y} \\ \tilde{Z} \end{bmatrix} = \boldsymbol{R}(\alpha, \beta, \gamma) \begin{bmatrix} X \\ Y \\ Z \end{bmatrix}. \tag{1.18}$$

For the definition of the rotation matrix $\boldsymbol{R}(\alpha, \beta, \gamma)$ and the meaning of its angles refer to Eq. 5.5 of Sect. 5.2.2. The selection of a suitable set of HRIRs is a question of directional discretization of the 3D directions, as addressed in the decoder above. Signals obtained for virtual loudspeakers are again to be convolved with the corresponding HRIRs for the left and the right ear.

1.4 Practical Free-Software Examples

The practical examples below show first-order Ambisonic panning a mono sound, decoded to simple loudspeaker layouts. These are either a square layout with 4 loudspeakers at the azimuth angles [0°, 90°, 180°, -90°] or an octahedral layout with 6 loudspeakers at azimuth [0°, 90°, 180°, -90°, 0°, 0°] and elevation [0°, 0°, 0°, 0°, 90°, -90°].

1.4.1 Pd with Iemmatrix, Iemlib, and Zexy

Pd is free and it can load and install its extensions from the internet. Required software components are:

- `pure-data` (free, http://puredata.info/downloads/pure-data)
- `iemmatrix` (free download within pure-data)
- `zexy` (free download within pure-data)
- `iemlib` (free download within pure-data)

Figure 1.13 gives an example for horizontal (2D) first-order Ambisonic panning, decoded to 4 loudspeaker and 2 headphone signals.

Figure 1.14 shows the processing inside the Pd abstraction `[FOA_binaural_decoder]` contained in the Fig. 1.13 example, which uses SADIE database[1] subject 1 (KU100 dummy head) HRIRs to render headphone signals.

Figure 1.15 sketches a first-order Ambisonic panning in 3D with decoding to an octahedral loudspeaker layout; master level `[multiline~]` and hardware outlets `[dac~]` were omitted for easier readability.

1.4.2 Ambix VST Plugins

This example uses a DAW and ready-to-use VST plug-ins to render first-order Ambisonics. As DAW, we recommend Reaper (reaper.fm) because it nicely facilitates higher-order Ambisonics by allowing tracks of up to 64 channels. Moreover, it is relatively low-priced and there is a fully functional free evaluation version available. You can also use any other DAW that supports VST and sufficiently many multi-track

[1] https://www.york.ac.uk/sadie-project/Resources/SADIEIIDatabase/D1/D1_HRIR_WAV.zip.

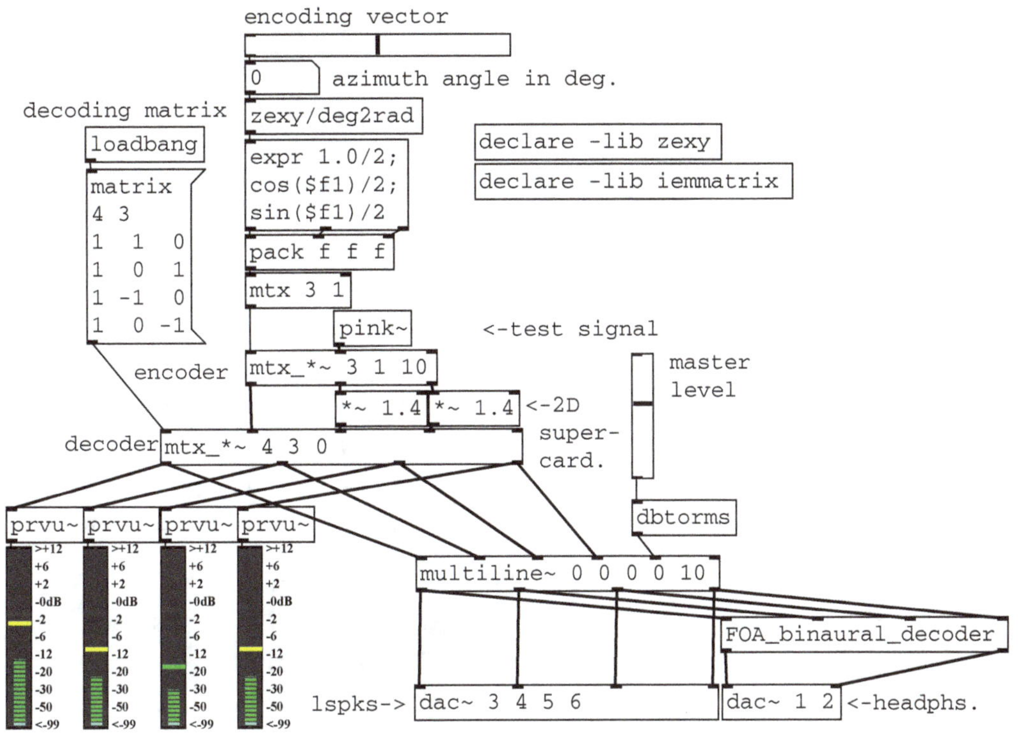

Fig. 1.13 First-order 2D encoding and decoding in pure data (Pd) for a square layout

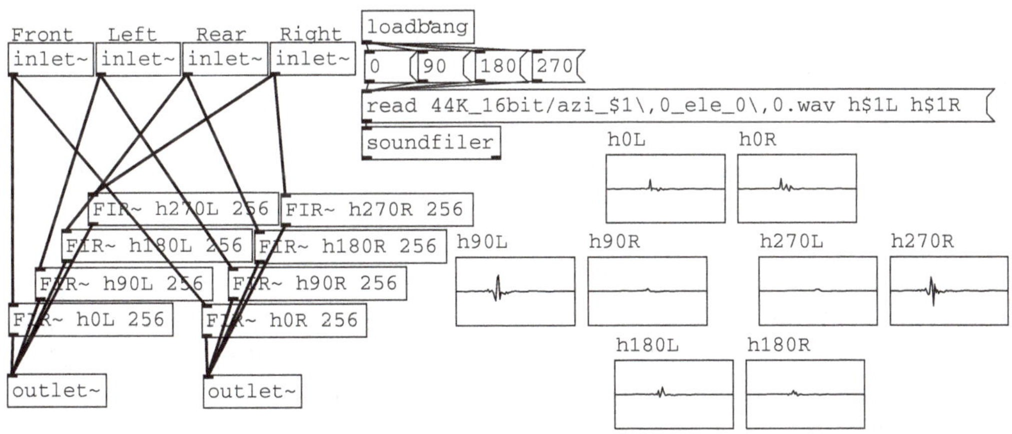

Fig. 1.14 Binaural rendering to headphones on pure data (Pd) by convolution with SADIE KU100 HRIRs

channels. The example employs the freely available ambiX plug-in suite (http://www.matthiaskronlachner.com/?p=2015), although there exist other Ambisonics plug-ins, especially for first-order.

Track Name	Ins	Outs	FX
Virtual source 1	1	4	ambix_encoder_o1
MASTER	4	6	ambix_decoder_o1

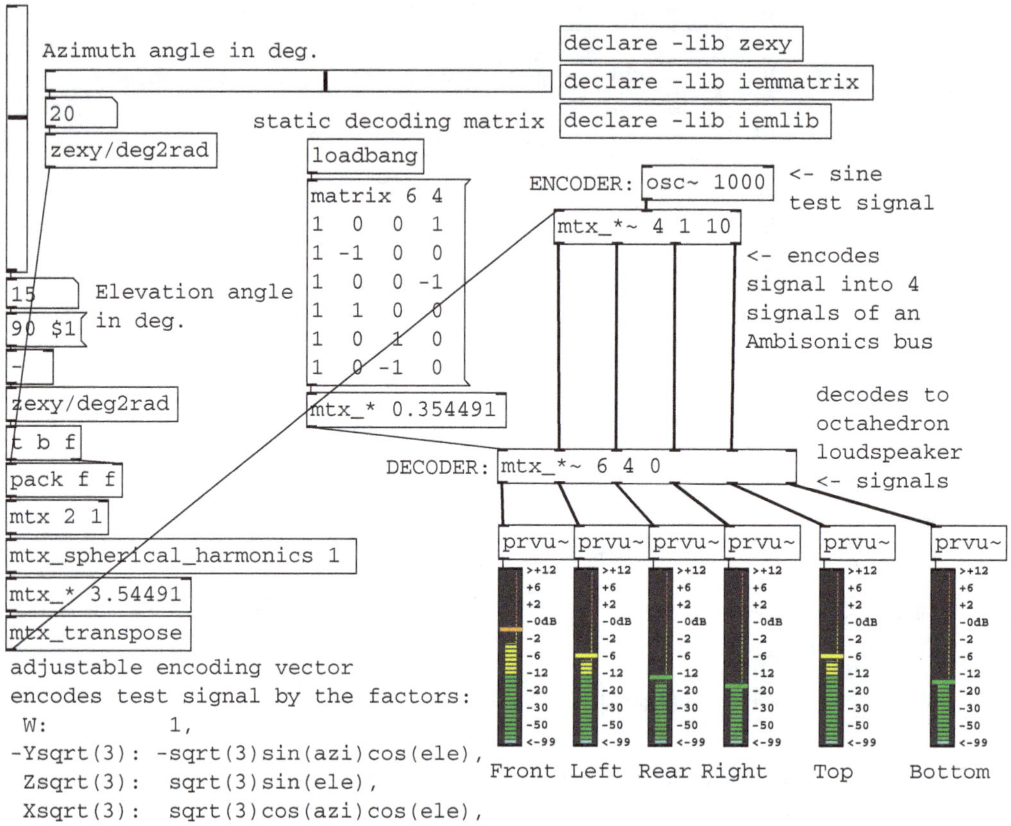

Fig. 1.15 First-order 3D encoding and decoding in pure data (Pd) using `[mtx_spherical_harmonics]` for an octahedral layout (`[dac~]` omitted for simplicity)

After creating the new track for the virtual source and importing a mono/stereo audio file (per drag-and-drop), the next step is the setup of the track channels. As shown in the table, the virtual source has a single-channel (mono) input and 4 output channels to send the 4 channels of first-order Ambisonics to the Master. The option to send to the Master is activated by default, cf. left in Fig. 1.16. The Master track itself requires 4 input channels and 6 output channels to feed the 6 loudspeakers (right). In Reaper, there is no separate adjustment for input and output channels, thus the Master track has to be set to 6 channels.

In the source track FX, the `ambix_encoder_o1` can be used to encode the virtual source signal at an arbitrary location on a sphere by inserting the plug-in into the track of the virtual source, cf. its panning GUI in Fig. 1.17. For adding more sources, the track of the virtual source can simply be copied or duplicated. All effects and routing options are maintained for the new tracks.

In order to decode the 4 first-order Ambisonics Master channels to the loudspeakers the `ambix_decoder_o1` plug-in is added to the Master track. The plug-in requires a preset that defines the decoding matrix and its channel sequence and normalization. For the exemplary octahedral setup with 6 loudspeakers, the following text can be copied to a text file and saved as config-file, e.g., "`octahedral.config`". The decoder

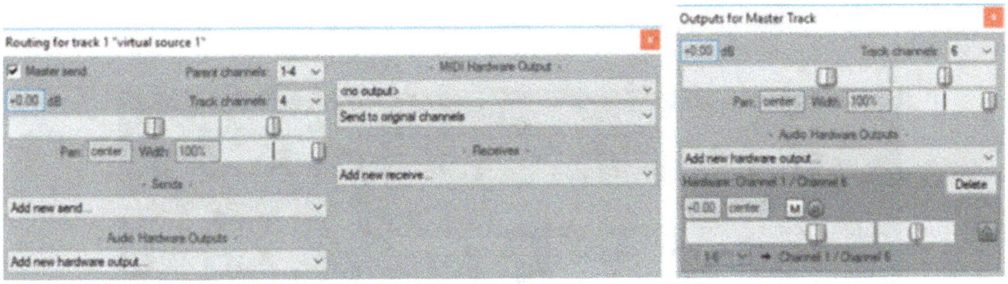

Fig. 1.16 1st-order example in Reaper DAW: routing

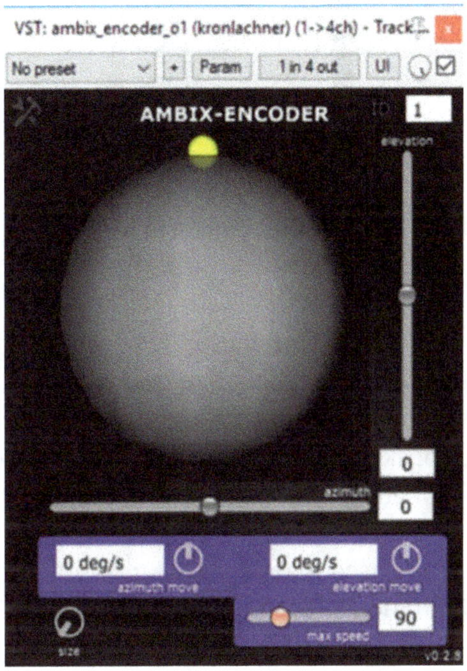

Fig. 1.17 1st-order example in Reaper DAW: encoder

matrix contains W, -Y, Z, X, with W as constant and -Y, Z, X refer to Cartesian coordinates of the octahedron.

```
#GLOBAL
/coeff_scale n3d
/coeff_seq acn
#END

#DECODERMATRIX
1  0  0  1
1  1  0  0
1 -1  0  0
1  0  0 -1
1  0  1  0
1  0 -1  0
#END
```

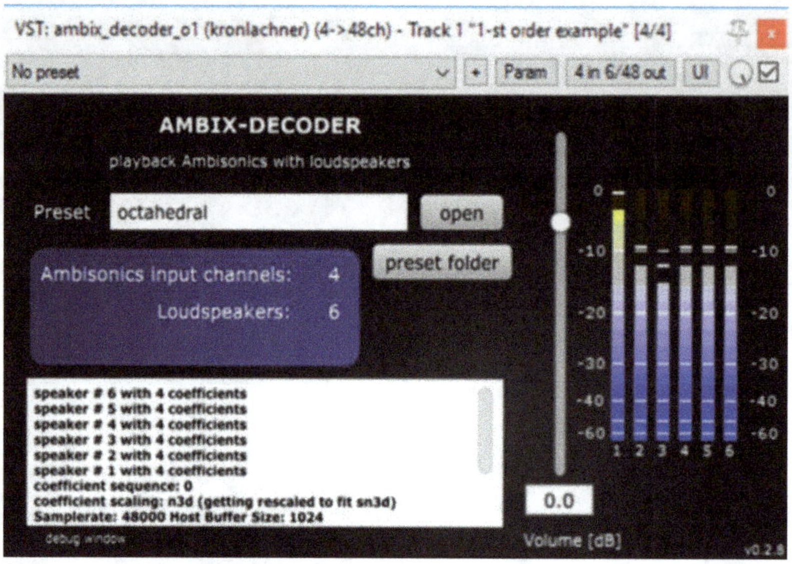

Fig. 1.18 1st-order example in Reaper DAW: decoder

After loading the preset into the decoder plug-in, the decoder can generate the loudspeaker signals as shown in Fig. 1.18. In the example, the virtual source is panned to the front, resulting in the highest level for loudspeaker 1 (front). The loudspeaker 3 (back) is 12dB quieter because of a side-lobe suppressing super cardioid weighting implied by the switch /coeff_scale n3d, as *a trick to keep things simple*.

As shown on the SADIE-II website,[2] the SADIE-II head-related impulse responses can be used to rendering Ambisonics to headphones. The listing below shows a configuration file to be used with `ambix_binaural`, cf. Fig. 1.19, again using the trick to select `n3d` to keep the numbers simple and super-cardioid weighting

```
#GLOBAL
/coeff_scale n3d
/coeff_seq acn
#END

#HRTF
44K_16bit/azi_0,0_ele_0,0.wav
44K_16bit/azi_90,0_ele_0,0.wav
44K_16bit/azi_180,0_ele_0,0.wav
44K_16bit/azi_270,0_ele_0,0.wav
44K_16bit/azi_0,0_ele_90,0.wav
44K_16bit/azi_0,0_ele_-90,0.wav
#END
```

[2]https://www.york.ac.uk/sadie-project/ambidec.html.

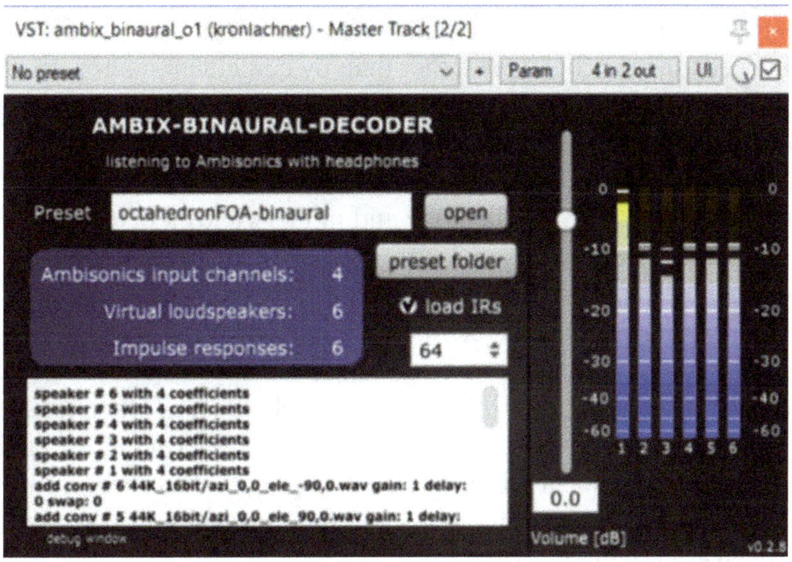

Fig. 1.19 1st-order example in Reaper DAW: binaural decoder

```
#DECODERMATRIX
1  0  0  1
1  1  0  0
1 -1  0  0
1  0  0 -1
1  0  1  0
1  0 -1  0
#END
```

For decoding to less regular loudspeaker layouts, the `IEM AllRADecoder`[3] permits editing loudspeaker coordinates and automatically calculating a decoder within the plugin. For decoding to headphones, the `IEM BinauralDecoder` offers a high-quality decoder. The technology behind both plugins is explained in Chap. 4.

In addition to the virtual sources, you can also add a 4-channel recording done with a B-format microphone by placing the 4-channel file in a new track. Reaper will automatically set the number of track channels to 4 and send the channels to the Master. Note that some B-format microphones use a different order and/or weighting of the Ambisonics channels. Simple conversion to the AmbiX-format can be done by inserting the ambix_converter_o1 plug-in into the microphone track.

1.5 Motivation of Higher-Order Ambisonics

Diffuseness, spaciousness, depth? Diffuse sound fields are typically characterized by sound arriving randomly from evenly distributed directions at evenly distributed delays. It is practical knowledge that the impression of diffuseness and spaciousness

[3] https://plugins.iem.at/.

requires benefits from decorrelated signals, which is typically achieved by large distances between the microphones rather than by coincident microphones.

Due to the evenness of diffuse sound fields, one would still hope that a low spatial resolution is sufficient to map diffuseness and spatial depth of a room, using coincident microphones or first-order Ambisonics. Nevertheless, high directional correlation during playback destroys this hope and in fact yields a perceptually impeded playback of diffuseness, spaciousness, and depth.

The technical advantages in interactivity and VR as well as the known shortcomings of first-order coincident recording techniques offer enough motivation to increase the directional resolution and go to higher-order Ambisonics, as presented in the subsequent chapters. For professional productions, it is often not sufficient to only rely on first-order coincident microphone recordings. By contrast, higher-order Ambisonics is able to drastically improve the mapping of diffuseness, spaciousness, and depth, as shown in the upcoming chapter about psychoacoustical properties of many-loudspeaker systems.

Recording with a higher-order main microphone array increases the required technological complexity. Nevertheless, digital signal processing and the theory presented in the later chapters is powerful nowadays to achieve this goal.

After all, it seems that delay-based stereophonic recording, such as AB, or equivalence-based recording, such as ORTF, INA5, etc., is often required and well-known in its mapping properties for spaciousness and diffuseness, correspondingly. What is nice about higher-order Ambisonics: it can make use of these benefits by embedding such recordings appropriately, see Fig. 1.20.

Facts about higher orders: Ambisonics extended to higher orders permits a refinement of the directional resolution and hereby improves the mapping of uncorrelated sounds in playback. Figure 1.21a shows the correlation introduced in two neighboring loudspeaker signals when using Ambisonics, given their spacing of 60°. Given the just noticeable difference (JND) of the inter-aural cross correlation, the figure indicates that an Ambisonic order of ≥3 might be necessary to perceptually preserve decorrelation.

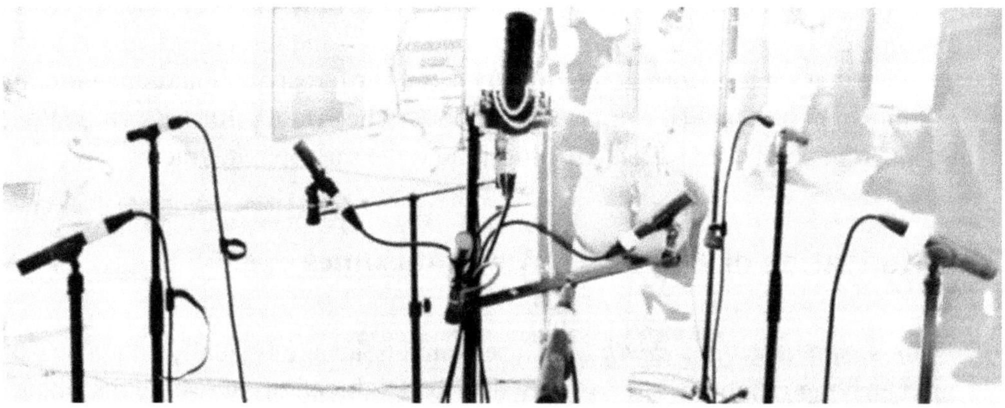

Fig. 1.20 How is a microphone tree represented in Ambisonics, when it consist of 6 cardioids spaced by 60 cm and 60° on a horizontal ring, and a ring of 4 super cardioids spaced by 40 cm and 90° as height layer, pointing upwards?

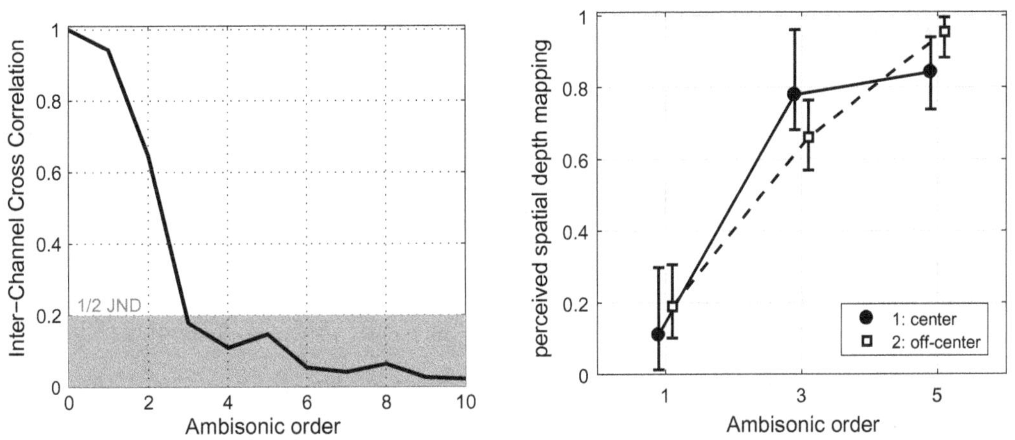

(a) Inter-channel cross correlation of two Ambisonically driven loudspeakers spaced by 60°

(b) Perceived depth in dependence of Ambisonics playback order for two listening positions

Fig. 1.21 Relation between the Ambisonic order, decorrelation, and perceived depth

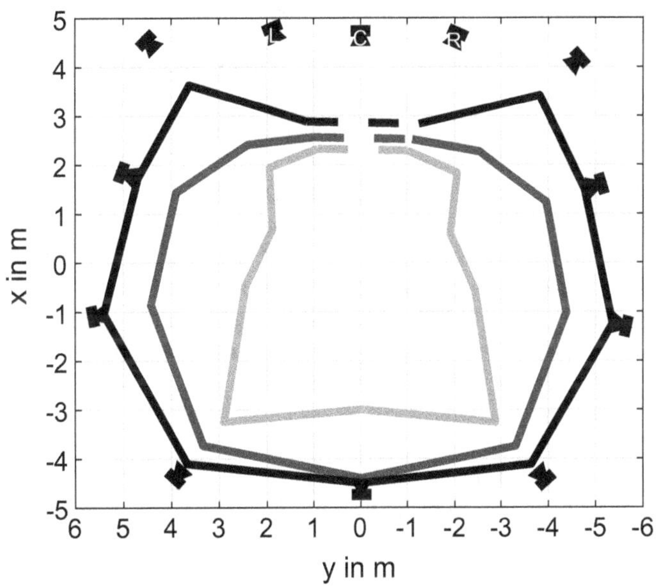

Fig. 1.22 Perceptual sweet area of Ambisonic playback from the front at first (light gray), third (dark gray), and fifth order (black). It marks the area in which the perceived direction is plausible, i.e., does not collapse into a single loudspeaker other than C

For this reason, the perception of spatial depth strongly improves when increasing the Ambisonic order from 1 up to 3, Fig. 1.21b. However, this is only the case when seated at the central listening position. Outside this sweet spot, higher orders than 3, e.g., 5, additionally improve the mapping of depth [19]. Therefore, higher-order Ambisonics is important for preserving spatial impressions and when supplying a large audience.

Figure 1.22 shows that the sweet area of perceptually plausible playback increases with the Ambisonic order [20]. With fifth-order Ambisonics, nearly all the area spanned by the horizontal loudspeakers at the IEM CUBE, the 12×10 m concert space at our lab, becomes a valid listening area.

References

1. A.D. Blumlein, Improvements in and relating to sound-transmission, sound-recording and sound-reproducing systems. GB patent 394325 (1933), http://www.richardbrice.net/blumlein_patent.htm
2. M.A. Gerzon, Why coincident microphones? in *Reproduced from Studio Sound, by permission of IPC Media Ltd, publishers of Hi-Fi News* (www.hifinews.co.uk), vol. 13 (1971), https://www.michaelgerzonphotos.org.uk/articles/Why
3. P.B. Vanderlyn, In search of Blumlein: the inventor incognito. J. Audio Eng. Soc. **26**(9) (1978)
4. S.P. Lipshitz, Stereo microphone techniques: Are the purists wrong? J. Audio Eng. Soc. **34**(9) (1986)
5. S. Weinzierl, *Handbuch der Audiotechnik* (Springer, Berlin, 2008)
6. A. Friesecke, *Die Audio-Enzyklopädie*. G. Sazur (2007)
7. M.A. Gerzon, The design of precisely coincident microphone arrays for stereo and surround sound, prepr. L-20 of 50th Audio Eng. Soc. Conv. (1975)
8. G. Bore, S.F. Temmer, M-s stereophony and compatibility, Audio Magazine (1958), http://www.americanradiohistory.com
9. M.A. Gerzon, Application of blumlein shuffling to stereo microphone techniques. J. Audio Eng. Soc. **42**(6) (1994)
10. D.H. Cooper, T. Shiga, Discrete-matrix multichannel stereo. J. Audio Eng. Soc. **20**(5), 346–360 (1972)
11. P. Felgett, Ambisonic reproduction of directionality in surround-sound systems. Nature **252**, 534–538 (1974)
12. P. Craven, M.A. Gerzon, Coincident microphone simulation covering three dimensional space and yielding various directional outputs, U.S. Patent, no. 4,042,779 (1977)
13. C. Faller and M. Kolundžija, "Design and limitations of non-coincidence correction filters forsoundfield microphones," in *prepr. 7766, 126th AES Conv*, Munich (2009)
14. J.-M. Batke, The b-format microphone revisited, in *1st Ambisonics Symposium*, Graz (2009)
15. A. Heller E. Benjamin, Calibration of soundfield microphones using the diffuse-field response, in prepr. 8711, 133rd AES Conv, San Francisco (2012)
16. M. Gerzon, General metatheory of auditory localization, in prepr. 3306, Conv. Audio Eng. Soc. (1992)
17. D.G. Malham, A. Myatt, 3D Sound spatialization using ambisonic techniques. Comput. Music. J. **19**(4), 58–70 (1995)
18. J. Daniel, J.-B. Rault, J.-D. Polack, Acoustic properties and perceptive implications of stereophonic phenomena, in *AES 6th International Conference: Spatial Sound Reproduction* (1999)
19. M. Frank, F. Zotter, Spatial impression and directional resolution in the reproduction of reverberation, in Fortschritte der Akustik - DEGA, Aachen (2016)
20. M. Frank, F. Zotter, Exploring the perceptual sweet area in ambisonics, in AES 142nd Conv. (2017)

Chapter 2
Auditory Events of Multi-loudspeaker Playback

It is evident that until one knows what information needs to be presented at the listener's ears, no rational system design can proceed.

Michael A. Gerzon 1976 and AES Vienna [1] 1992.

Abstract This chapter describes the perceptual properties of auditory events, the sound images that we localize in terms of direction and width, when distributing a signal with different amplitudes to one or a couple of loudspeakers. These amplitude differences are what methods for amplitude panning implement, and they are also what mapping of any coincident-microphone recording implies when reproduced over the directions of a loudspeaker layout. Therefore several listening experiments on localization are described and analyzed that are essential to understand and model the psychoacoustical properties of amplitude panning on multiple loudspeakers of a 3D audio system. For delay-based recordings or diffuse sounds, there is some relation, however, it is found to be less stable for the desired applications. Moreover, amplitude panning is not only about consistent directional localization. Loudness, spectrum, temporal structure, or the perceived width should be panning-invariant. The chapter also shows experiments and models required to understand and provide those panning-invariant aspects, especially for moving sounds. It concludes with openly-available response data of most of the presented listening experiments.

Starting from classic listening experiments on stereo panning by Leakey [2], Wendt [3], and pairwise horizontal panning by Theile [4], this chapter explores the relevant perceptual properties for 3D amplitude panning and their models. Important experimental studies considered here are for instance those by Simon [5], Kimura [6], F. Wendt [7], Lee [8], Helm [9], and Frank [10, 11]. By the experimental results, it

F. Zotter and M. Frank, *Ambisonics*, Springer Topics in Signal Processing 19,
https://doi.org/10.1007/978-3-030-17207-7_2

is possible to firmly establish Gerzon's [1] E, $\boldsymbol{r}_\mathrm{E}$ and $\|\boldsymbol{r}_\mathrm{E}\|$ estimators for perceived loudness, direction, and width that apply to most stationary sounds in typical studio and performance environments.

2.1 Loudness

At a measurement point in the free field, the same signal fed to equalized loudspeakers of exactly the same acoustic distance would superimpose constructively (+6 dB).

In a room with early reflections and a less strict equality of the incoming pair of sounds (typical, slight inaccuracy in loudspeaker/listener position, different mounting situations, different directions in the directivities of ears and loudspeakers), the superposition can be regarded as stochastically constructive (+3 dB) in particular at frequencies that aren't very low.

For the above reasoning, typical amplitude panning rules try to keep the weights distributing the signal to the loudspeakers normalized by root of squares instead of normalizing to the linear sum, in order to obtain constant loudness ([12], VBAP):

$$g_l \leftarrow \frac{g_l}{\sqrt{\sum_{l=1}^{\mathrm{L}} g_l^2}}. \tag{2.1}$$

Loudness Model. If all loudspeakers are equalized, located at the same distance to the listener, and fed by the same signal with different amplitude gains g_l, a constructive interference could be expected so that the amplitude becomes [1]

$$P = \sum_{l=1}^{\mathrm{L}} g_l. \tag{2.2}$$

However, the interference stops to be strictly constructive as soon as the room is not entirely anechoic, the sitting position is not exactly centered, or even for anechoic and centered conditions at high frequencies, when the superposition at the ears cannot be assumed to be purely constructive anymore. Then it is better to assume a less well-defined, stochastic superposition in which a squared amplitude is determined by the sum of the squared weights [1]:

$$E = \sum_{l=1}^{\mathrm{L}} g_l^2. \tag{2.3}$$

Therefore, the most common amplitude panning rules use root-squares normalization to obtain a loudness impression that is as constant as possible.

The measure E seems to be most useful when designing and evaluating amplitude-panning or coincident microphone techniques. It is not surprising that the ITU-R

BS.1770-4[1] uses the Leq(RLB) measure as a loudness model: it is essentially the RMS level after high-pass filtering, cf. [13], which is closely related to the E measure detected from loudspeaker signals.

An interesting refinement was proposed by Laitinen et al. [14], which uses a measure $\sqrt[p]{\sum_{l=1}^{\mathrm{L}} g_l^p}$ in which the exponent p is close to 1 at low frequencies under anechoic conditions and close to 2 at high frequencies/under reverberant conditions.

2.2 Direction

In the early years of stereophony, researchers investigated the differences in delay times and amplitudes required to control the perceived direction. Below, only experiments are considered that did not use fixation of the listener's head.

2.2.1 *Time Differences on Frontal, Horizontal Loudspeaker Pair*

The dissertation of K. Wendt in 1963 [3] shows notably accurate listening experiments done on ±30° two-channel stereophony using time delays, in which listeners indicated from where they heard the sounds for each of the tested time differences. H. Lee revisited the properties in 2013 [8], but with musical sound material and an experiment, in which the listener adjusted the time differences until the perceived direction matched the one of a corresponding fixed reference loudspeaker, Fig. 2.1.

The time differences are seldom applicable to reliable angular auditory event placement: auditory images are strongly frequency-dependent (not shown here) and therefore unstable for narrow-band sounds. Leakey and Cherry showed 1957 [2] that time-delay stereophony loses its effect under the presence of background noise.

2.2.2 *Level Differences on Frontal, Horizontal Loudspeaker Pair*

K. Wendt's [3] and H. Lee's [8] experiments deliver insights in sound source positioning with ±30° two-channel stereophony, however this time with level differences.

As opposed to Fig. 2.1, in which auditory image panning with time differences were characterized by statistical spreads of up to 15°, level-difference-based panning is clearly smaller in the spread of perceived directions than 10°, Fig. 2.2.

Signal dependency. Wendt [3] described the signal dependency of panning curves on various transient and band-limited sounds, and Lee [8] for musical sounds. A new

[1] https://www.itu.int/rec/R-REC-BS.1770-4-201510-I/en Algorithms to measure audio programme loudness and true-peak audio level (10/2015).

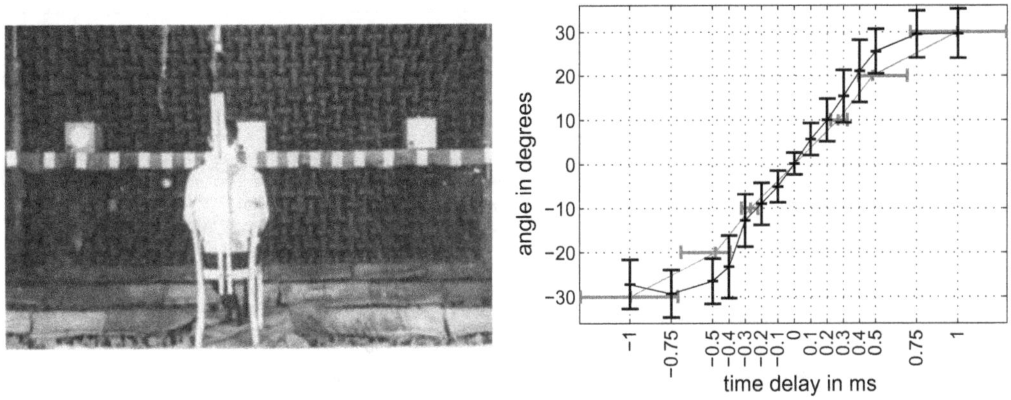

Fig. 2.1 K. Wendt's experiment [3] used an angular marks helping to specify the localized direction (left). Right shows results for time differences between impulse signals fed to loudspeakers, no head fixation (diagram shows means and standard deviation; the standard deviation was interpolated for the figure). In gray: Results of the time-difference adjustment experiment of Lee [8] using musical material (25, 50, 75% quartiles, symmetrized diagram)

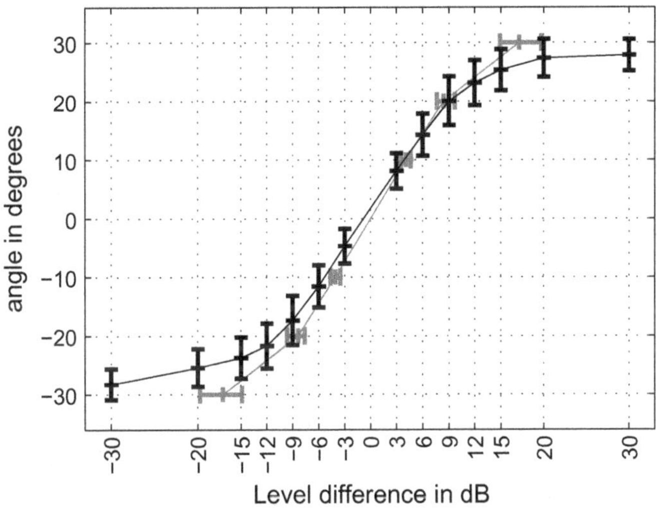

Fig. 2.2 Wendt's [3] results to crack (impulsive) signals with level differences and without head fixation (the figure shows means and standard deviation; standard deviation was interpolated to plot this figure). In gray: Results of Lee's [8] level-difference adjustment experiment with musical sounds (25, 50, 75% quartiles, symmetrized diagram)

comprehensive investigation on frequency dependency was carried out by Helm and Kurz [9]. With level differences {0, 3, 6, 9, 12} dB and third-octave filtered pulsed pink noise at {125, 250, 500, 1k, 2k, 4k} Hz, they showed that the perceived angle pointed at by the listeners using a motion-tracked pointer was similar between the broad-band case and third-octave bands below 2 kHz. In bands above 2 kHz, smaller level differences cause a larger lateralization, see interpolated curves in Fig. 2.3.

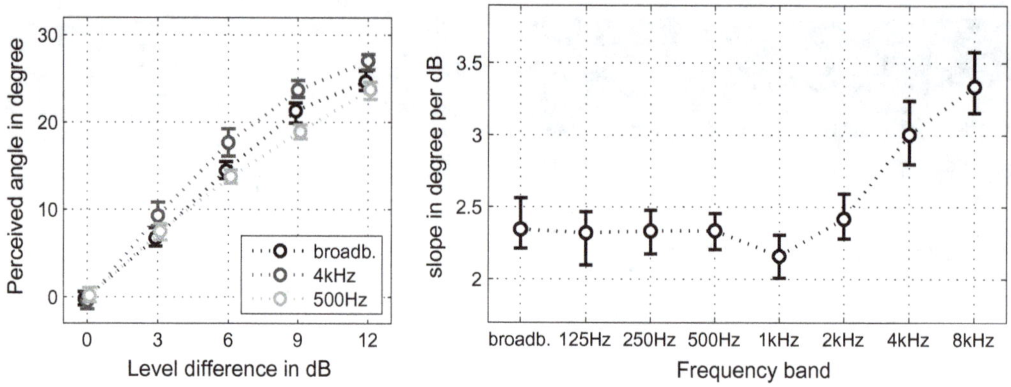

(a) Third-octave vs. broad-band

(b) Slope of different bands

Fig. 2.3 Panning curve for frontal ±30° loudspeaker pair from [9] on the example of the 500 and 4 kHz third-octave band and the slopes for different bands, based on the 3 and 6 dB conditions

2.2.3 *Level Differences on Horizontally Surrounding Pairs*

Successive pairwise panning on neighboring loudspeaker pairs is typically used to pan auditory events freely along the loudspeakers of a horizontally surrounding loudspeaker ring. The classical research done specifically targeted at such applications was contributed by Theile and Plenge 1977 [4]. They used a mobile reference loudspeaker with some reference sound that could be moved to match the perceived direction of a loudspeaker pair playing pink noise with level differences at different orientations with respect to the listener's head. There is also the experiment of Pulkki [15] using a level-adjustment task, in which levels were adjusted as to match the auditory event to one of a reference loudspeaker at three different reference directions and for different head orientations. A comprehensive experiment was done by Simon et al. [5], who used a graphical user interface displaying the floor plan of a 45°-spaced loudspeaker ring to have the listeners specify the perceived direction. Martin et al. in 1999 [16] used a graphical user interface showing the floorplan of a 5.1 ring in their experiment, and last but not least, Matthias Frank used a direct pointing method to enter the perceived direction [10] in one of his experiments.

As the experiments did not seem to yield consistent results, a comprehensive level-difference adjustment experiment with 24 loudspeakers arranged as a horizontal ring was done in [17] and partially repeated later in [11], see results in Fig. 2.4. In the repeated experiment [11] it became clear that in the anechoic room, a large amount of the differently pronounced localization biases can be avoided by encouraging the listeners to do front-back and left-right head motion by a few of centimeters, whenever there is doubt.

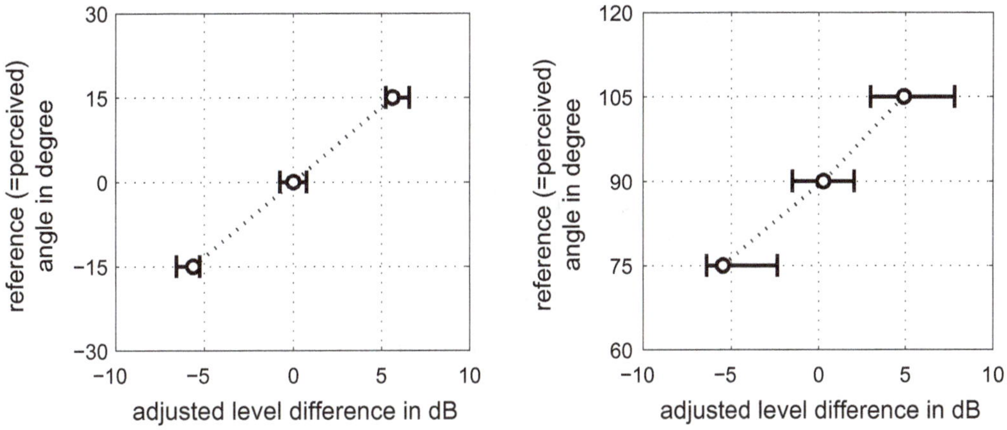

(a) frontal 60° stereo pair (b) lateral 60° stereo pair

Fig. 2.4 Medians and 95% confidence intervals for adjusted level differences to align amplitude-panned pink-noise with harmonic complex tone from $\{\pm 15°, 0°\}$, for **a** frontal and **b** lateral 60° stereo pair; **a** uses data from [17] with 4 responses per direction from 5 listeners; **b** used data from [11] with 20 responses per direction. Despite the considerably different spread, frontal and lateral stereo pairs seem to yield pretty much the same tendency

2.2.4 *Level Differences on Frontal, Horizontal to Vertical Pairs*

Quite extensively, T. Kimura investigates the localization of auditory events between frontal, vertical $\pm 13.5°$ loudspeaker pairs in 2012 [6, 18]. The work of F. Wendt in 2013 [7, 19] also investigates a slant and vertical loudspeaker pair, Fig. 2.5. Kimura uses pulsed white noise, Wendt uses pulsed pink noise.

Obviously, the horizontal spread is always smaller than the vertical spread and the spread does not align with the direction of the loudspeaker pair. The largest vertical spread appears for the vertical loudspeaker pair.

2.2.5 *Vector Models for Horizontal Loudspeaker Pairs*

A weighted sum of the loudspeakers' direction vectors $\boldsymbol{\theta}_1$, $\boldsymbol{\theta}_2$ could be conceived as simple linear model of the perceived direction, using a linear blending parameter $0 \leq q \leq 1$

$$\boldsymbol{r} = (1-q)\,\boldsymbol{\theta}_1 + q\,\boldsymbol{\theta}_2. \tag{2.4}$$

The parameter q adjusts where the resulting vector $\boldsymbol{r}$ is located on the connecting line between $\boldsymbol{\theta}_1$ and $\boldsymbol{\theta}_2$. On frontal loudspeaker pairs, localization curves typically run through the middle direction $q = \frac{1}{2}$ for level differences of 0 dB. If only one

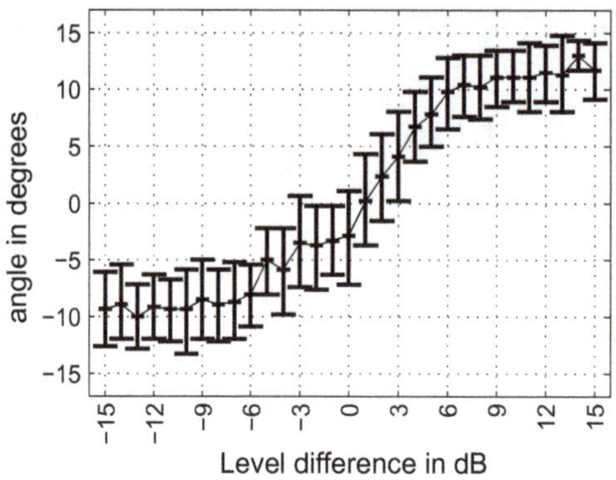

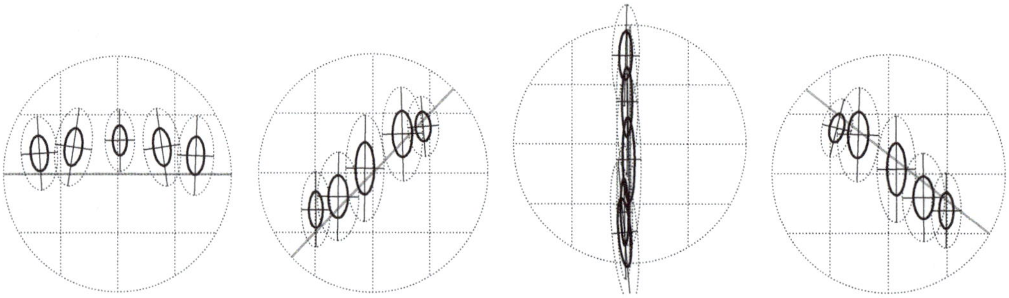

Fig. 2.5 Mean values and 95% confidence intervals of the direct-pointing experiments of Kimura (top) with level differences on a vertical ±13.5° loudspeaker pair and results of F. Wendt (bottom) on frontally arranged horizontal, slant, and vertical ±20° loudspeaker pairs showing two-dimensional 95% confidence (solid) and standard deviation ellipses (dotted)

loudspeakers is active, the result is either of the loudspeaker directions, thus the parameter is $q = 0$ or $q = 1$.

Classical definitions. As the simplest choice for q, one could insert $q = \frac{g_2}{g_1+g_2}$ or $q = \frac{g_2^2}{g_1^2+g_2^2}$ to get the vector definitions as weighted average using either the linear or squared gains according to [1]:

$$\boldsymbol{r}_\mathrm{V} = \frac{g_1\,\boldsymbol{\theta}_1 + g_2\,\boldsymbol{\theta}_2}{g_1 + g_2}, \qquad \boldsymbol{r}_\mathrm{E} = \frac{g_1^2\,\boldsymbol{\theta}_1 + g_2^2\,\boldsymbol{\theta}_2}{g_1^2 + g_2^2}. \tag{2.5}$$

For both models, equal gains $g_1 = g_2$ yield $q = \frac{1}{2}$, and also the endpoints with $g_2 = 0$ or $g_1 = 0$ correspond to $q = 0$ or $q = 1$, respectively. However, the slope of the $\boldsymbol{r}_\mathrm{E}$ vector is steeper than the one of the $\boldsymbol{r}_\mathrm{V}$. For instance, if $g_2 = 2\,g_1$, the vector $\boldsymbol{r}_\mathrm{V}$ lies on $q = 2/3$ of the line between $\boldsymbol{\theta}_1$ and $\boldsymbol{\theta}_2$, while $\boldsymbol{r}_\mathrm{E}$ lies at $q = 4/5$ of the connecting line.

The $\boldsymbol{r}_V$ vector for the $\pm\alpha$ loudspeaker pair at the directions $\boldsymbol{\theta}_{1,2}^\mathrm{T} = (\cos\alpha,\ \pm\sin\alpha)$ corresponds to the tangent law [20], whose formal origin lies in a model of summing localization based on a simple model of the ear signals, cf. Appendix A.7. The equivalence of this law to the vector model follows from the tangent $\tan\varphi$ as ratio of the y divided by x component of the $\boldsymbol{r}_V$ vector, $\tan\varphi = \frac{g_1\,\sin(\alpha)+g_2\,\sin(-\alpha)}{g_1\,\cos(\alpha)+g_2\,\cos(\alpha)} = \frac{g_1-g_2}{g_1+g_2}\tan\alpha$.

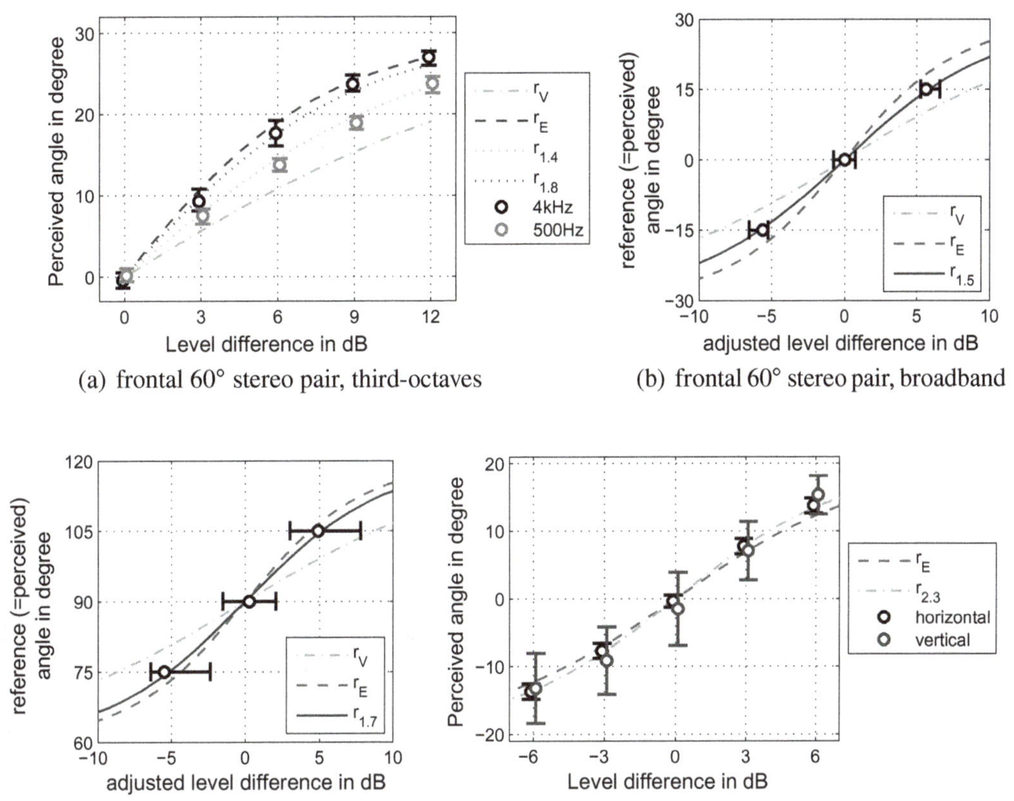

(a) frontal 60° stereo pair, third-octaves (b) frontal 60° stereo pair, broadband

(c) lateral 60° stereo pair, broadband (d) frontal horizontal and vertical ±20° pair, broadband

Fig. 2.6 Fit of the $\boldsymbol{r}_{\mathrm{V}}$, $\boldsymbol{r}_{\mathrm{E}}$, and $\boldsymbol{r}_{\gamma}$ models for **a** third-octave noise on a frontal stereo pair using data from [9], and with data from [11]: **b** pink noise frontal and **c** lateral, cf. Figs. 2.3 and 2.4; **d** horizontal and vertical from [7], Fig. 2.5

Adjusted slope. Differently steep curves were fitted by an adjustable-slope model [17]

$$\boldsymbol{r}_{\gamma} = \frac{|g_1|^{\gamma}\,\boldsymbol{\theta}_1 + |g_2|^{\gamma}\,\boldsymbol{\theta}_2}{|g_1|^{\gamma} + |g_2|^{\gamma}}, \tag{2.6}$$

which uses $\gamma = 1$ for $\boldsymbol{r}_{\mathrm{V}}$ and $\gamma = 2$ for $\boldsymbol{r}_{\mathrm{E}}$. Figure 2.6 compares the prediction by $\boldsymbol{r}_{\mathrm{V}}$, $\boldsymbol{r}_{\mathrm{E}}$, and $\boldsymbol{r}_{\gamma}$ to frequency-dependently perceived directions in frontal horizontal pairs, to perceived directions in a lateral stereo pair, and to perceived directions in a frontal pair that is either horizontal or vertical, using various studies mentioned above.

Practical choice $\boldsymbol{r}_{\mathrm{E}}$. While a specific exponent γ closely fitting the experimental data may vary, a constant value is preferable. Figure 2.6 indicates that in most cases focusing on $\boldsymbol{r}_{\mathrm{E}}$ is reasonable and sufficiently precise, see also [11].

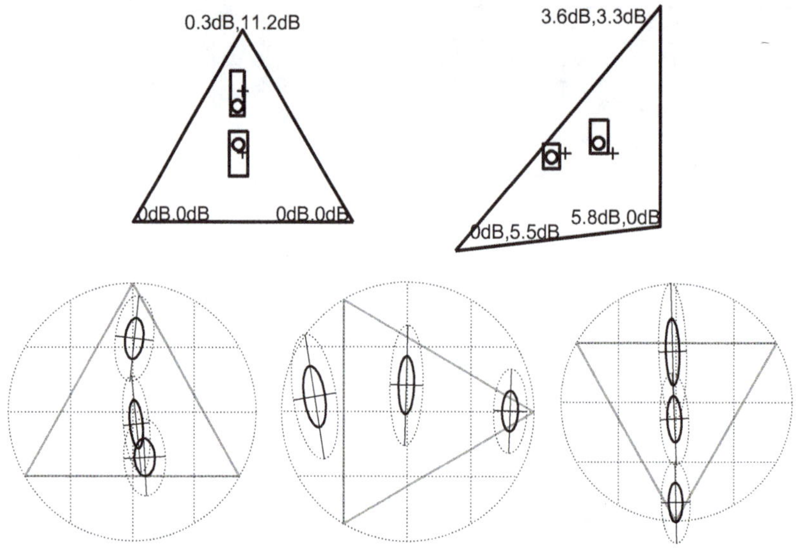

Fig. 2.7 Indirect level-adjustment experiment of Pulkki [21] shows the spread and mean of the adjusted VBAP angles for frontal loudspeaker triplets, and the experiments of F. Wendt [7, 19] use a direct pointing method to obtain results in the shape of two-dimensional 95% confidence (solid) and standard deviation ellipses (dotted) for $\{-\infty,\ 0,\ +11.71\}$ dB for the top loudspeaker (left diagram), or the right loudspeaker (center diagram) respectively, or $\{-\infty,\ 0,\ +11.51\}$ dB for the bottom loudspeaker (right)

2.2.6 Level Differences on Frontal Loudspeaker Triangles

V. Pulkki [21] and F. Wendt [7, 19] investigated localization properties for frontal loudspeaker triplets with level differences, see Fig. 2.7. Both used pulsed pink noise in their experiments.

While V. Pulkki used an indirect adjustment task to evaluate VBAP control angles to obtain auditory events directionally matching the respective reference loudspeakers, F. Wendt uses a direct pointing method. Wendt's experiments indicate that loudspeaker triplets with three different azimuthal positions yield a smaller spread in the indicated direction than such with vertical loudspeaker pairs (not the case in Pulkki's experiments).

2.2.7 Level Differences on Frontal Loudspeaker Rectangles

F. Wendt [7, 19] moreover presents experiments about frontal loudspeaker rectangles, again using a pointer method and pulsed pink noise, Fig. 2.8.

Again it seems that arrangements avoiding vertical loudspeaker pairs exhibit a smaller statistical spread in the responses.

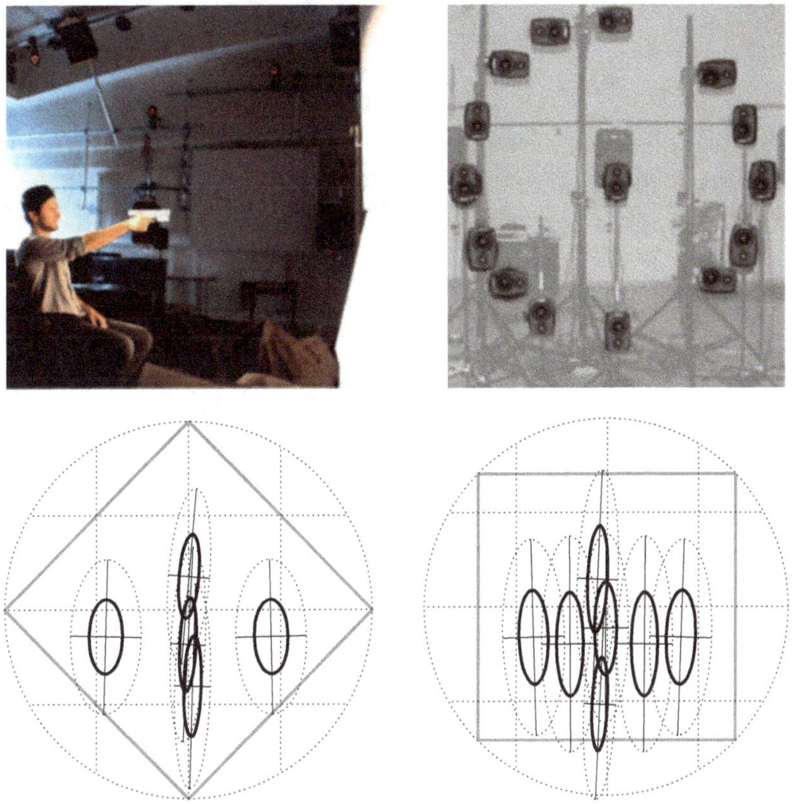

Fig. 2.8 Wendt's experiments about frontal loudspeaker rectangles showing two-dimensional 95% confidence (solid) and standard deviation ellipses (dotted). The experimental setup of this and above-mentioned experiments is shown. Left: each of the corner loudspeakers is raised once by +6 dB in level, right: both left/right loudspeaker levels are raised once by {+3, +6} dB, and both top/bottom pairs are once raised by +6 dB

2.2.8 Vector Model for More than 2 Loudspeakers

For more than two active loudspeakers and in 3D, a vector model based on the exponent $\gamma = 2$ yields the $\boldsymbol{r}_\mathrm{E}$ vector [1]

$$\boldsymbol{r}_\mathrm{E} = \frac{\sum_{l=1}^{\mathrm{L}} g_l^2\, \boldsymbol{\theta}_l}{\sum_{l=1}^{\mathrm{L}} g_l^2}. \tag{2.7}$$

2.2.9 Vector Model for Off-Center Listening Positions

At off-center listening positions, the distances to the loudspeakers are not equal any-more, resulting in additional attenuation and delay for each loudspeaker depending on the position. For stationary sounds, this effect can be incorporated into the energy vector by additional weights $w_{\mathrm{r},i}$ and $w_{\tau,i}$

$$\boldsymbol{r}_\mathrm{E} = \frac{\sum_{l=1}^{\mathrm{L}} (w_{\mathrm{r},l}\, w_{\tau,l}\, g_l)^2\, \boldsymbol{\theta}_l}{\sum_{l=1}^{\mathrm{L}} (w_{\mathrm{r},l}\, w_{\tau,l}\, g_l)^2}. \quad (2.8)$$

The weight $w_{\mathrm{r},l}$ models the attenuation of a point-source-like propagation $\frac{1}{r}$. The reference distance is the distance to the closest loudspeaker at the evaluated listening position, thus the weight of each loudspeaker results in

$$w_{\mathrm{r},l} = \frac{1}{r_l}. \quad (2.9)$$

The incorporation of delays into the energy vector requires a transformation that yields the weights $w_{\tau,l}$ for each loudspeaker. It is reasonable that these weights attenuate the lagging signals in order to reduce their influence on the predicted direction. An attenuation of $\frac{1}{4}\frac{\mathrm{dB}}{\mathrm{ms}}$ is known from the echo threshold in [22], similarly [23], and has successfully been applied for the prediction of localization in rooms [24]. The weight of each loudspeaker is calculated as $\tau_l = \frac{c}{r_l}$ in seconds at the listening position under test

$$w_{\tau,l} = 10^{\frac{-1000}{4\cdot 20}\tau_l}. \quad (2.10)$$

Further weights can be applied in order to model the precedence effect in more detail, as proposed by Stitt [25, 26]. Listening test results in [27] compared the differently complex extensions of the energy vector and revealed that the simple weighting with $w_{\mathrm{r},i}$ and $w_{\tau,i}$ is sufficient for a rough prediction of the perceived direction in typical playback scenarios.

The left side of Fig. 2.9 shows the predicted directions by the energy vector for various listening positions when playing back the same signal on a standard stereo loudspeaker pair with a radius of 2.5 m. The absolute localization error can be calculated from the difference of the predicted direction and the desired panning direction. The right side of Fig. 2.9 depicts areas with localization errors within 4 ranges: 0° ... 10° (white, perfect localization), 10° ... 30° (light gray, plausible localization), 30° ... 90° (gray, rough localization), and >90° (dark gray, poor localization).

Concerning a single playback scenario, i.e. a single panning direction on a loudspeaker setup, the perceptual sweet area for plausible playback can be estimated by the area with localization errors below 30°. For the prediction of a more general sweet area, the absolute localization errors can be computed for all possible panning directions in a fine grid of 1° and averaged at each listening position as shown in Fig. 2.10.

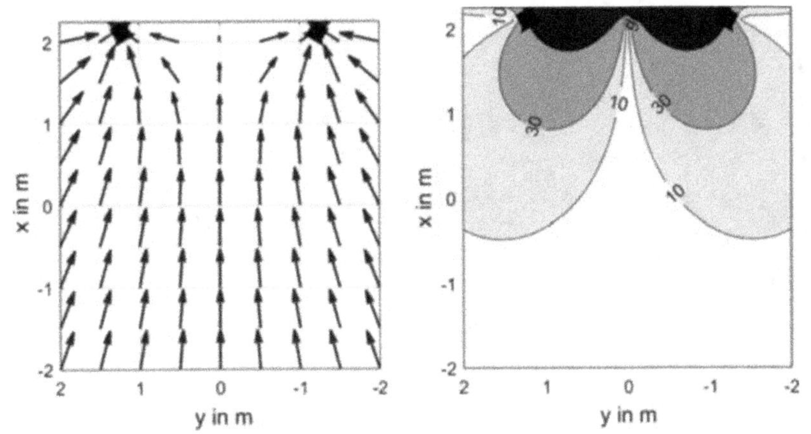

Fig. 2.9 Predictions of perceived directions by the energy vector for different listening positions in a standard stereo setup with two loudspeakers playing the same signal. Gray-scale areas on the right indicate listening areas with predicted absolute localization errors within different angular ranges

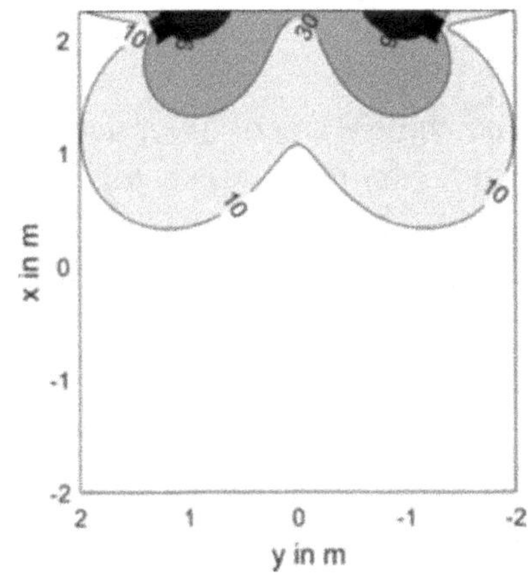

Fig. 2.10 Predictions of mean absolute localization errors by the energy vector in a standard stereo setup for panning directions between −30° and 30°

2.3 Width

M. Frank [10] investigated the auditory source width for frontal loudspeaker pairs with 0 dB level difference and various aperture angles, as well as the influence of an additional center loudspeaker on the auditory source width. The response was given by reading numbers off a left-right symmetric scale written on the loudspeaker arrangement (Fig. 2.11).

Figure 2.11 (right) shows the statistical analysis of the responses. Obviously the additional center loudspeaker decreases the auditory source width.

Auditory source with is difficult to compare for different directions and also single loudspeakers yield auditory source widths that vary with direction. Still, a relatively constant auditory source width is desirable for moving auditory events. For static auditory events, the narrowest-possible extent can be desirable.

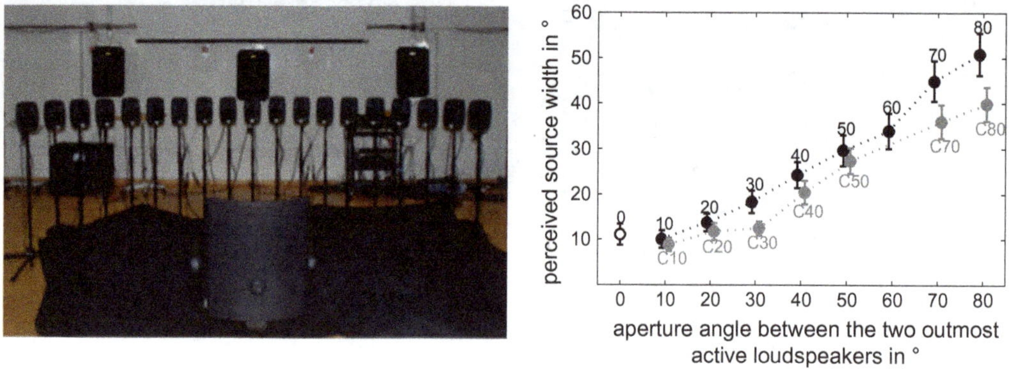

Fig. 2.11 Experimental setup and results of experiments of M. Frank (confidence intervals) about auditory source width of frontal stereo pairs of the angles $\pm5°, \ldots, \pm40°$ and with an additional center loudspeaker (C)

2.3.1 Model of the Perceived Width

The angle $2 \arccos \|\boldsymbol{r}_\mathrm{E}\|$ describes the aperture of a cap cut off the unit sphere perpendicular to the $\boldsymbol{r}_\mathrm{E}$ vector, at its tip, from the origin, see Fig. 2.12. As the $\boldsymbol{r}_\mathrm{E}$ vector length is between 0 (unclear direction) and 1 (only one loudspeaker active), this angle stays between 180° and 0°.

M. Frank's experiments about the auditory source width [10, 28] showed that stereo pairs of larger half angles α were also heard as wider. The length of the $\boldsymbol{r}_\mathrm{E}$ vector gets shorter with the half angle α. In a symmetrical loudspeaker pair $\boldsymbol{\theta}_{12}^\mathrm{T} = (\cos\alpha, \pm\sin\alpha)$ with $g_1 = g_2 = 1$, the y coordinate of the $\boldsymbol{r}_\mathrm{E}$ vector cancels and its length is

$$\|\boldsymbol{r}_\mathrm{E}\| = r_{\mathrm{E,x}} = \cos\alpha.$$

The corresponding spherical cap is same size as the loudspeaker pair $2 \arccos \|\boldsymbol{r}_\mathrm{E}\| = 2\alpha$. However, only $\frac{5}{8}$ of the size was indicated by the listeners of the experiments, which yields the following estimator of the perceived width:

$$ASW = \tfrac{5}{8} \cdot \tfrac{180°}{\pi} \cdot 2 \arccos \|\boldsymbol{r}_\mathrm{E}\|. \tag{2.11}$$

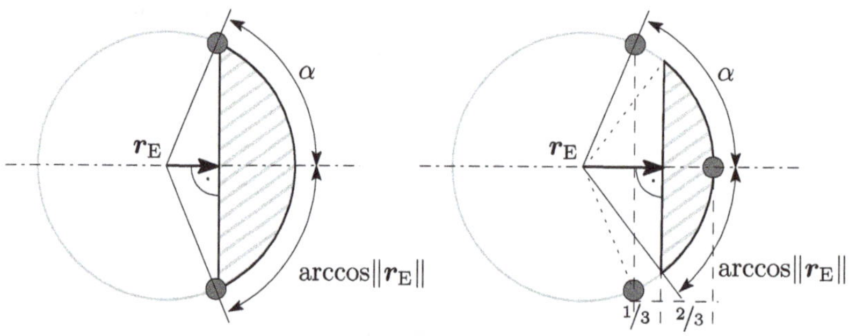

Fig. 2.12 Cap size associated with $\boldsymbol{r}_\mathrm{E}$ length model for L+R (left plot) and L+R+C (right plot)

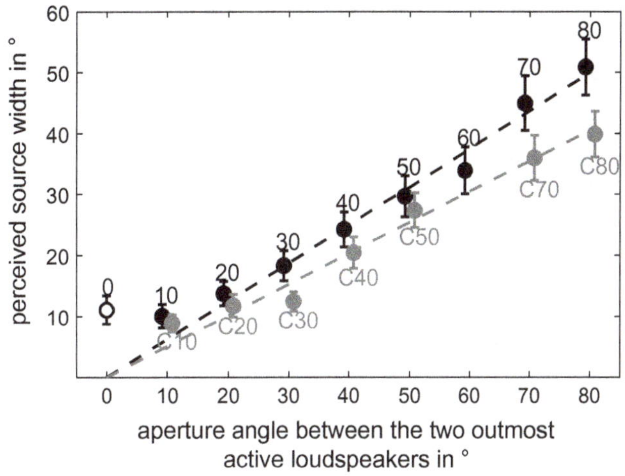

Fig. 2.13 Model of the perceived width as $\frac{5}{8}$ of the half-angle $\arccos \|\boldsymbol{r}_\mathrm{E}\|$ matches the half-angle of the experiment. Except for a lower limit, which is determined by the apparent source width (ASW) due to the room acoustical setting

For an additional center loudspeaker $g_3 = 1$, $\boldsymbol{\theta}^\mathrm{T} = (1, 0)$, the estimator yields

$$\|\boldsymbol{r}_\mathrm{E}\| = r_{\mathrm{E,x}} = \frac{1}{3} + \frac{2}{3}\cos\alpha,$$

an increase matching the experiments as $\arccos \|\boldsymbol{r}_\mathrm{E}\| < \alpha$, see Figs. 2.13 and 2.12.

2.4 Coloration

Despite research primarily focuses on the spatial fidelity of multi-loudspeaker playback, the overall quality of surround sound playback was found to be largely determined by timbral fidelity (70%) [29]. Loudspeakers in a studio or performance space are often characterized by different colorations that are caused by different reflection patterns (most often the wall behind the loudspeaker). When changing the active loudspeakers, or their number, these differences become audible. On the one hand, static coloration, e.g. the frequency responses of the loudspeakers, can typically be equalized. On the other hand, changes in coloration during the movement of a source cannot be equalized easily and yield annoying comb filters.

Although coloration is often assessed verbally [30], we employ a simple technical predictor based on the composite loudness level (CLL) by Ono [31, 32]. The CLL spectrum predicts the perceived coloration and is calculated from the sum of the loudnesses of both ears in each third-octave band. Studies about loudspeaker and headphone equalization show that differences in third-octave band levels of less than 1dB are inaudible by most listeners [33, 34]. This criterion can also be applied for the perception of coloration, i.e., differences between CLL spectra of less than 1dB are assumed to be inaudible.

Pairwise panning between loudspeakers results in a single active loudspeaker for source directions that coincide with the direction of a loudspeaker and two equally loud loudspeakers for source directions exactly between two neighboring loudspeakers, cf. Fig. 2.14. In the second case, the different propagation paths from the two loudspeakers to the ears create a comb filter. This comb filter is not present for sources

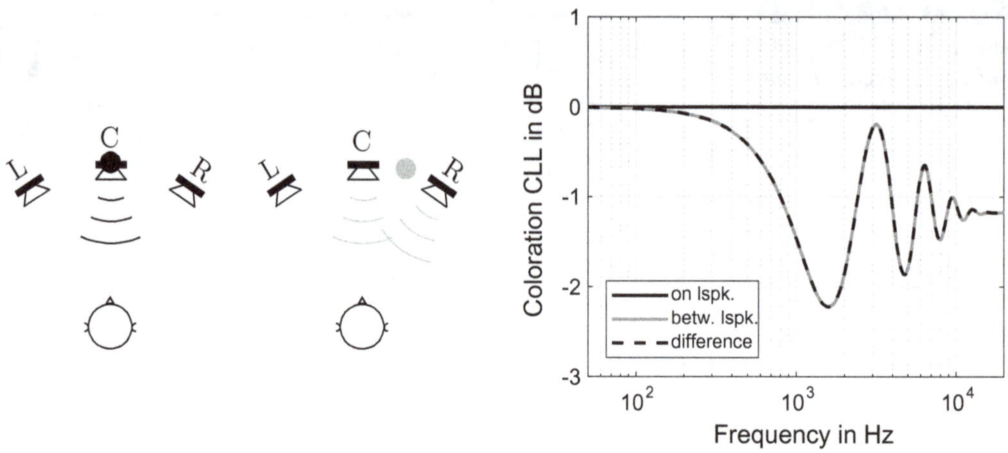

Fig. 2.14 Coloration predicted by composite loudness levels for a single loudspeaker C (black), two equally loud loudspeakers C and R (light gray), and their difference (dashed dark gray)

played from a single loudspeaker. Thus, moving a source between the two directions yields noticeable coloration. This is in contrast to static sources, for which Theile's experiments [35] indicated that they are perceived without coloration.

The actual shape of the afore-mentioned comb filter depends on the angular distance between the loudspeakers. The first notch and its depth decreases with the distance. This implies that coloration increases for playback with higher loudspeaker densities.

A similar comb filter is created when using a triplet of loudspeakers with the same loudspeaker density as the pair, e.g. L, C, R compared to C, R. In order to avoid a strong increase in source width or annoying phasing effects, the outmost loudspeakers L and R are strongly reduced in their level, typically around -12dB compared to loudspeaker C. In doing so, the similarity of the comb filters yields barely any coloration when moving a source between the two directions, cf. Fig. 2.15.

Judging from what is shown above, it appears beneficial to activate always a few loudspeaker to stabilize the coloration, as opposed to using just one loudspeaker and moving the playback to another one. Keeping the number of simultaneously active loudspeakers more or less constant does not only prevent coloration of source movements, it also yields a more constant source width. Because of this relation between coloration and source width, the fluctuation of $\|\boldsymbol{r}_\mathrm{E}\|$ is also a simple predictor of panning-dependent coloration.

In general, the strongest coloration is perceived under anechoic listening conditions. In reverberant rooms, the additional comb filters introduced by reflections help to conceal the comb filters due to multi-loudspeaker playback.

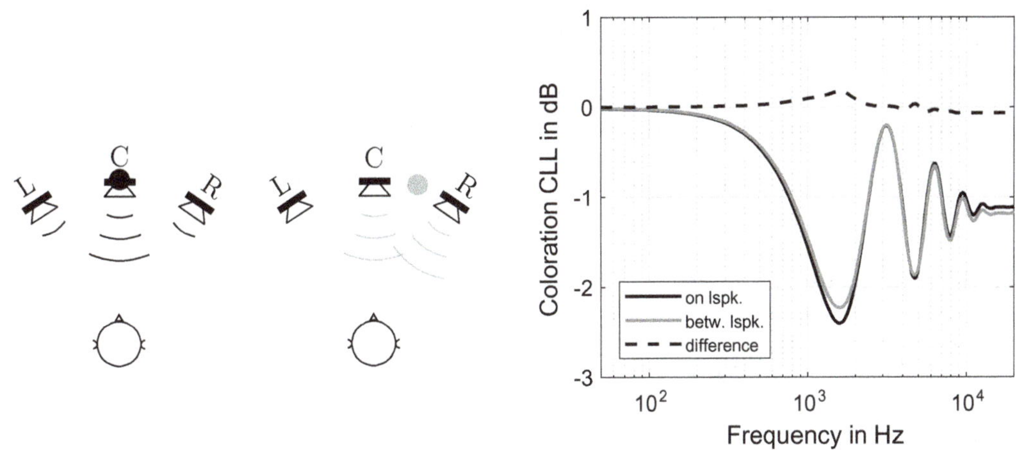

Fig. 2.15 Coloration predicted by composite loudness levels for loudspeaker C with additional −12 dB from L and R (black), two equally loud loudspeakers C and R (light gray), and their difference (dashed dark gray)

2.5 Open Listening Experiment Data

Experimental data from azimuthal localization in frontal and lateral loudspeaker pairs Figs. 2.3 and 2.4, azimuthal/elevational localization in horizontal, skew, and vertical frontal pairs Fig. 2.5, triangles Fig. 2.7, and quadrilaterals Fig. 2.8 are available online at https://opendata.iem.at in the listening experiment data project, as well as the data to the width experiment in Fig. 2.11.

The opendata.iem.at listening experiment data project contains evaluation routines to analyze the 95%-confidence intervals symmetrically based on means, standard deviations and the inverse Student's t-distribution `CIMEAN.m`, or more robustly based on median and inter-quartile ranges `CI2.m` and Student's t-distribution, or for two-dimensional data analysis `robust_multivariate_confidence_region.m`. The MATLAB script `plot_gathered_data.m` reads the formatted listening experiment data and its exemplary code generates figures like the above.

In order to support others providing own listening experiment data, the MATLAB functions `write_experimental_data.m` `read_experimental_data.m` are provided on the website.

References

1. M. Gerzon, General metatheory of auditory localization, in *prepr. 3306, Convention Audio Engineering Society* (1992)
2. D.M. Leakey E.C. Cherry, Influence of noise upon the equivalence of intensity differences and small time delays in two-loudspeaker systems. J. Acoust. Soc. Am. **29**, 284–286 (1957)
3. K. Wendt, Das Richtungshören bei Überlagerung zweier Schallfelder bei Intensitäts- und Laufzeitsterophonie, Ph.D. Thesis, RWTH-Aachen (1963)

4. G. Theile, G. Plenge, Localization of lateral phantom sources. J. Audio Eng. Soc. **25**(4), 96–200 (1977)
5. L. Simon, R. Mason, F. Rumsey, Localisation curves for a regularly-spaced octagon loudspeaker array, in *prepr. 7015, Convention Audio Engineering Society* (2009)
6. T. Kimura, H. Ando, 3D audio system using multiple vertical panning for large-screen multiview 3D video display. ITE Trans. Media Technol. Appl. **2**(1), 33–45 (2014)
7. F. Wendt, M. Frank, F. Zotter, Panning with height on 2, 3, and 4 loudspeakers, in *Proceedings of the ICSA* (Erlangen, 2013)
8. H. Lee, F. Rumsey, Level and time panning of phantom images for musical sources. J. Audio Eng. Soc. **61**(12), 978–988 (2013)
9. J.M. Helm, E. Kurz, M. Frank, Frequency-dependent amplitude-panning curves, in *29th Tonmeistertagung* (Köln, 2016)
10. M. Frank, Phantom sources using multiple loudspeakers in the horizontal plane, Ph.D. Thesis, Kunstuni Graz (2013)
11. M. Frank F. Zotter, Extension of the generalized tangent law for multiple loudspeakers, in *Fortschritte der Akustik - DAGA* (Kiel, 2017)
12. V. Pulkki, Virtual sound source positioning using vector base amplitude panning. J. Audio Eng. Soc. **45**(6), 456–466 (1997)
13. G. Soulodre, Evaluation of objective measures of loudness, in *prepr. 6161, 116th AES Convention* (Berlin, 2004)
14. M.-V. Laitinen, J. Vilkamo, K. Jussila, A. Politis, V. Pulkki, Gain normalization in amplitude panning as a function of frequency and room reverberance, in *Proceedings 55th AES Conference* (Helsinki, 2014)
15. V. Pulkki, Compensating displacement of amplitude-panned virtual sources, in *22nd Conference Audio Engineering Society* (2003)
16. G. Martin, W. Woszczyk, J. Corey, R. Quesnel, Sound source localization in a five-channel surround sound reproduction system, in *prepr. 4994, Convention Audio Engineering Society* (1999)
17. F. Zotter, M. Frank, Generalized tangent law for horizontal pairwise amplitude panning, in *Proceedings of the 3rd ICSA, Graz* (2015)
18. T. Kimura, H. Ando, Listening test for three-dimensional audio system based on multiple vertical panning, in *Aoustics-12, Hong Kong* (2012)
19. F. Wendt, Untersuchung von phantomschallquellen vertikaler lautsprecheranordnungen, M. Thesis, Kunstuni Graz (2013)
20. H. Clark, G. Dutton, P. Vanerlyn, The 'stereosonic' recording and reproduction system. *Repr. J. Audio Eng. Soc. Proc. Inst. Electr. Eng. 1957* **6**(2), 102–117 (1958)
21. V. Pulkki, Localization of amplitude-panned virtual sources ii: two- and three-dimensional panning. J. Audio Eng. Soc. **49**(9), 753–767 (2001)
22. B. Rakerd, W.M. Hartmann, J. Hsu, Echo suppression in the horizontal and median sagittal planes. J. Acoust. Soc. Am. **107**(2) (2000)
23. P.W. Robinson, A. Walther, C. Faller, J. Braasch, Echo thresholds for reflections from acoustically diffusive architectural surfaces. J. Acoust. Soc. Am. **134**(4) (2013)
24. F. Zotter, M. Frank, M. Kronlachner, J. Choi, Efficient phantom source widening and diffuseness in ambisonics, in *EAA Symposium on Auralization and Ambisonics* (Berlin, 2014)
25. P. Stitt, S. Bertet, M. van Walstijn, Extended energy vector prediction of ambisonically reproduced image direction at off-center listening positions. J. Audio Eng. Soc. **64**(5), 299–310 (2016)
26. P. Stitt, S. Bertet, M. Van Walstijn, Off-center listening with third-order ambisonics: dependence of perceived source direction on signal type. J. Audio Eng. Soc. **65**(3), 188–197 (2017)
27. E. Kurz, M. Frank, Prediction of the listening area based on the energy vector, in *Proceedings of the 4th International Conference on Spatial Audio (ICSA)* (Graz, 2017)
28. M. Frank, Source width of frontal phantom sources: perception, measurement, and modeling. Arch. Acoust. **38**(3), 311–319 (2013)

29. F. Rumsey, S. Zieliński, R. Kassier, S. Bech, On the relative importance of spatial and timbral fidelities in judgments of degraded multichannel audio quality. J. Acoust. Soc. Am. **118**(2), 968–976 (2005), http://link.aip.org/link/?JAS/118/968/1
30. S. Choisel, F. Wickelmaier, Evaluation of multichannel reproduced sound: scaling auditory attributes underlying listener preference. J. Acoust. Soc. Am. **121**(1), 388–400 (2007)
31. K. Ono, V. Pulkki, M. Karjalainen, Binaural modeling of multiple sound source perception: methodology and coloration experiments. Audio Eng. Soc. Conv. 111, **11** (2001), http://www.aes.org/e-lib/browse.cfm?elib=9884
32. K. Ono, V. Pulkki, M. Karjalainen, Binaural modeling of multiple sound source perception: Coloration of wideband sound. *Audio Eng. Soc. Conv. 112* **4** (2002), http://www.aes.org/e-lib/browse.cfm?elib=11331
33. A. Ramsteiner, G. Spikofski, *Ermittlung von Wahrnehmbarkeitsschwellen für Klangfarbenunterschiede unter Verwendung eines diffusfeldentzerrten Kopfhörers* (Fortschritte der Akustik, DAGA, 1987)
34. M. Karjalainen, E. Piirilä, A. Järvinen, J. Huopaniemi, Comparison of loudspeaker equalization methods based on DSP techniques. J. Audio Eng. Soc. **47**(1/2), 14–31 (1999), http://www.aes.org/e-lib/browse.cfm?elib=12117
35. G. Theile, Über die Lokalisation im überlagerten Schallfeld, Ph.D. dissertation, Technische Universität (Berlin, 1980)

Chapter 3
Amplitude Panning Using Vector Bases

The method is straightforward and can be used on many occasions succesfully.

Ville Pulkki [1], Ph.D. Thesis, 2001.

Abstract This chapter describes Ville Pulkki's famous vector-base amplitude panning (VBAP) as the most robust and generic algorithm of amplitude panning that works on nearly any surrounding loudspeaker layout. VBAP activates the smallest-possible number of loudspeakers, which gives a directionally robust auditory event localization for virtual sound sources, but it can also cause fluctuations in width and coloration for moving sources. Multiple-direction amplitude panning (MDAP) proposed by Pulkki is a modification that increases the number of activated loudspeakers. In this way, more direction-independence is achieved at the cost of an increased perceived source width and reduced localization accuracy at off-center positions. As vector-base panning methods rely on convex hull triangulation, irregular loudspeaker layouts yielding degenerate vector bases can become a problem. Imaginary loudspeaker insertion and downmix is shown as robust method improving the behavior, in particular for smaller surround-with-height loudspeaker layouts. The chapter concludes with some practical examples using free software tools that accomplish amplitude panning on vector bases.

Vector-base amplitude panning (VBAP) was extensively described and investigated in [2], alongside with the stabilization of moving sources by adding spread with multiple-direction amplitude panning (MDPA) [3]. Since then, VBAP and MDAP have been becoming the most common and popular amplitude panning techniques, which is particularly robust and can automatically adapt to specific playback layouts.

F. Zotter and M. Frank, *Ambisonics*, Springer Topics in Signal Processing 19,
https://doi.org/10.1007/978-3-030-17207-7_3

3.1 Vector-Base Amplitude Panning (VBAP)

Assuming the $\boldsymbol{r}_\mathrm{V}$ model to predict the perceived direction, an intended auditory event at a panning direction $\boldsymbol{\theta}$, we call it the *virtual source*, can theoretically be controlled by the criterion according to V. Pulkki [2]

$$\boldsymbol{\theta} = \sum_{l=1}^{\mathrm{L}} \tilde{g}_l\, \boldsymbol{\theta}_l. \tag{3.1}$$

Here, $\boldsymbol{\theta}_l$ are the direction vectors of the loudspeakers involved and the amplitude weights $\tilde{g}_l$ need to be normalized for constant loudness

$$g_l = \frac{\tilde{g}_l}{\sqrt{\sum_{l=1}^{\mathrm{L}} \tilde{g}_l^2}}. \tag{3.2}$$

Moreover, the weights g_l should always stay positive to avoid in-head localization or other irritating listening experiences. For loudspeaker rings around the horizon, always 1 or 2 loudspeakers will be contributing to the auditory event, for loudspeakers arranged on a surrounding sphere, always 1 up to 3 loudspeakers will be used, whose directions must enclose the direction of the desired auditory event, the virtal source. For the directional stability of the auditory event, the angle enclosed between the loudspeakers should stay smaller than 90°.

The system of equations for VBAP [2] uses 3 loudspeaker directions and gains to model the panning direction $\boldsymbol{\theta}$

$$\boldsymbol{\theta} = [\boldsymbol{\theta}_1,\, \boldsymbol{\theta}_2,\, \boldsymbol{\theta}_3] \begin{bmatrix} \tilde{g}_1 \\ \tilde{g}_2 \\ \tilde{g}_3 \end{bmatrix} = \mathbf{L} \cdot \tilde{\boldsymbol{g}} \qquad \Rightarrow \tilde{\boldsymbol{g}} = \mathbf{L}^{-1}\boldsymbol{\theta}, \qquad \boldsymbol{g} = \frac{\tilde{\boldsymbol{g}}}{\|\tilde{\boldsymbol{g}}\|}. \tag{3.3}$$

The selection of the activated loudspeaker triplet is preceded by forming all triplets of the convex hull spanned by all the given playback loudspeakers. To find the loudspeaker triplet that needs to be activated, the list of all triplets is being searched for the one with all-positive weights, $g_1 \geq 0$, $g_2 \geq 0$, $g_3 \geq 0$.

Figure 3.1 shows the localization curve for VBAP between a loudspeaker at 0° and 45° for a centrally seated listener and one shifted to the left. The experiment is described in [4] and results were gathered by a 1.8 m circle of 8 loudspeakers, and listeners indicated the perceived direction by naming numbers from a 5° scale mounted on the loudspeaker setup. Black whiskers of the results (95% confidence intervals and medians) for the centrally seated listener indicate a mismatch between slope of the perceived angles with VBAP; the ideal curve is represented by the dashed line and the mismatch can be understood by a better match of other exponents γ in Fig. 2.6. The directional spread is quite narrow. For an off-center left-shifted listening position the perceived directions is shown in terms of a 5° histogram (gray bubbles) in Fig. 3.1. For this off-center position, it becomes clear that the closest loudspeaker dominates localization within a third of the panning directions. Still, the directional

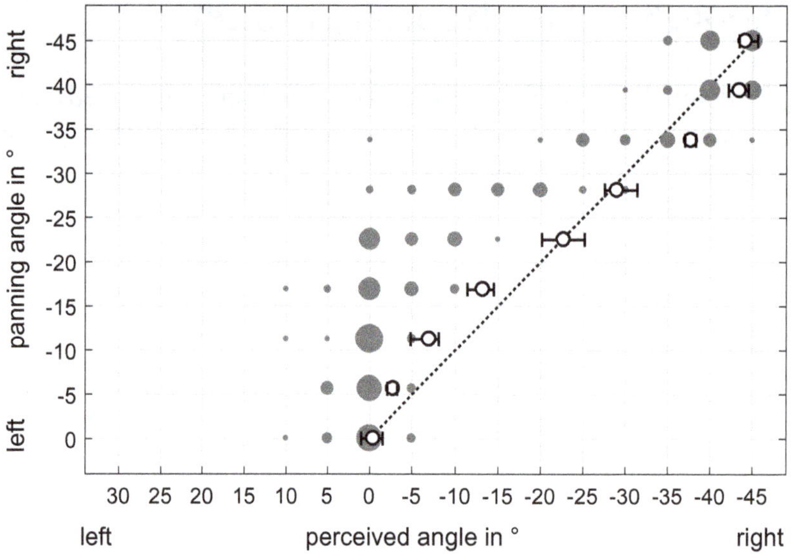

Fig. 3.1 Perceived directions for VBAP between loudspeakers at 0° and 45° from [4]. 95% confidence intervals and medians (black) are for a centrally seated listener in a circle of 2.5 m radius. Localization for left-shifted listener (1.25 m) can become bi-modal, so that 5° bubble histogram is shown (gray)

mapping seems to be monotonic with the panning angle, and the perceived direction stays within the loudspeaker pair, which is a robust result, at least.

In Fig. 3.2 we see that responses from [5] in which the panning angle was adjusted to match reference loudspeakers set up in steps of 15° on amplitude-panned lateral loudspeaker pairs fairly match the reference directions using VBAP. The $\boldsymbol{r}_\mathrm{E}$ vector model (black curve) delivers a better match with only one exception at 105°. This motivates VBIP as alternative strategy.

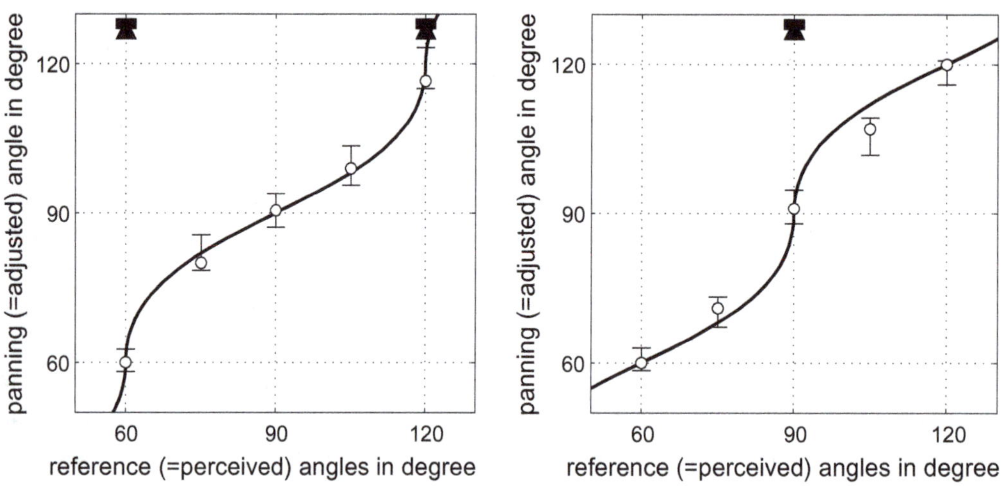

(a) Ring starting at 0°

(b) Ring starting at 30°

Fig. 3.2 VBAP angles on a 60°-spaced horizontal loudspeaker ring starting at 0° (**a**) or 30° (**b**), perceptually adjusted to match panned pink noise with harmonic-complex acoustic reference in 15° steps, from [5]; black curve shows $\boldsymbol{r}_\mathrm{E}$ model prediction

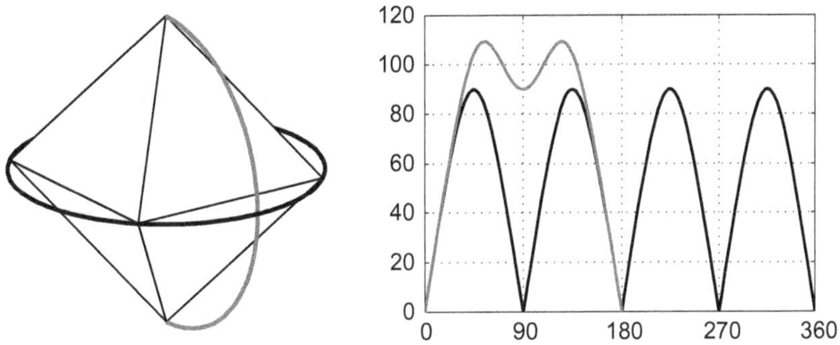

Fig. 3.3 The width measure $2\arccos\|\boldsymbol{r}_\mathrm{E}\|$ for a virtual source on a horizontal and a vertical trajectory (45° azimuth) using VBAP on an octahedral arrangement

Vector-Base Intensity Panning (VBIP). With nearly the same set of equations, but improving the perceptual mapping by the squares of the weights, the auditory event can be controlled corresponding to the direction of the $\boldsymbol{r}_\mathrm{E}$ vector

$$\boldsymbol{\theta} = [\boldsymbol{\theta}_1,\, \boldsymbol{\theta}_2,\, \boldsymbol{\theta}_3] \begin{bmatrix} \tilde{g}_1^2 \\ \tilde{g}_2^2 \\ \tilde{g}_3^2 \end{bmatrix} = \mathbf{L} \cdot \tilde{\boldsymbol{g}}_\mathrm{sq} \quad \Rightarrow \tilde{\boldsymbol{g}}_\mathrm{sq} = \mathbf{L}^{-1}\boldsymbol{\theta}, \quad \tilde{\boldsymbol{g}} = \begin{bmatrix} \sqrt{\tilde{g}_\mathrm{sq1}} \\ \sqrt{\tilde{g}_\mathrm{sq2}} \\ \sqrt{\tilde{g}_\mathrm{sq3}} \end{bmatrix}, \quad \boldsymbol{g} = \frac{\tilde{\boldsymbol{g}}}{\|\tilde{\boldsymbol{g}}\|}. \tag{3.4}$$

This formulation appears more contemporary due to the excellent match of the $\boldsymbol{r}_\mathrm{E}$ model to predict experimental results, as shown earlier.

Non-smooth VBAP/VBIP width. If one of the loudspeakers is exactly aligned with the virtual source for either VBAP or VBIP, e.g. $\boldsymbol{\theta}_1 = \boldsymbol{\theta}$, the resulting gains are $g_{1,2,3} = (1, 0, 0)$, and therefore only 1 loudspeaker will be activated. For a virtual source between the 2 loudspeakers, e.g. $\boldsymbol{\theta}_1 + \boldsymbol{\theta}_2 \propto \boldsymbol{\theta}$, then we obtain $g_{1,2,3} = (1, 1, 0)/\sqrt{2}$, and hereby only 2 loudspeakers will be active. This behavior in particular yields audible variation in the perceived width and coloration. For virtual source movements that cross a common edge of neighboring loudspeaker triplets, there will often be unexpectedly intense jumps that are quite pronounced.

Figure 3.3 illustrates the variation of the perceived width with VBAP on an octahedral arrangements of loudspeakers in the directions $\boldsymbol{\theta}_l^\mathrm{T} \in \{[\pm 1, 0, 0], [0, \pm 1, 0], [0, 0, \pm 1]\}$.

3.2 Multiple-Direction Amplitude Panning (MDAP)

In order to adjust the $\boldsymbol{r}_\mathrm{E}$ or $\boldsymbol{r}_\mathrm{V}$ vector not only directionally but also in length, and thus to control the number of active loudspeakers for moving sound objects, Pulkki extended VBAP to multiple-direction amplitude panning (MDAP [3]). Hereby not only the perceived width but also the coloration can be held constant.

Direction spread in MDAP. MDAP employs more than one virtual source distributed around the panning direction as a directional spreading strategy. For horizontal loudspeaker rings, MDAP can consist of a pair of virtual VBAP sources at the angle $\pm\alpha$ around the panning direction $\varphi_s \pm \alpha$. In a ring of L loudspeakers with uniform angular spacing of $\frac{360^\circ}{L}$, the angle $\alpha = 90\% \frac{180^\circ}{L}$ yields optimally flat width for all panning directions, as shown for L = 6 in comparison between MDAP and VBAP in Fig. 3.4. Moreover, MDAP seems to equalize the aiming of the $\boldsymbol{r}_E$ measure to the aiming of the $\boldsymbol{r}_V$ measure, which is the one controlled by VBAP and MDAP.

Listening experiment results. Experiments from [4] in Fig. 3.5 investigate the perceived width for two possible horizontal loudspeaker ring layouts, both with 45° spacings, but one starting at 0° ("0") the other at 22.5° ("1/2"). Widths of MDAP with a direction spread of $\alpha = 22.5^\circ$ are perceived as significantly similar on both

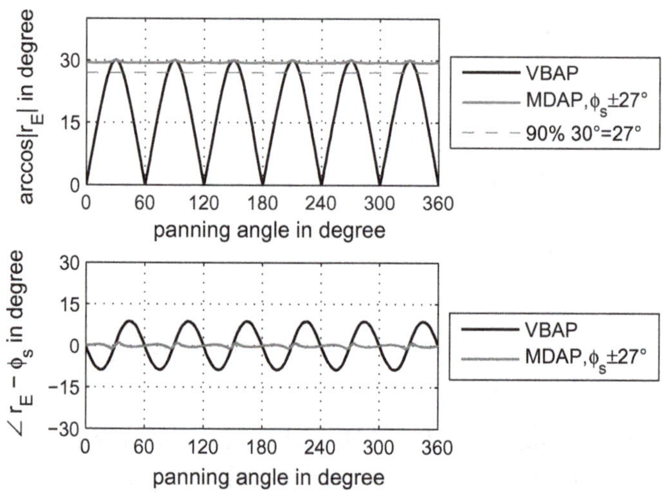

Fig. 3.4 Width-related angle $\arccos \|\boldsymbol{r}_E\|$ and angular error $\angle\boldsymbol{r}_E - \varphi_s$ for VBAP and MDAP, with MDAP using two virtual sources at $\pm 90\% \frac{180^\circ}{L}$ around the intended panning direction. This not only achieves minimal angular errors but also panning-invariant width

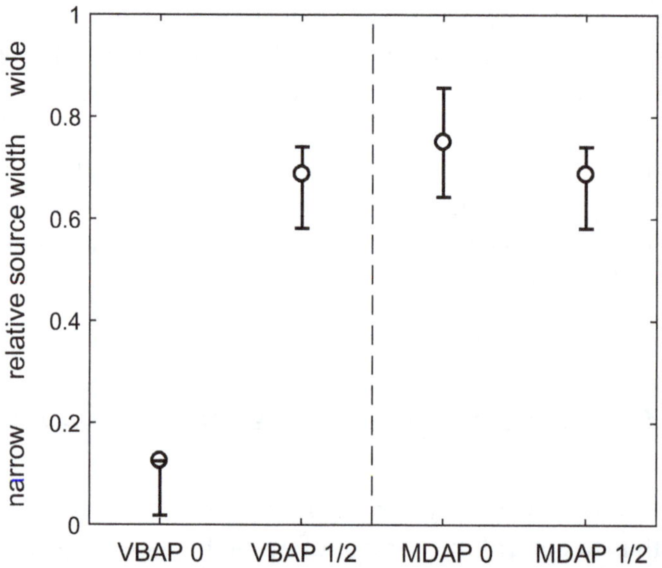

Fig. 3.5 Perceived width differences when using VBAP and MDAP for panning onto or between loudspeaker directions from listening experiment in [4]

ring layouts, while VBAP yields significantly narrower results for panning onto the frontal loudspeaker in the "0" layout, which activates a single loudspeaker, only. Note that VBAP1/2 and MDAP1/2 are identical with $\alpha = 22.5^\circ$ and were treated as one condition.

Moreover, a more constant width measure also describes a more constant number of activated loudspeakers while panning. Figure 3.6 shows that listeners can hear the difference in coloration changes with rotatory panning using pink noise and a constant speed. The figure shows that coloration fluctuations of MDAP are always clearly smaller than with VBAP on similar loudspeaker rings. Moreover, coloration changes are more pronounced on rings of 16 loudspeakers than with 8 loudspeakers, which is explained by their faster fluctuation.

Figure 3.7 shows the results from [6] for a central and left-shifted off-center listening position when using MDAP on an 8-channel ring of loudspeakers. At the central listening position, the perceived directional spread around the loudspeaker positions 0° and 45°, obviously increases as expected, as indicated by the whiskers (95% confidence intervals and medians). Moreover, the spread of MDAP seems to slightly decrease the slope mismatch between the underlying VBAP algorithm and the perceptual curve around the 22.5° direction.

Despite MDAP enforces a larger number of active loudspeakers, its localization is still similarly robust as the one of VBAP, also at on off-center listening positions. The perceived direction can be assumed to stay at least confined within a strictly directionally limited activation of loudspeakers. Correspondingly, the perceived directions shown in the gray 5°-histogram bubbles of Fig. 3.7 indicate the perceived directions when the listener is located left-shifted off-center. While localization is slightly attracted by the closer loudspeaker at 0°, the larger spread causes a more monotonic outcome that is less split than with VBAP in Fig. 3.1.

For a more exhaustive study, Frank used 6 loudspeakers on the horizon and gave the task to his listeners to align an MDAP pink-noise direction to match acoustical references every 15° (harmonic complex) by adjusting the panning direction [5]. The results in Fig. 3.8 contain 24 answers from 6 subjects responding four times

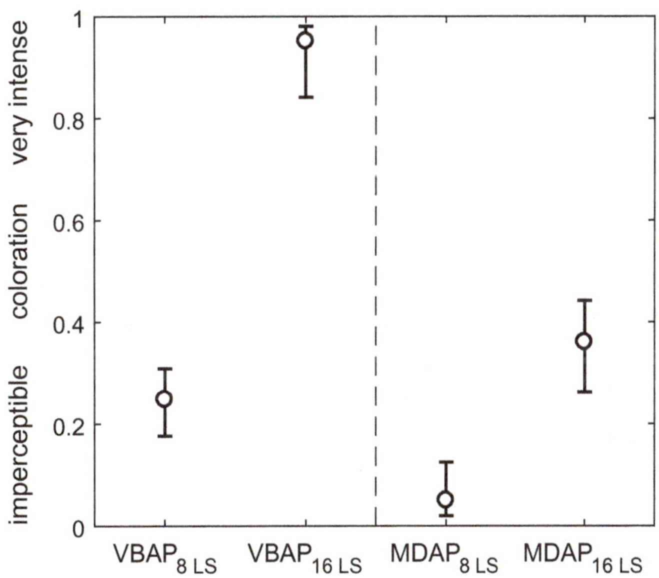

Fig. 3.6 Coloration change of horizontally moving sources with VBAP and MDAP on different loudspeaker aperture angles from listening experiment in [4]

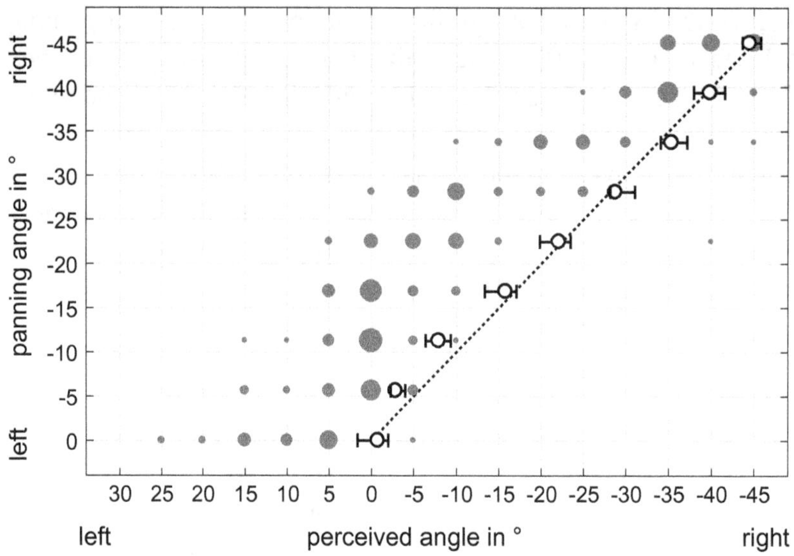

Fig. 3.7 Perceived directions for MDAP panning on an 8-channel 2.5 m radius loudspeaker ring within the interval [0°, 45°] at a central (black medians and 95%- confidence whiskers) and 1.25 m left-shifted off-center listening position (gray 5° bubble histogram); dashed line indicates ideal panning curve

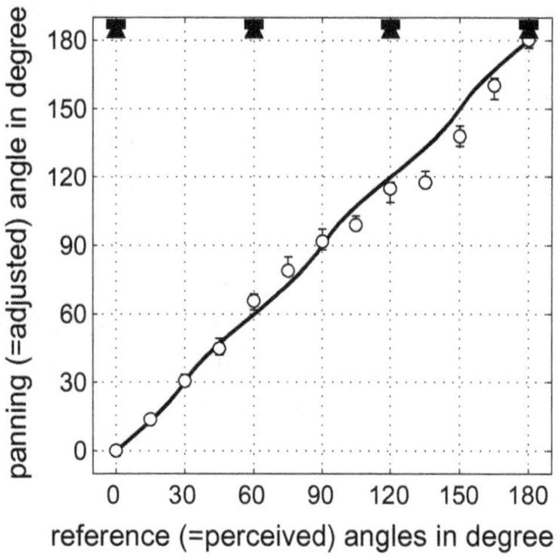

(a) Ring starting at 0°

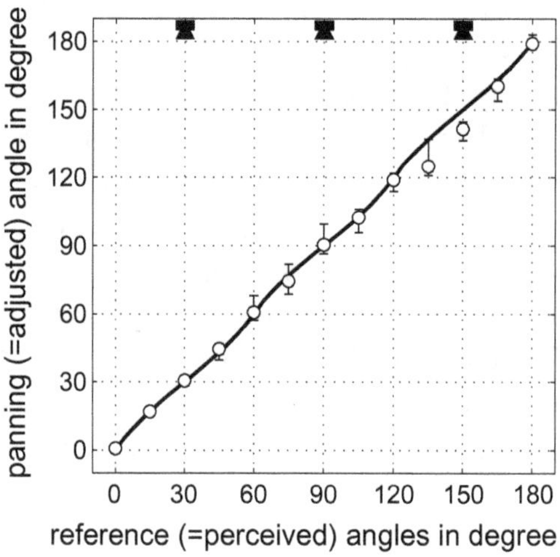

(b) Ring starting at 30°

Fig. 3.8 MDAP pink-noise directions on horizontal rings of 60°-spaced loudspeakers adjusted to perceptually match reference loudspeaker directions (harmonic complex) every 15°. Markers and whiskers indicate 95% confidence interals and medians, black curve the $\boldsymbol{r}_{\mathrm{E}}$ vector model

(by repetition and symmetrization). The black line shows directions indicated by the $\boldsymbol{r}_{\mathrm{E}}$ vector model for the tested conditions. Obviously, the confidence intervals of the adjusted MDAP angles match quite well both the reference directions and predictions by the $\boldsymbol{r}_{\mathrm{E}}$ vector model, in particular for angles between 0° and 90° (except 75°) for the ring starting at 0°, and from 0° to 120° for the 30°-rotated ring. The mismatch is much less than 4° for panning angles $\leq 90°$.

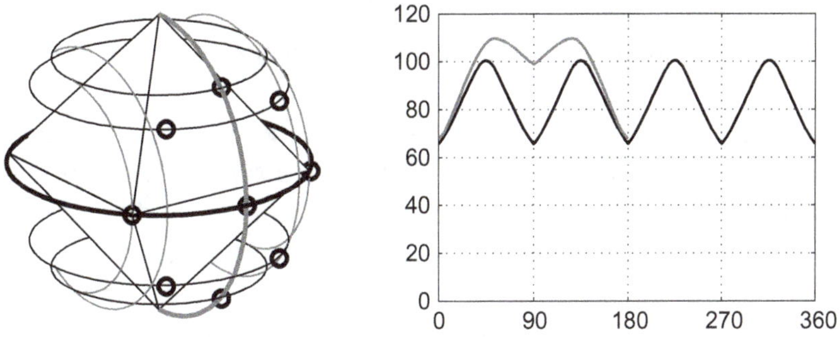

Fig. 3.9 The width measure 2 arccos $\|\boldsymbol{r}_{\mathrm{E}}\|$ for virtual sources on a horizontal and vertical path on an octahedron setup using MDAP with additional 8 half-amplitude virtual sources at 45° distance to the main virtual source

MDAP with 3D loudspeaker layouts. For more arbitrary 3D loudspeaker arrangements, multiple-directions could be arranged ring-like, see Fig. 3.9. This arrangement uses 8 additional virtual sources inclined by 45° wrt. the main virtual source.

At least mathematically, however, it requires to post optimize the amplitudes and angles of the virtual sources in order to accurately match the desired $\boldsymbol{r}_{\mathrm{V}}$ or $\boldsymbol{r}_{\mathrm{E}}$ vector in direction and length on irregular loudspeaker arrangements, cf. [7]. Non-uniform $\boldsymbol{r}_{\mathrm{V}}$ vector lengths of the individual virtual sources involved cause a distorted resultant vector. In particular, their superposition is distorted towards those of the multiple virtual source directions with the longest $\boldsymbol{r}_{\mathrm{V}}$ vectors. Epain's article [7] proposes optimization retrieving optimal orientation and weighting of the multiple virtual sources for every panning direction.

3.3 Challenges in 3D Triangulation: Imaginary Loudspeaker Insertion and Downmix

Surrounding loudspeaker hemispheres typically exhibit the following two problems, in most cases:

- Loudspeaker rectangles at the sides of standard setups with height (ITU-R BS.2051-0 [8]) can be decomposed ambiguously into triangles at the sides, back, and top. This can yield noticeable ranges within loudspeaker quadrilaterals, in which auditory events are unexpectedly created by just two of the loudspeakers.
- Signals of virtual sources below the horizon usually get lost.

The problem of unfavorable or ambiguous triangulations into loudspeaker triplets appears subtle, however, it can cause clearly audible deficiencies. Especially when ambiguous triangulation yields asymmetric behavior between left and right, e.g., for the top, rear, and lateral directions, where we would manually define loudspeaker quadrilaterals instead of triangles, see [9].

As surrounding loudspeaker hemispheres are typically open by 180° towards below, VBAP/ VBIP/ MDAP is numerically unstable and theoretically useless for any panning direction below. Despite the absence of loudspeakers below renders downwards amplitude panning theroretically infeasible, it is still reasonable to preserve signals of virtual sources that are meant for playback on spherically surrounding setups.

In the case of the asymmetric loudspeaker rectangles, see Fig. 3.10, and a missing lower hemisphere of surrounding loudspeakers, the insertion of one or more *imaginary loudspeakers* in the vertical direction (nadir) or in the middle of the rectangle (the average direction vector) has proven to be a useful strategy, e.g. in [10]. Any imaginary loudspeaker aims at either extending the admissible triangulation towards open parts of the surround loudspeaker setup, or to cover for parts with potential asymmetry, see [9].

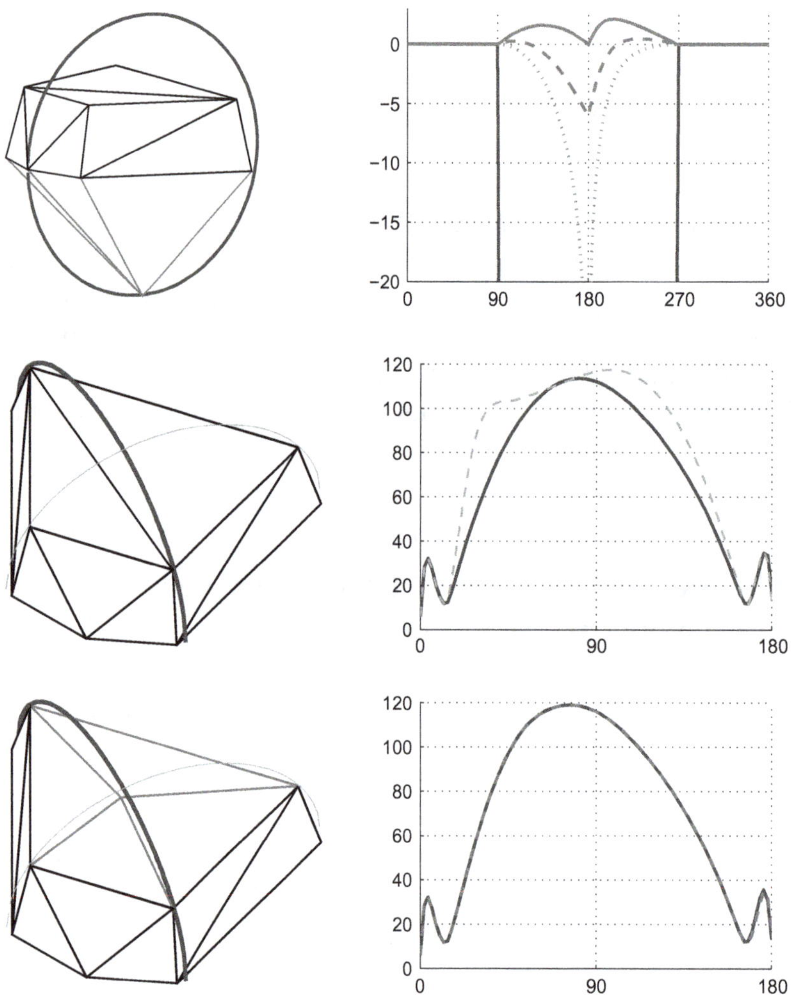

Fig. 3.10 VBAP on the ITU D (4 + 5 + 0) setup [8]. *Top row:* Insertion of imaginary loudspeaker at nadir preserves loudness of downward-panned signals, shown for vertical path and E values in dB for factors $\{\frac{1}{\sqrt{5}}, \frac{1}{2\sqrt{5}}, 0\}$ to re-distribute the signal to the 5 existing horizontal loudspeakers. *Middle row:* Due to typical triangulation, two left-right mirrored vertical paths (45° azimuth) yield asymmetric behavior, as shown by the $2 \arccos \|\boldsymbol{r}_\mathrm{E}\|$ measure. *Bottom row:* Insertion of imaginary loudspeaker at 65° fixes symmetry and feeds the signal with a factor $\frac{1}{\sqrt{4}}$ to the 4 existing neighbor loudspeakers

The signal of the imaginary loudspeaker can be dealt with in two ways

- it can be dismissed, e.g., for loudspeaker below at nadir, this would still yield a signal near the closest horizontal pair of loudspeakers for virtual sources panned to below-horizontal directions unless panned exactly to nadir
- it can be down-mixed to the neighboring M loudspeakers by a factor of $\frac{1}{\sqrt{M}}$, or less as in Fig. 3.10; alternatively for control yielding perfectly flat E measures, the resulting down-mixed gain vector can be re-normalized by Eq. (3.2).

3.4 Practical Free-Software Examples

3.4.1 VBAP/MDAP Object for Pd

There is a classic VBAP/MDAP implementation by Ville Pulkki that is available as external in pure data (Pd). The example in Fig. 3.11 illustrates its use together with some other useful externals in Pd. Software requirements are:

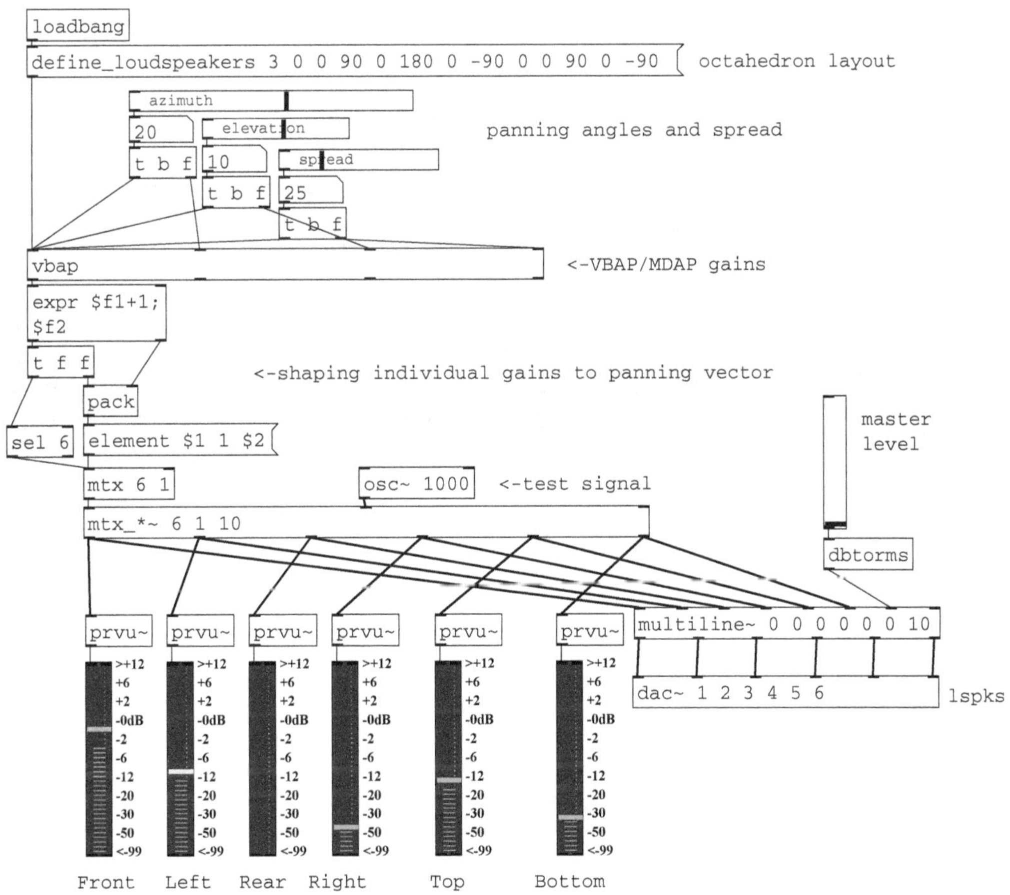

Fig. 3.11 Vector-Base/Multi-Direction Amplitude Panning (VBAP/MDAP) example in pure data (Pd) using Pulkki's [vbap] external for an octahedral layout

- `pure-data` (free, http://puredata.info/downloads/pure-data)
- `iemmatrix` (free download within pure-data)
- `zexy` (free download within pure-data)
- `vbap` (free download within pure-data).

3.4.2 SPARTA Panner Plugin

The `SPARTA Panner` under http://research.spa.aalto.fi/projects/sparta_vsts/plugins.html provides a vector-base amplitude panning interface (VBAP) and multiple-direction amplitude panning (MDAP), see Fig. 3.12, with frequency-dependent loudness normalization by $\sqrt[p]{\sum_{l=0}^{L} g_l^p}$ adjustable to the listening conditions, see Laitinen [11].

The parameter DTT can be varied between 0 (standard, frequency-independent VBAP normalization, i.e. diffuse-field normalization), 0.5 for typical listening environments, and 1 for the anechoic chamber. The plugin allows to either manually enter the azimuth and elevation angles of multiple panning directions (if more than one input signal is used) and for the playback loudspeakers, or import/export from/to preset files. Of course all panning directions can be time-varying and be moved per mouse, automations, or controls.

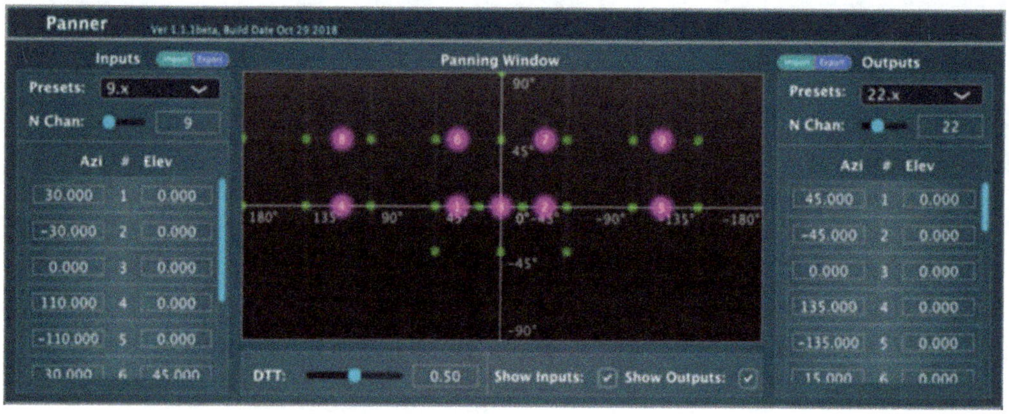

Fig. 3.12 The `Panner` VST plug-in from Aalto University's SPARTA plug-in suite manages Vector-Base Amplitude Panning within sequencers supporting VST

References

1. V. Pulkki, Spatial sound generation and perception by amplitude panning techniques, Ph.D. dissertation, Helsinki University of Technology (2001)
2. V. Pulkki, Virtual sound source positioning using vector base amplitude panning. J. Audio Eng. Soc. **45**(6), 456–466 (1997)
3. V. Pulkki, Uniform spreading of amplitude panned virtual sources, in *Proceedings of the Applications of Signal Processing to Audio and Acoustics (WASPAA), IEEE* (New Paltz, NY, 1999)
4. M. Frank, Phantom sources using multiple loudspeakers in the horizontal plane, Ph.D. Thesis, Kunstuni Graz (2013)
5. M. Frank, F. Zotter, Extension of the generalized tangent law for multiple loudspeakers, in *Fortschritte der Akustik - DAGA* (Kiel, 2017)
6. M. Frank, Source width of frontal phantom sources: perception, measurement, and modeling. Arch. Acoust. **38**(3), 311–319 (2013)
7. N. Epain, C.T. Jin, F. Zotter, Ambisonic decoding with constant angular spread. Acta Acust. United Acust. **100**(5) (2014)
8. ITU, *Recommendation BS.2051: Advanced sound system for programme production* (ITU, 2018)
9. M. Romanov, M. Frank, F. Zotter, T. Nixon, Manipulations improving amplitude panning on small standard loudspeaker arrangements for surround with height, in *Tonmeistertagung* (2016)
10. F. Zotter, M. Frank, All-round ambisonic panning and decoding. J. Audio Eng. Soc. (2012)
11. M.-V. Laitinen, J. Vilkamo, K. Jussila, A. Politis, V. Pulkki, Gain normalization in amplitude panning as a function of frequency and room reverberance, in *Proceedings 55th AES Conference* (Helsinki, 2014)

Chapter 4
Ambisonic Amplitude Panning and Decoding in Higher Orders

...the second-order Ambisonic system offers improved imaging over a wider area than the first-order system and is suitable for larger rooms.

Jeffrey S. Bamford [1], Canadian Acoustics, 1994.

Abstract Already in the 1970s, the idea of using continuous harmonic functions of scalable resolution was described by Cooper and then Gerzon, who introduced the name Ambisonics. This chapter starts by reviewing properties of first-order horizontal Ambisonics, using an interpretation in terms of panning functions. And the required mathematical formulations for 3D higher-order Ambisonics are developed here, with the idea to improve the directional resolution. Based on this formalism, ideal loudspeaker layouts can be defined for constant loudness, localization, and width, according to the previous models. The chapter discusses how Ambisonics can be decoded to less ideal, typical loudspeaker setups for studios, concerts, sound-reinforcement systems, and to headphones. The behavior is analyzed by a rich variety of listening experiments and for various decoding applications. The chapter concludes with example applications using free software tools.

Cooper [2] used higher-order angular harmonics to formulate circular panning of auditory events. Due to the work of Felgett [3], Gerzon [4], and Craven [5], the term Ambisonics became common for technology using spherical harmonic functions. Around the early 2000s, most notably Bamford [6], Malham [7], Poletti [8], Jot [9], and Daniel [10] pioneered the development of higher-order Ambisonic panning and decoding, Ward and Abhayapala [11], Dickens [12], and at the lab of the authors Sontacchi [13].

Another leap happened around 2010, when Ambisonic decoding to loudspeakers could be largely improved by considering regularization methods [14], singular-value decomposition [15], and all-round Ambisonic decoding (AllRAD) [15, 16],

F. Zotter and M. Frank, *Ambisonics*, Springer Topics in Signal Processing 19,
https://doi.org/10.1007/978-3-030-17207-7_4

a combination of vector-base panning techniques with Ambisonics, yielding the most robust and flexible higher-order decoding method known today.

For headphones, after the work of Jot [9] that outlined the basic problems of binaural decoding in the 1990s, Sun, Bernschütz, Ben-Hur, and Brinkmann [17–19] made important contributions to binaural decoding, and we consider TAC and MagLS decoders by Zaunschirm and Schörkhuber [20, 21] as the essential binaural decoders. Both remove HRTF delays or optimize HRTF phases at high frequencies to avoid spectral artifacts. By interaural covariance correction, MagLS/TAC manage to play back diffuse fields consistently, using the formalism of Vilkamo et al [22].

4.1 Direction Spread in First-Order 2D Ambisonics

In 2D first-order Ambisonics as discussed in Chap. 1, the directional mapping of a single sound source from the angle φ_s to the direction of each loudspeaker φ is described by the shape of panning function (or direction-spread function) in Eq. (1.17). The directional spreading is not infinitely narrow, but determined by what can be represented by first-order directivity patterns. Consequently, sound from the angle φ_s will be mapped by a dipole pattern aligned with the source and an additional omnidirectional pattern. We can involve a spread parameter a to make the directional spread to the loudspeakers system adjustable and either cardioid-shaped $a = 1$, 2D-supercardioid-shaped $a = \sqrt{2}$, or 2D-hypercardioid-shaped $a = 2$, using:

$$g(\varphi) = 1 + a \cos(\varphi - \varphi_S). \tag{4.1}$$

This function represents how first-order Ambisonic panning would distribute a mono signal to loudspeakers. With the loudspeaker positions described by the set of angles $\{\varphi_l\}$, a vector of amplitude-panning gains with an entry for each loudspeaker could be determined by sampling the direction-spread function:

$$\boldsymbol{g} = \begin{bmatrix} g_1 \\ \vdots \\ g_L \end{bmatrix} = 1 + a \begin{bmatrix} \cos(\varphi_1 - \varphi_S) \\ \vdots \\ \cos(\varphi_L - \varphi_S) \end{bmatrix}. \tag{4.2}$$

With these gain values, we evaluate models of perceived loudness, direction, and width, as introduced in Chap. 2, in order to enter a discussion of perceptual goals.

If the loudspeaker directions $\{\boldsymbol{\theta}_l\}$ are chosen suitably, it is possible to obtain panning-independent loudness, direction, and width measures $E = \sum_l g_l^2$, $\boldsymbol{r}_E = \frac{1}{E}\sum_l g_l^2 \boldsymbol{\theta}_l$, and $\frac{5}{8}\frac{180^\circ}{\pi}\, 2 \arccos \|\boldsymbol{r}_E\|$. How is it done?

For first-order 2D Ambisonics, it is theoretically optimal to use at least a ring of 4 loudspeakers with uniform angular spacing and $a = \sqrt{2}$, which is easily checked with the aid of a computer, cf. Fig. 4.1, and explained below and in Sect. 4.4.

Direction spread in FOA. The panning-function interpretation with its directional spread has some similarity to MDAP, with its attempt to directionally spread an amplitude-panned signal. Similar to the discrete virtual spread by $\pm\alpha = \arccos \|\boldsymbol{r}_E\|$

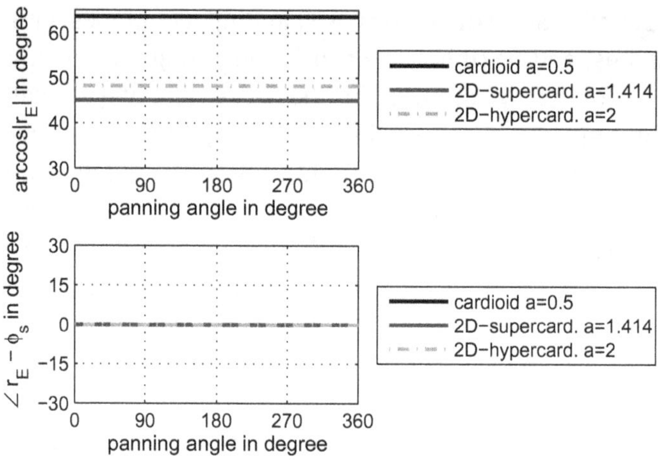

Fig. 4.1 Width-related angle $\arccos \|\boldsymbol{r}_\mathrm{E}\|$ and angular error $\angle \boldsymbol{r}_\mathrm{E} - \varphi_\mathrm{s}$ for first-order Ambisonics on a ring with 90° spaced loudspeakers using a cardioid, or 2D super-/hyper-cardioid patterns; the measures are all panning-invariant and the super-cardioid weighting is the 2D max-$\boldsymbol{r}_\mathrm{E}$ weighting with $\arccos \|\boldsymbol{r}_\mathrm{E}\| = 45°$

around the panning direction. The virtual direction spread of first-order Ambisonics is described by its continuous panning function $g(\varphi)$ in Eq. (4.1). To inspect the continuous function by the $\boldsymbol{r}_\mathrm{E}$ measure defined in Eq. (2.7), we may evaluate an integral over the panning function instead of the sum. Because of the symmetry around φ_s, we may set for convenience $\varphi_\mathrm{s} = 0$, which knowingly causes $r_{\mathrm{E,y}} = 0$, and evaluate

$$r_{\mathrm{E,x}} = \frac{\int_0^{2\pi} g^2(\varphi)\, \cos\varphi\, \mathrm{d}\varphi}{\int_0^{2\pi} g^2(\varphi)\, \mathrm{d}\varphi} = \frac{\int_0^{\pi} [1 + 2a\cos\varphi + a^2\cos^2\varphi]\, \cos\varphi\, \mathrm{d}\varphi}{\int_0^{\pi} [1 + 2a\cos\varphi + a^2\cos^2\varphi]\, \mathrm{d}\varphi} = \frac{a}{1 + \frac{a^2}{2}}. \tag{4.3}$$

The maximum of $r_{\mathrm{E,x}} = \frac{2a}{2+a^2}$ is found by $\frac{\mathrm{d}}{\mathrm{d}a} r_{\mathrm{E,x}} = \frac{4+2a^2-4a^2}{2+a^2} = 0$, hence at $a = \sqrt{2}$. Consequently, the 2D max-$\boldsymbol{r}_\mathrm{E}$ weight is $r_{\mathrm{E,x}} = \frac{\sqrt{2}}{2} = \frac{1}{\sqrt{2}}$ and yields the angle $\arccos \|\boldsymbol{r}_\mathrm{E}\| = 45°$. This would resemble a 2D-MDAP-equivalent source spread to ±45°. Note that first-order Ambisonics cannot map to a smaller spread than this. Only higher orders permit to further reduce this spread to a desired angle below 90°.

Ideal loudspeaker layouts. Not only is the directional aiming of the virtual, continuous first-order Ambisonic panning function ideal and its width panning-invariant, also its loudness measure is panning-invariant. However, decoding to a physical loudspeaker setup can degrade the ideal behavior. For which loudspeaker layout are these properties preserved by sampling decoding?

The 2D first-order Ambisonic components (W, X, Y) correspond to $\{1, \cos\varphi, \sin\varphi\}$ patterns, a first-order Fourier series in the angle. Sampling the playback directions by $\mathrm{L} = 3$ uniformly spaced loudspeakers on the horizon, the sampling theorem for this series is already fulfilled. Accordingly, Parseval's theorem ensures panning-invariant loudness E for any panning direction.

For an ideal $\boldsymbol{r}_\mathrm{E}$ measure, however, one more loudspeaker is required $\mathrm{L} \geq 4$ for a uniformly spaced horizontal ring. To explain this increase exhaustively, the concept of circular/spherical polynomials and t-designs will be introduced in this chapter. For a brief explanation, $g^2(\varphi)$ is a second-order expression and therefore to represent the ideal constant loudness $E = \int g^2(\varphi)\, \mathrm{d}\varphi$ of the continuous panning function consistently after discretization $E = \frac{2\pi}{\mathrm{L}} \sum_l g_l^2$, it requires $\mathrm{L} = 3$ uniformly spaced loudspeakers, as argued before. By contrast, the expressions $g^2(\varphi)\cos\varphi$ and $g^2(\varphi)\sin\varphi$ are third-order and appear in $\boldsymbol{r}_\mathrm{E}\cdot E=\int g^2(\varphi)\,[\cos\varphi,\ \sin\varphi]^\mathrm{T}\mathrm{d}\varphi$. Consequently, ideal mapping of $\boldsymbol{r}_\mathrm{E}$ (direction and width) requires at least one more loudspeaker $\mathrm{L}=4$ for a uniformly spaced arrangement to make the continuous and the discretized form $\boldsymbol{r}_\mathrm{E}\,E=\frac{2\pi}{\mathrm{L}}\sum_l g_l^2\,[\cos\varphi_l,\ \sin\varphi_l]^\mathrm{T}$ perfectly equal.

Towards a higher-order panning function. An Nth-order cardioid pattern is obtained from the cardioid pattern by taking its Nth power

$$g_\mathrm{N}(\varphi)=\frac{1}{2^\mathrm{N}}(1+\cos\varphi)^\mathrm{N},$$

which makes it narrower. With $\mathrm{N}=2$, this becomes, using $\cos^2\varphi=\frac{1}{2}(1+\cos 2\varphi)$,

$$g_2(\varphi)=\frac{1}{4}(1+2\cos\varphi+\cos^2\varphi)=\frac{1}{8}(3+4\cos\varphi+\cos 2\varphi).$$

More generally, Chebyshev polynomials $T_m(\cos\varphi)=\cos m\varphi$, cf. [23, Eq. 3.11.6] can be used to argue that there is always a fully equivalent cosine series describing the higher-order 2D panning function in the azimuth angle

$$g(\varphi)=\sum_{m=0}^{\mathrm{N}}a_m\,\cos m\varphi. \tag{4.4}$$

Rotated panning function. In first-order Ambisonics, panning functions consist of an omnidirectional part, $\cos(0\varphi)=1$, and a figure-of-eight to x, $\cos\varphi$, but that was not all: Recording and playback also required a figure-of-eight pattern to y, $\sin\varphi$. The additional component allows to express rotated first-order directivities by a basis set of fixed directivities. For higher orders, a panning function rotated to a non-zero aiming $\varphi_\mathrm{s}\neq 0$

$$g(\varphi-\varphi_s)=\sum_{m=0}^{\mathrm{N}}a_m\,\cos[m(\varphi-\varphi_\mathrm{s})] \tag{4.5}$$

can be re-expressed by the addition theorem $\cos(\alpha+\beta)=\cos\alpha\,\cos\beta-\sin\alpha\sin\beta$ into a series involving the sinusoids (odd symmetric part of a Fourier series),

$$\begin{aligned} g(\varphi-\varphi_s)&=\sum_{m=0}^{\mathrm{N}}a_m\,[\cos m\varphi_\mathrm{s}\,\cos m\varphi+\sin m\varphi_\mathrm{s}\,\sin m\varphi] \\ &=\sum_{n=0}^{\mathrm{N}}a_m^{(c)}\,\cos m\varphi+\sum_{m=0}^{\mathrm{N}}a_m^{(s)}\,\sin m\varphi. \end{aligned} \tag{4.6}$$

We conclude: *Higher-order Ambisonics in 2D (and the associated set of theoretical microphone directivities) is based on the Fourier series in the azimuth angle* φ.

4.2 Higher-Order Polynomials and Harmonics

The previous section required that direction and length of the $\boldsymbol{r}_{\mathrm{E}}$ vector resulting from amplitude panning on loudspeakers matched the desired auditory event direction and width. Harmonic functions with strict symmetry around a panning direction $\boldsymbol{\theta}_{\mathrm{s}}$ will help us in achieving this goal and in defining good sampling.

Regardless of the dimensions, be it in 2D or 3D, we desire to define continuous and resolution-limited axisymmetric functions around the panning direction $\boldsymbol{\theta}_{\mathrm{s}}$ to fulfill our perceptual goals of a panning-invariant loudness E, width $\|\boldsymbol{r}_{\mathrm{E}}\|$, and perfect alignment between panning direction $\boldsymbol{\theta}_{\mathrm{s}}$ and localized direction $\boldsymbol{r}_{\mathrm{E}}$. Then we hope to find suitable directional discretization schemes for ideal loudspeaker layouts, so that the measures E and $\boldsymbol{r}_{\mathrm{E}}$ are perfectly reconstructed in playback.

The projection of a variable direction vector $\boldsymbol{\theta}$ onto the panning direction $\boldsymbol{\theta}_{\mathrm{s}}$ always yields the cosine of the enclosed angle $\boldsymbol{\theta}_{\mathrm{s}}^{\mathrm{T}}\boldsymbol{\theta} = \cos\phi$, no matter whether it is in two or three dimensions. Hereby constructing the panning function based on this projection readily meets the desired goals. The mth power thereof, $(\boldsymbol{\theta}_{\mathrm{s}}^{\mathrm{T}}\boldsymbol{\theta})^m = \cos^m\phi$ helps to build an Nth-order power series $g = \sum_{m=0}^{\mathrm{N}} a_m (\boldsymbol{\theta}_{\mathrm{s}}^{\mathrm{T}}\boldsymbol{\theta})^m$ to describe a virtual Ambisonic panning function.

For 2D, such a *circular polynomial* $g = \sum_{m=0}^{\mathrm{N}} a_m (\boldsymbol{\theta}_{\mathrm{s}}^{\mathrm{T}}\boldsymbol{\theta})^m$ contains all $(\mathrm{N}+1)(\mathrm{N}+2)/2$ mixed powers by $(\boldsymbol{\theta}_{\mathrm{s}}^{\mathrm{T}}\boldsymbol{\theta})^m = (\theta_{\mathrm{xs}}\theta_{\mathrm{x}} + \theta_{\mathrm{ys}}\theta_{\mathrm{y}})^m = \sum_{k=0}^{m} \binom{m}{k} (\theta_{\mathrm{xs}}\theta_{\mathrm{x}})^k (\theta_{\mathrm{ys}}\theta_{\mathrm{y}})^{m-k}$ of the direction vectors' entries $\boldsymbol{\theta}_{\mathrm{s}}^{\mathrm{T}} = [\theta_{\mathrm{xs}}, \theta_{\mathrm{ys}}]$ and $\boldsymbol{\theta} = [\theta_{\mathrm{x}}, \theta_{\mathrm{y}}]^{\mathrm{T}}$. However, we could already recognize that it only takes $2\mathrm{N}+1$ functions to express $g = \sum_{m=0}^{\mathrm{N}} a_m (\boldsymbol{\theta}_{\mathrm{s}}^{\mathrm{T}}\boldsymbol{\theta})^m = \sum_{m=0}^{\mathrm{N}} a_m \cos^m\phi$: First an initial polynomial with relative azimuth $\phi = \varphi - \varphi_{\mathrm{s}}$ relating to a harmonic series of $\mathrm{N}+1$ cosines or Chebyshev-polynomials $g = \sum_{m=0}^{\mathrm{N}} b_m \cos m\phi = \sum_{m=0}^{\mathrm{N}} b_m\, T_m(\boldsymbol{\theta}_{\mathrm{s}}\boldsymbol{\theta})$. Then, in terms of absolute azimuth φ, the trigonometric addition theorem re-expresses the series into one of $\mathrm{N}+1$ cosines and N sines, with $T_m(\boldsymbol{\theta}_{\mathrm{s}}\boldsymbol{\theta}) = \cos[m(\varphi-\varphi_{\mathrm{s}})] = \cos m\varphi_{\mathrm{s}} \cos m\varphi + \sin m\varphi_{\mathrm{s}} \sin m\varphi$. As shown in the upcoming section, we can alternatively obtain such orthonormal harmonic functions by solving a second-order differential equation that is generally used to define harmonics, which bears the later benefit that we can use the approach to define spherical harmonics in three space dimensions.

Spherical polynomials are similar, $g = \sum_{n=0}^{\mathrm{N}} a_n\, (\boldsymbol{\theta}_{\mathrm{s}}^{\mathrm{T}}\boldsymbol{\theta})^n$, involving the expressions $(\boldsymbol{\theta}_{\mathrm{s}}^{\mathrm{T}}\boldsymbol{\theta})^n = (\theta_{\mathrm{xs}}\theta_{\mathrm{x}} + \theta_{\mathrm{ys}}\theta_{\mathrm{y}} + \theta_{\mathrm{zs}}\theta_{\mathrm{z}})^n = \sum_{k=0}^{n} \sum_{l=0}^{n-k} \binom{n}{k}\binom{l}{n-k} (\theta_{\mathrm{zs}}\theta_{\mathrm{z}})^k (\theta_{\mathrm{xs}}\theta_{\mathrm{x}})^l (\theta_{\mathrm{ys}}\theta_{\mathrm{y}})^{n-k-l}$. Again, all these $(\mathrm{N}+1)(\mathrm{N}+2)(\mathrm{N}+3)/6$ combinations would be too many to form an orthogonal set of basis functions. Moreover, while the different cosine harmonics are orthogonal axisymmetric functions in 2D, they are not in 3D. On the sphere, the $\mathrm{N}+1$ orthogonal Legendre polynomials $P_n(\cos\phi)$ replace the cosine series as a basis for $g = \sum_{n=0}^{\mathrm{N}} c_n\, P_n(\cos\phi)$, as shown below. All mathematical derivations for the sphere rely on the definition of harmonics. They result in $(\mathrm{N}+1)^2$ spherical harmonics and their addition theorem as a basis in terms of absolute directions $\frac{2n+1}{4\pi} P_n(\boldsymbol{\theta}_{\mathrm{s}}^{\mathrm{T}}\boldsymbol{\theta}) = \sum_{m=-n}^{n} Y_n^m(\boldsymbol{\theta}_{\mathrm{s}}) Y_n^m(\boldsymbol{\theta})$. Dickins' thesis is interesting for further reading [12].

In both regimes, 2D and 3D, the circular or spherical polynomials concept will be used to determine optimal layouts, so-called t-designs. Such t-designs are directional sampling grids that are able to keep the information about the constant part of any either circular (2D) or spherical (3D) polynomials up to the order $\mathrm{N} \leq t$. This will be a mathematical key property exploited to determine requirements for preserving E and $\boldsymbol{r}_\mathrm{E}$ measures during Ambisonic playback with optimal loudspeaker setups, but not only. Also t-designs simplify numerical integration of circular or spherical harmonics to define state-of-the-art Ambisonic decoders or mapping effects.

4.3 Angular/Directional Harmonics in 2D and 3D

The Laplacian is defined in the D-dimensional Cartesian space as

$$\triangle = \sum_{j=1}^{\mathrm{D}} \frac{\partial^2}{\partial x_j^2}, \tag{4.7}$$

and for any function f, the Laplacian $\triangle f$ describes the curvature. Any harmonic function is proportional to its curvature by an eigenvalue λ,

$$\triangle f = -\lambda\, f, \tag{4.8}$$

and therefore is an oscillatory function. Generally, eigensolutions $\triangle f = -\lambda\, f$ to the Laplacian are called *harmonics*. For suitable eigenvalues λ, harmonics span an orthogonal set of basis functions that are typically used for Fourier expansion on a finite interval. It seems desirable to find such harmonics for functions only exhibiting directional dependencies, i.e. in the azimuth angle φ in 2D, and azimuth and zenith angle φ, ϑ in 3D.

4.4 Panning with Circular Harmonics in 2D

For 2 dimensions Appendix A.3.2 uses the generalized chain rule to convert the Laplacian of a 2D coordinate system $\triangle = \frac{\partial^2}{\partial x^2} + \frac{\partial^2}{\partial y^2}$ to a polar coordinate system with the radius r and the angle φ to the x axis, $\triangle = \frac{1}{r}\frac{\partial}{\partial r} + \frac{\partial^2}{\partial r^2} + \frac{1}{r^2}\frac{\partial^2}{\partial \varphi^2}$. And for functions $\Phi = \Phi(\varphi)$ purely in the angle φ, the radial derivatives of $\triangle\Phi$ all vanish and it remains ($\partial \rightarrow \mathrm{d}$)

$$\frac{\mathrm{d}^2}{\mathrm{d}\varphi^2}\Phi = -\lambda\, r^2\, \Phi. \tag{4.9}$$

It is only yielding useful solutions with $\lambda\, r^2 = m^2$, $m \in \mathbb{Z}$, cf. Appendix A.3.4, Fig. 4.2,

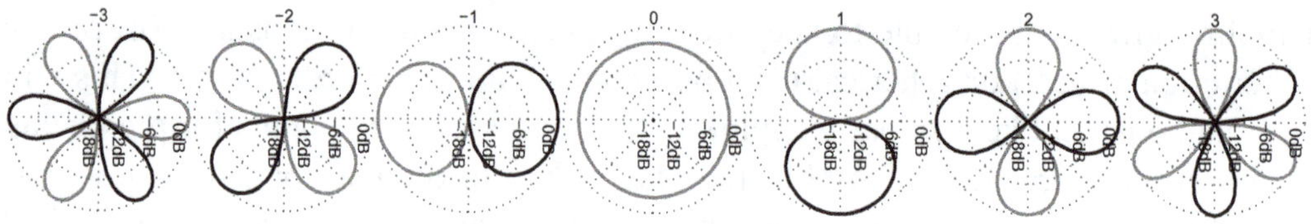

Fig. 4.2 Circular harmonics with $m = -3, \dots, 3$ plotted as polar diagram using the radius $R = 20\lg|\sqrt{\pi}\Phi_m|$ and grayscale to distinguish between positive (gray) and negative (black) signs

$$\Phi_m = \frac{1}{\sqrt{2\pi}} \begin{cases} \sqrt{2}\sin(|m|\varphi), & \text{for } m < 0, \\ 1, & \text{for } m = 0, \\ \sqrt{2}\cos(m\varphi), & \text{for } m > 0, \end{cases} \tag{4.10}$$

which defines how to decompose panning functions of limited order $|m| < \mathrm{N}$. The harmonics are periodic in azimuth, orthogonal and normalized (orthonormal) on the period $-\pi \leq \varphi \leq \pi$. Due to their completeness, any square-integrable function $g(\varphi)$ can be expanded into a series of the harmonics using coefficients γ_m

$$g(\varphi) = \sum_{m=-\infty}^{\infty} \gamma_m \, \Phi_m(\varphi). \tag{4.11}$$

For a known function $g(\varphi)$, the coefficients γ_m are obtained by the transformation integral

$$\gamma_m = \int_{-\pi}^{\pi} g(\varphi)\, \Phi_m(\varphi)\, \mathrm{d}\varphi, \tag{4.12}$$

as shown in Appendix Eq. (A.14).

2D panning function. An infinitely narrow angular range around a desired direction $|\varphi - \varphi_\mathrm{s}| < \varepsilon \to 0$ is represented by the transformation integral over a Dirac delta distribution $\delta(\varphi - \varphi_s)$, cf. Appendix Eq. (A.16), so that the coefficients of such a panning function are

$$\gamma_m = \Phi_m(\varphi_\mathrm{s}). \tag{4.13}$$

As the infinite circular harmonic series is complete, the panning function is

$$g(\varphi) = \sum_{m=-\infty}^{\infty} \Phi_m(\varphi_\mathrm{s})\, \Phi_m(\varphi) = \delta(\varphi - \varphi_\mathrm{s}), \tag{4.14}$$

and in practice we resolution-limit it to the Nth Ambisonic order, $|m| \leq \mathrm{N}$, and use an additional weight a_m that allows us to design its side lobes

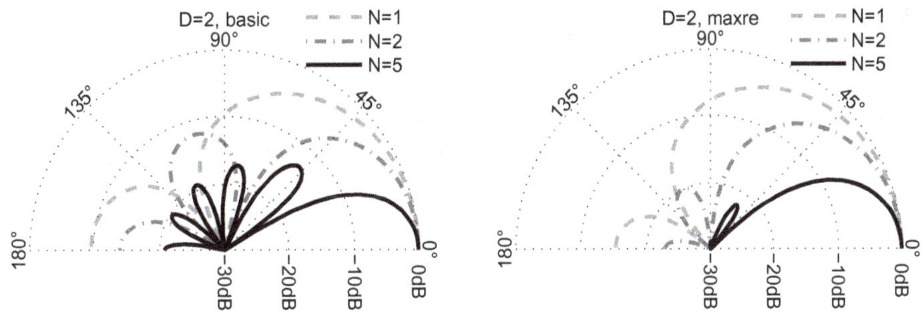

(a) basic $a_n = 1$ (b) max-$\boldsymbol{r}_\mathrm{E}$

Fig. 4.3 2D unweighted $a_n = 1$ basic and weighted max-$\boldsymbol{r}_\mathrm{E}$ Ambisonic panning functions for the orders $\mathrm{N} = 1, 2, 5$

$$g_\mathrm{N}(\varphi) = \sum_{m=-\mathrm{N}}^{\mathrm{N}} a_m\, \Phi_m(\varphi_\mathrm{s})\, \Phi_m(\varphi). \tag{4.15}$$

The max-$\boldsymbol{r}_\mathrm{E}$ panning function [24] uses the weights $a_m = \cos(\frac{\pi\, m}{2(\mathrm{N}+1)})$, as derived in Appendix Eq. (A.20). The spread is now adjustable by the order to $\pm\frac{90°}{\mathrm{N}+1}$. The result is shown in Fig. 4.3, compared with no side-lobe suppression when $a_n = 1$ (basic).

It is easy to recognize: $\Phi_m(\varphi_s)$ represents the recorded or encoded directions, and $\Phi_m(\varphi)$ represents the decoded playback directions.

Optimal sampling of the 2D panning function. In the theory of circular/spherical polynomials in the variable $\zeta = \cos(\varphi - \varphi_\mathrm{s})$, so-called t-designs in 2D are optimal point sets of given angles $\{\varphi_l\}$ with $l = 1, \ldots, \mathrm{L}$ and size L. A t-design allows to perfectly compute the integral (constant part) over the polynomials $\mathcal{P}_m(\zeta)$ of limited degree $m \le t$ by discrete summation

$$\int_{-\pi}^{\pi} \mathcal{P}_m(\cos\phi)\, \mathrm{d}\phi = \sum_{l=1}^{\mathrm{L}} \mathcal{P}_m[\cos(\varphi_l - \varphi_\mathrm{s})]\, \tfrac{2\pi}{\mathrm{L}}, \tag{4.16}$$

regardless of any angular shift φ_s. In 2D, Chebyshev polynomials $T_m(\cos\phi) = \cos(m\phi)$ are orthogonal polynomials, therefore an Nth-order panning function composed out of $\cos(m\phi)$ is always a polynomial of Nth degree. Knowing this, it is clear that the integral over g_N^2 required to evaluate the loudness measure E is a polynomial of the order 2N. The integral to calculate $\boldsymbol{r}_\mathrm{E}$ is over $g_\mathrm{N}^2\,\cos(\phi)$ and thus of the order $2\mathrm{N} + 1$. In playback, to get a perfectly panning-invariant loudness measure E of the continuous panning function and also the perfectly oriented $\boldsymbol{r}_\mathrm{E}$ vector of constant spread $\arccos\|\boldsymbol{r}_\mathrm{E}\|$, the parameter t must be $t \ge 2\mathrm{N} + 1$. In 2D, all regular polygons are t-designs with $\mathrm{L} = t + 1$ points

$$\varphi_l = \frac{2\pi}{t+1}\,(l-1). \tag{4.17}$$

We can use the smallest set of $2\mathrm{N} + 2$ angles $\varphi_l = \frac{180°}{\mathrm{N}+1}\,(l-1)$ as optimal 2D layout.

4.5 Ambisonics Encoding and Optimal Decoding in 2D

To encode a signal s into Ambisonic signals χ_m, we multiply the signal with the encoder representing the direction of the signal at the angle φ_s by the weights $\Phi_m(\varphi_\mathrm{s})$

$$\chi_m(t) = \Phi_m(\varphi_\mathrm{s})\, s(t), \tag{4.18}$$

or in vector notation

$$\boldsymbol{\chi}_\mathrm{N} = \boldsymbol{y}_\mathrm{N}(\varphi_\mathrm{s})\, s, \tag{4.19}$$

using the column vector $\boldsymbol{y}_\mathrm{N} = [\Phi_{-\mathrm{N}}(\varphi_\mathrm{s}), \ldots, \Phi_\mathrm{N}(\varphi_\mathrm{s})]^\mathrm{T}$ of $2\mathrm{N}+1$ components. The Ambisonic signals in $\boldsymbol{\chi}_\mathrm{N}$ are weighted by side-lobe suppressing weights $\boldsymbol{a}_\mathrm{N} = [a_{|-\mathrm{N}|}, \ldots, a_\mathrm{N}]^\mathrm{T}$, expressed by the multiplication with a diagonal matrix $\mathrm{diag}\{\boldsymbol{a}_\mathrm{N}\}$, and then decoded to the L loudspeaker signals $\boldsymbol{x}$ by a sampling decoder

$$\boldsymbol{D} = \sqrt{\tfrac{2\pi}{\mathrm{L}}}\left[\boldsymbol{y}_\mathrm{N}(\varphi_1), \ldots, \boldsymbol{y}_\mathrm{N}(\varphi_\mathrm{L})\right]^\mathrm{T} = \sqrt{\tfrac{2\pi}{\mathrm{L}}}\,\boldsymbol{Y}_\mathrm{N}^\mathrm{T}, \tag{4.20}$$

using

$$\boldsymbol{x} = \boldsymbol{D}\,\mathrm{diag}\{\boldsymbol{a}_\mathrm{N}\}\,\boldsymbol{\chi}_\mathrm{N}. \tag{4.21}$$

In total, the system for encoding and decoding can also be written to yield a set of loudspeaker gains for one virtual source

$$\boldsymbol{g} = \boldsymbol{D}\,\mathrm{diag}\{\boldsymbol{a}_\mathrm{N}\}\,\boldsymbol{y}_\mathrm{N}(\varphi_\mathrm{s}), \tag{4.22}$$

or in particular for the 2D sampling decoder $\boldsymbol{g} = \sqrt{\frac{2\pi}{\mathrm{L}}}\,\boldsymbol{Y}_\mathrm{N}^\mathrm{T}\,\mathrm{diag}\{\boldsymbol{a}_\mathrm{N}\}\,\boldsymbol{y}_\mathrm{N}(\varphi_\mathrm{s})$.

4.6 Listening Experiments on 2D Ambisonics

There are several listening experiments discussing the features of Ambisonics, most of which are summarized in [25], which will be discussed complemented with those from [26] below.

The perceptually adjusted panning angle of 2nd-order max-$\boldsymbol{r}_\mathrm{E}$ Ambisonics panning on 6 horizontal loudspeakers matches quite well the acoustic reference direction as shown in Fig. 4.4, similar to MDAP in Fig. 3.8, but with a slightly more

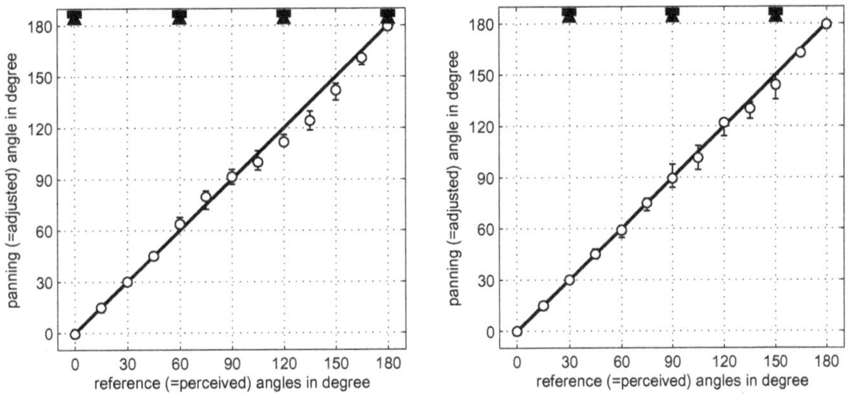

(a) Ring starting at 0°

(b) Ring starting at 30°

Fig. 4.4 2nd-order max-$\boldsymbol{r}_\mathrm{E}$-weighted Ambisonic panning with pink-noise on horizontal rings of 60°-spaced loudspeakers adjusted to perceptually match reference loudspeaker directions (harmonic complex) every 15°. Markers and whiskers indicate 95% confidence intervals and medians, black curve the $\boldsymbol{r}_\mathrm{E}$ vector model

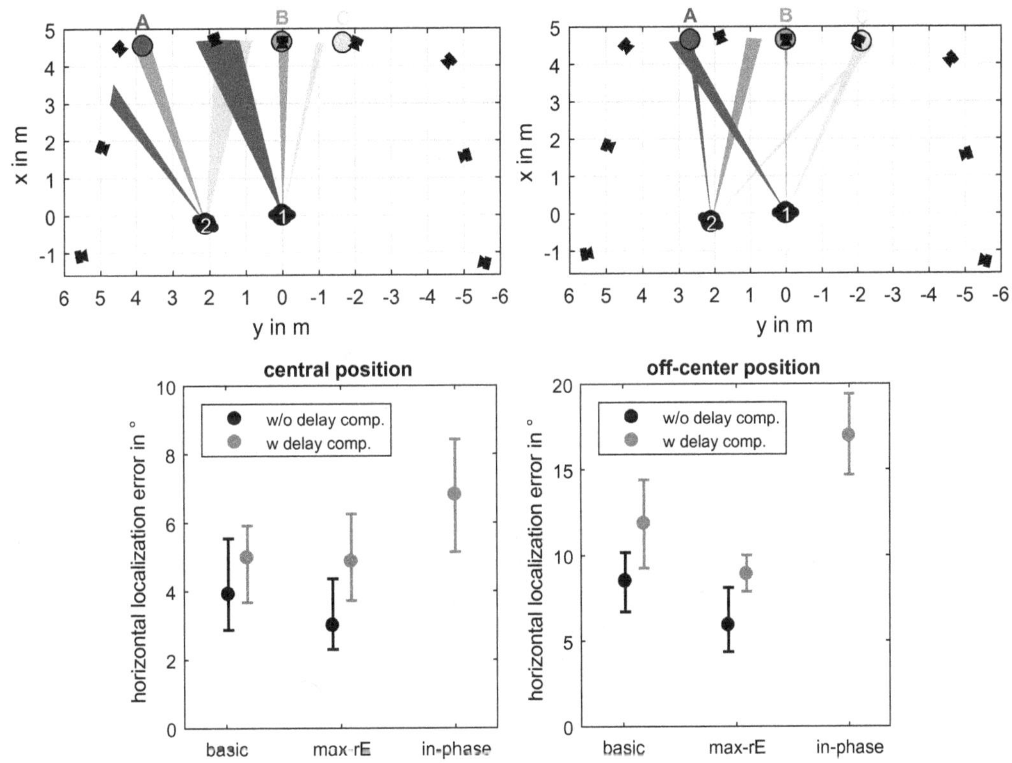

Fig. 4.5 In Frank's 2008 pointing experiment [27] on center and off-center listening seats for 3 virtual sources (*A, B, C*) using 1st-order (left) and 5th-order (right) Ambisonics on 12 horizontal loudspeakers (IEM CUBE) indicate a more stable localization with high orders. Moreover, for 5th-order, max-$\boldsymbol{r}_\mathrm{E}$ weighting and omission of delay compensation were preferred. Omission of max-$\boldsymbol{r}_\mathrm{E}$ weights ("basic") or alternative "in-phase" weights that entirely suppresses any side lobe yield less precise localization at off-center listening positions

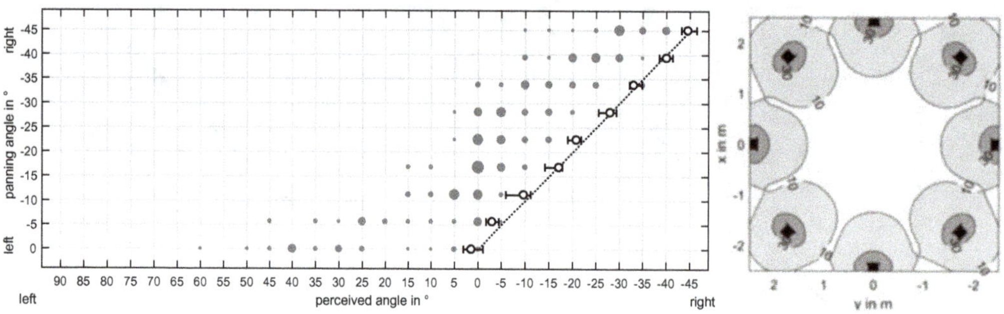

(a) Frank 2013 off-center experiment - max $\boldsymbol{r}_{\mathrm{E}}$

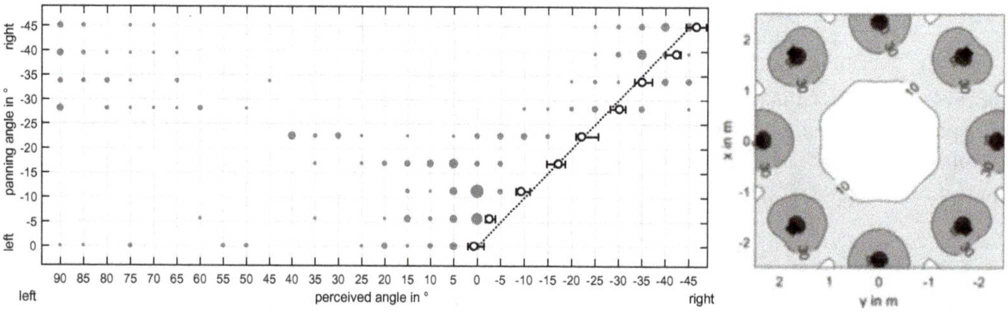

(b) Frank 2013 off-center experiment - basic (no sidelobe suppression)

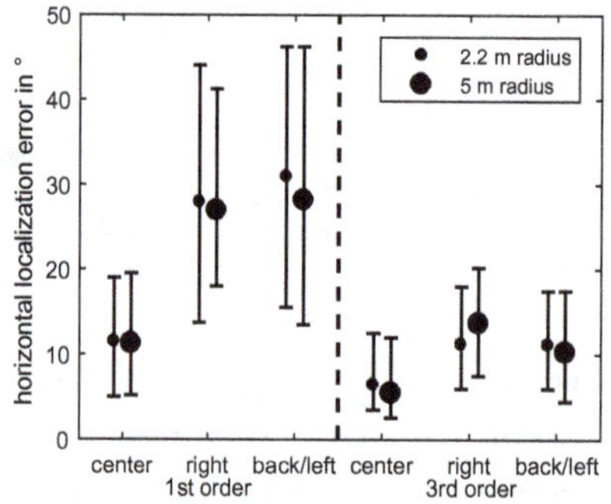

(c) Stitt 2014 off-center experiments

Fig. 4.6 Experiments on an off-center position in **a** show that max-$\boldsymbol{r}_{\mathrm{E}}$ outperforms the basic, rectangularly truncated Fourier series at off-center listening positions, **b** where it can avoid splitting of the auditory event. Stitt's experiments **c** imply that localization with higher orders is more stable and that the localization deficiency at off-center listening seats seems to be proportional to the ratio between distance to the center divided by radius of the loudspeaker ring, and not the specific time-delays that are larger for large loudspeaker rings, cf. [28]

accurate median by 0.5° on average, and in particular at side and back panning directions.

Another aspect to investigate is how stable the results are for center and off-center listening seats as shown in Fig. 4.5. It illustrates that max-$\boldsymbol{r}_{\mathrm{E}}$ with the highest order achieves the best stability with regard to localization at off-center listening seats. Astonishingly, the delay compensation for non-uniform delay times to the center deteriorated the results, most probably because of the nearly linear frontal

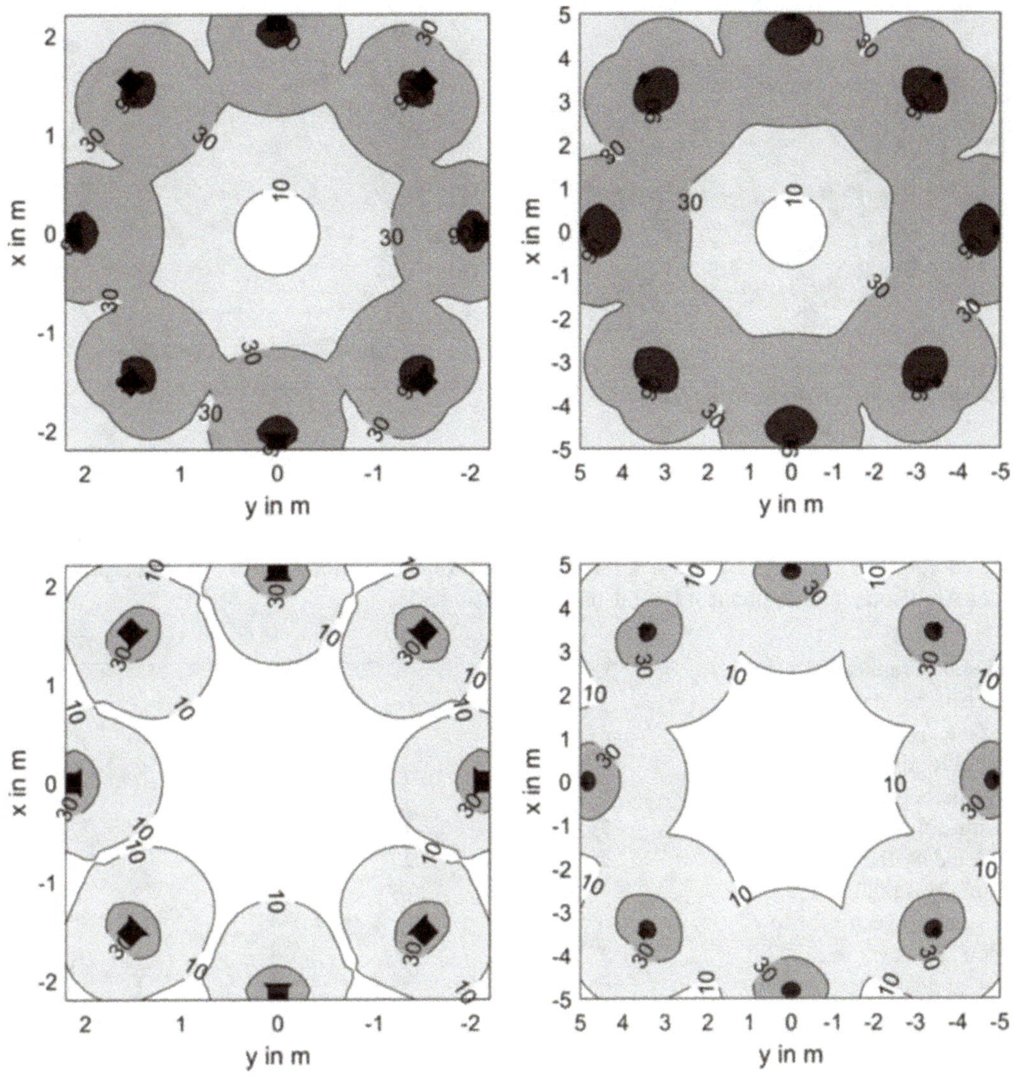

Fig. 4.7 Predicted sweet area sizes using the $\boldsymbol{r}_{\mathrm{E}}$ model Sect. 2.2.9 for loudspeaker layouts and playback orders used in Stitt's experiments [28]: first order (top), third order (bottom), small (left), and large (right)

arrangement of loudspeakers that is more robust to lateral shifts of the listening positions than a circular arrangement.

Figure 4.6a, b shows the direction histogram for two different weightings a_m, and it illustrates that proper sidelobe suppression of the panning function by using max-$\boldsymbol{r}_{\mathrm{E}}$ weights is decisive at shifted listening positions to avoid splitting of the auditory image, as it appears in Fig. 4.6b without the weights (basic).

Peter Stitt's work shows that the localization offsets at off-center listening seats do not increase with the radius of the loudspeaker arrangement as long as the off-center seat stays in proportion to the radius, Fig. 4.6c. The result are predicted by the sweet area model from Sect. 2.2.9 for the first order (top row) and third order (bottom row) in Fig. 4.7, with both sizes small setup (left) and large setup (right).

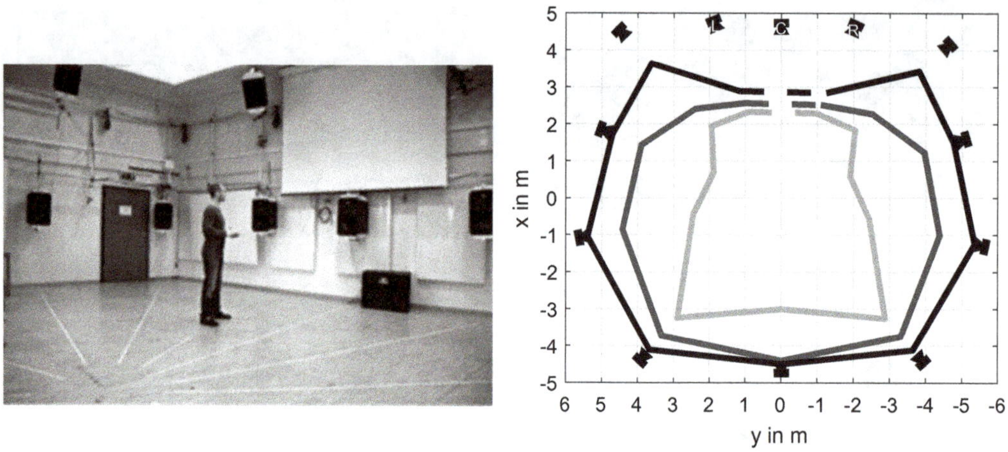

(a) Listening area size experiment (b) Sweet area for frontal sound

Fig. 4.8 The perceptual sweet spot size as investigated by Frank [29] is nearly covering the entire area enclosed by the IEM CUBE as a playback setup (black = 5th, gray = 3rd, light gray = 1st order Ambisonics). It is smallest for 1st-order Ambisonics

Fig. 4.9 Frank's 2013 experiments showed for max-$\boldsymbol{r}_{\mathrm{E}}$ Ambisonics on three frontal loudspeakers that large perceived widths are not entirely accurately modeled by the optimally-sampled, therefore constant $\boldsymbol{r}_{\mathrm{E}}$ vector for low orders, however reasonably well, and significantly for high orders/narrow width

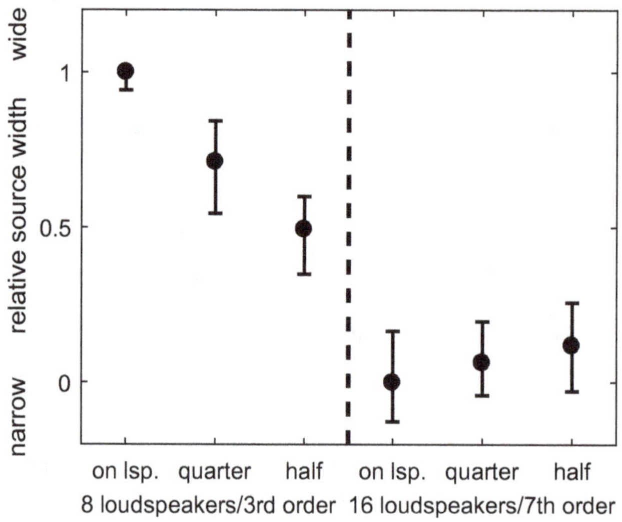

Frank's 2016 experiments [29] used scales on the floor from which listeners read off where the sweet area ends in every radial direction, cf. Fig. 4.8a. For Fig. 4.8b, the criterion for listeners to indicate leaving the sweet area was when the frontally panned sound was mapped outside the loudspeaker pairs L, C, and R. It showed that a sweet area providing perceptually plausible playback measures at least $\frac{2}{3}$ of the radius of the loudspeaker setup if the order is high enough.

The perceived width of auditory events is investigated in the experimental results of Fig. 4.9, [25], in which pink noise was frontally panned in different orientations of the loudspeaker ring (with one loudspeaker in front, with front direction lying quarter- and half-spaced wrt. loudspeaker spacing). Listeners compared the width of multiple stimuli, and the results were expected to indicate constant width for the differently rotated loudspeaker ring, as the optimal arrangement with $\mathrm{L} = 2\mathrm{N} + 2$

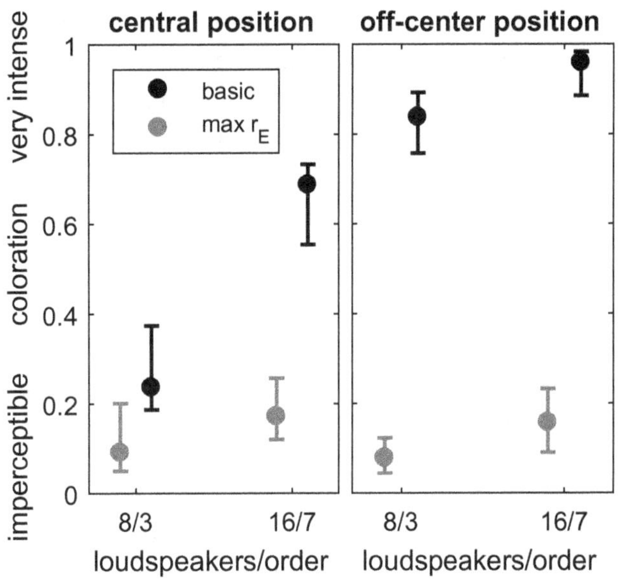

Fig. 4.10 Frank's 2013 experiments on the variation of the sound coloration of virtual sources rotating at a speed of 100°/s imply that max-$\boldsymbol{r}_{\mathrm{E}}$ weighting outperforms the "basic" weighting with $a_m = 1$, and that adjusting the number of loudspeakers to the Ambisonic order seems to be reasonable

provides constant $\boldsymbol{r}_{\mathrm{E}}$ length. The panning-invariant length is not perfectly reflected in the perceived widths with 3rd order on 8 loudspeakers, for which the on-loudspeaker position is perceived as being significantly wider. By contrast, the high-order experiment with 7th order on 16 loudspeakers would perfectly validate the model.

Figure 4.10 shows experiments investigating the time-variant change in sound coloration for a pink-noise virtual source rotating at a speed of 100°/s, and for different Ambisonic panning setups. There is an obvious advantage of a reduced fluctuation in coloration at both listening positions, centered and off-center, when using the side-lobe-suppressing "max-$\boldsymbol{r}_{\mathrm{E}}$" weighting instead of the "basic" rectangular truncation of the Fourier series. At the off-center listening position, max-$\boldsymbol{r}_{\mathrm{E}}$ weights achieve good results with regard to constant coloration for both 3rd and 7th order arrangements with 8 and 16 loudspeakers that were investigated.

How well would diffuse signals be preserved played back? All the above experiments deal with how non-diffuse signals are presented. To complement what is shown in Fig. 1.21 of Chap. 1 with an explanation, the relation between Ambisonic order and its ability to preserve diffuse fields is estimated here by the covariance between uncorrelated directions. Assume a max-$\boldsymbol{r}_{\mathrm{E}}$-weighted Nth-order Ambisonic panning function $g(\boldsymbol{\theta}_{\mathrm{s}}^{\mathrm{T}}\boldsymbol{\theta})$ that is normalized to $g(1) = 1$, encodes two sounds $s_{1,2}$ from two directions $\boldsymbol{\theta}_1$ and $\boldsymbol{\theta}_2$, with the sounds being uncorrelated and unit-variance $E\{s_1 s_2\} = \delta_{1,2}$. We can find that the Ambisonic representation mixes the sounds at their respective mapped directions and yields an increase of their correlation $x_1 = s_1 + g_{12}\, s_2$ and $x_2 = s_2 + g_{12}\, s_1$, using $g_{12} = g(\cos\phi)$,

$$R\{x_1\, x_2\} = \frac{E\{x_1\, x_2\}}{\sqrt{E\{x_1^2\}E\{x_2^2\}}} \tag{4.23}$$

$$= \frac{E\{(1 + g_{12}^2)\, s_1\, s_2 + g_{12}(s_1^2 + s_2^2)\}}{\sqrt{E\{s_1^2 + 2\, g_{12}\, s_1\, s_2 + g_{12}^2 s_2^2\}E\{s_2^2 + 2\, g_{12}\, s_1\, s_2 + g_{12}^2 s_1^2\}}} = \frac{2\, g_{12}}{1 + g_{12}^2}.$$

This result was presented in Fig. 1.21 and was used to argue that the directional separation of first-order Ambisonics by its high crosstalk term g_{12} might be too weak. Higher-order Ambisonics decreases this directional crosstalk and therefore improves the representation of diffuse sound fields.

4.7 Panning with Spherical Harmonics in 3D

In three space dimensions, the spherical coordinate system has a radius r and two angles, azimuth φ indicating the polar angle of the orthogonal projection to the xy plane, and the zenith angle ϑ indicating the angle to the z axis, according to the right-handed spherical coordinate systems in ISO31-11, ISO80000-2, [30, 31], Fig. 4.11.

By the generalized chain rule, Appendix A.3 re-writes the Laplacian to spherical coordinates in 3D with r signifying the radius, φ the azimuth angle, and the zenith angle ϑ re-expressed as $\zeta = \frac{z}{r} = \cos\vartheta$, yielding the operator $\triangle = \frac{2}{r}\frac{\partial}{\partial r} + \frac{\partial^2}{\partial r^2} + \frac{1}{r^2(1-\zeta^2)}\frac{\partial^2}{\partial\varphi^2} - \frac{2}{r^2}\zeta\frac{\partial}{\partial\zeta} + \frac{1-\zeta^2}{r^2}\frac{\partial^2}{\partial\zeta^2}$. Any radius-dependent part is removed to define an eigenproblem yielding the basis for panning functions, taking only $r^2\triangle_{\varphi,\zeta,\mathrm{3D}}$,

$$\left[\frac{1}{1-\zeta^2}\frac{\partial^2}{\partial\varphi^2} - 2\zeta\frac{\partial}{\partial\zeta} + (1-\zeta^2)\frac{\partial^2}{\partial\zeta^2}\right] Y = -\lambda\, Y \tag{4.24}$$

whose solution with $\lambda = n(n+1)$ defines the spherical harmonics

$$Y_n^m(\boldsymbol{\theta}) = Y_n^m(\varphi, \vartheta) = \Theta_n^m(\vartheta)\,\Phi_m(\varphi). \tag{4.25}$$

The pre-requisites are (i) periodicity in φ and (ii) that the function Y_n^m is finite on the sphere. In addition to the circular harmonics Φ_m expressing the dependency on azimuth φ according to Eq. (4.10), the spherical harmonics contain the associated Legendre functions P_n^m and their normalization term

$$\Theta_n^m(\vartheta) = N_n^{|m|}\, P_n^{|m|}(\cos\vartheta) \tag{4.26}$$

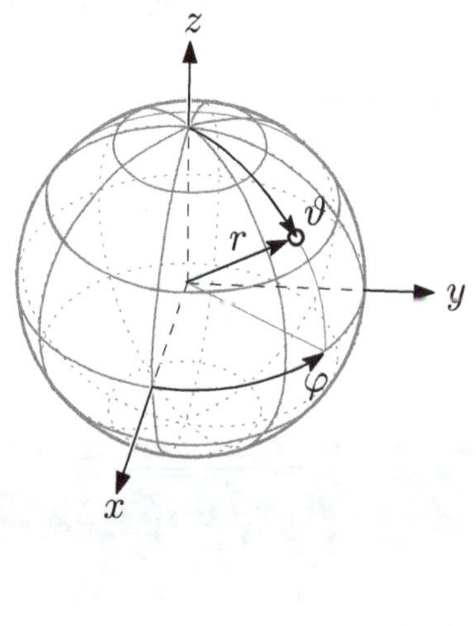

Fig. 4.11 The spherical coordinate system

to express the dependency on the zenith angle ϑ. The index $n \geq 0$ expresses the order and the directional resolution can be limited by requiring $0 \leq n \leq \mathrm{N}$. The index m is the degree and for each n it is limited by $-n \leq 0 \leq n$.

The spherical harmonics, Fig. 4.12, are orthonormal on the sphere $-\pi \leq \varphi \leq \pi$ and $0 \leq \vartheta \leq \pi$, and for unbounded order $\mathrm{N} \to \infty$ they are complete; see also Appendix A.3.7.

The spherical harmonics permit a series representation of square-integrable 3D directional functions by the coefficients γ_{nm},

$$g(\boldsymbol{\theta}) = \sum_{n=0}^{\infty} \sum_{m=-n}^{n} \gamma_{nm}\, Y_n^m(\boldsymbol{\theta}). \tag{4.27}$$

From a known function $g(\boldsymbol{\theta})$, the coefficients are obtained by the transformation integral over the unit sphere $\mathbb{S}^2$, cf. appendix Eq. (A.38)

$$\gamma_{nm} = \int_{\mathbb{S}^2} g(\boldsymbol{\theta})\, Y_n^m(\boldsymbol{\theta})\, \mathrm{d}\boldsymbol{\theta}. \tag{4.28}$$

Note that the above N3D normalization $\int_{\boldsymbol{\theta}\in\mathbb{S}^2} |Y_n^m(\boldsymbol{\theta})|^2\, \mathrm{d}\boldsymbol{\theta} = 1$ *defines each spherical harmonic except for an arbitrary-phase it might be multiplied with. Legendre functions for the zenith dependency might be defined differently in literature, and for azimuth, some implementations use* $\sin(m\varphi)$ *instead of* $\sin(|m|\varphi)$*. In Ambisonics, real-valued functions and the SN3D normalization* $\sqrt{\frac{1}{2}\frac{(n-|m|)!}{(n+|m|)!}}$ *are preferred, and positive signs of the first-order dipole components in the directions of the respective coordinate axes,* x, y, z, *are preferred. This might require to involve*

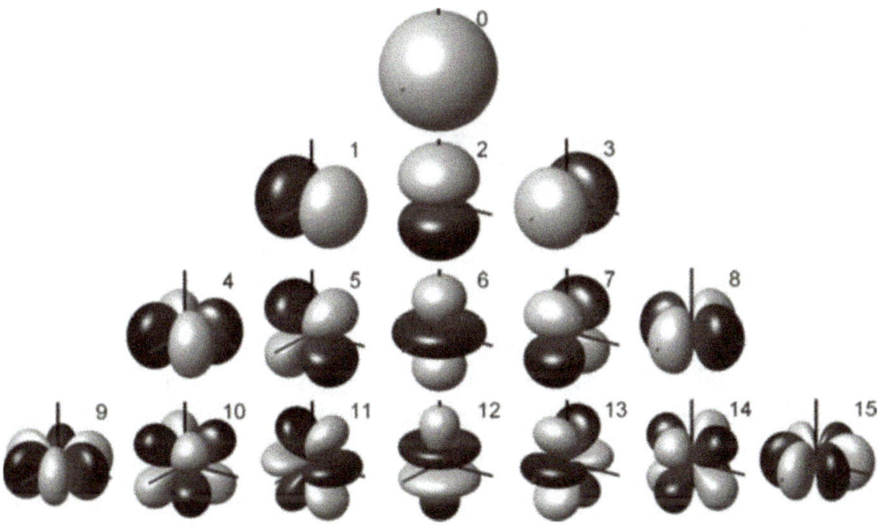

Fig. 4.12 Spherical harmonics indexed by Ambisonic channel number $ACN = n^2 + n + m$; rows show spherical harmonics for the order $0 \leq n \leq 3$ with the $2n + 1$ harmonics of the degree $-n \leq m \leq n$. What is plotted is a polar diagram with the radius $R = 20 \lg |Y_n^m|$ normalized to the upper 30 dB of each pattern, with positive (gray) and negative (black) color indicating the sign. The order n counts the circular zero crossings, and $|m|$ counts those running through zenith and nadir

the Condon-Shortley phase $(-1)^m$ to correct the signs of the Legendre functions, or -1 for $m < 0$ to correct the sign of azimuthal sinusoids, depending on the implementation of the respective functions. It is often helpful to employ converters and directional checks to ensure compatibility!

3D panning function. An infinitely narrow direction range around a desired direction $\boldsymbol{\theta}_{\mathrm{s}}^{\mathrm{T}}\boldsymbol{\theta} > \cos\varepsilon \rightarrow 1$ is represented by the transformation integral over the Dirac delta $\delta(1-\boldsymbol{\theta}_{\mathrm{s}}^{\mathrm{T}}\boldsymbol{\theta})$, cf. Eq.(A.41), so that the coefficients of the panning function are

$$\gamma_{nm} = Y_n^m(\boldsymbol{\theta}_{\mathrm{s}}). \tag{4.29}$$

As infinitely many spherical harmonics are complete, the panning function is

$$g(\boldsymbol{\theta}) = \sum_{n=0}^{\infty}\sum_{m=-n}^{n} Y_n^m(\boldsymbol{\theta}_{\mathrm{s}})\, Y_n^m(\boldsymbol{\theta}) = \delta(1-\boldsymbol{\theta}_{\mathrm{s}}^{\mathrm{T}}\boldsymbol{\theta}), \tag{4.30}$$

and in practice, the finite-resolution Nth-order panning function with $n \leq \mathrm{N}$ employs a weight a_n to reduce side lobes and optimize the spread

$$g_{\mathrm{N}}(\boldsymbol{\theta}) = \sum_{n=0}^{\mathrm{N}}\sum_{m=-n}^{n} a_n\, Y_n^m(\boldsymbol{\theta}_{\mathrm{s}})\, Y_n^m(\boldsymbol{\theta}). \tag{4.31}$$

The max-$\boldsymbol{r}_{\mathrm{E}}$ panning function uses the weights $a_n = P_n\big[\cos(\frac{137.9°}{\mathrm{N}+1.51})\big]$, as derived in Appendix Eq. (A.46). The spread is now adjustable by the order to $\pm\frac{137.9°}{\mathrm{N}+1.51}$. Figure 4.13 shows a comparison to the basic weighting $a_n = 1$. An alternative expression that uses Legendre polynomials P_n and only depends on the angle ϕ to the panning direction $\boldsymbol{\theta}_{\mathrm{s}}$ is obtained by replacing the sum over m by the *spherical harmonics addition theorem* $\sum_{m=-n}^{n} Y_n^m(\boldsymbol{\theta}_{\mathrm{s}})\, Y_n^m(\boldsymbol{\theta}) = \frac{2n+1}{4\pi}\, P_n(\cos\phi)$,

$$g_{\mathrm{N}}(\phi) = \sum_{n=0}^{\mathrm{N}} \tfrac{2n+1}{4\pi}\, a_n\, P_n(\cos\phi). \tag{4.32}$$

Comparison to first-order Ambisonics shows: now $Y_n^m(\boldsymbol{\theta}_s)$ represents the recorded or encoded directions, and $Y_n^m(\boldsymbol{\theta})$ represents the decoded playback directions.

Optimal sampling of the 3D panning function. In the theory of spherical polynomials in the variable $\zeta = \boldsymbol{\theta}_s^{\mathrm{T}}\boldsymbol{\theta}$, so-called t-designs describe point sets of given directions $\{\boldsymbol{\theta}_l\}$ with $l = 1, \ldots, \mathrm{L}$ and size L that allow to perfectly compute the integral (constant part) over the polynomials $\mathcal{P}_n(\zeta)$ of limited order $n \leq t$ by discrete summation

$$\int_{-\pi}^{\pi} \mathrm{d}\varphi \int_{-1}^{1} \mathcal{P}_n(\zeta)\, \mathrm{d}\zeta = \sum_{l=1}^{\mathrm{L}} \mathcal{P}_n(\boldsymbol{\theta}_s^{\mathrm{T}}\boldsymbol{\theta}_l)\, \tfrac{4\pi}{\mathrm{L}}, \tag{4.33}$$

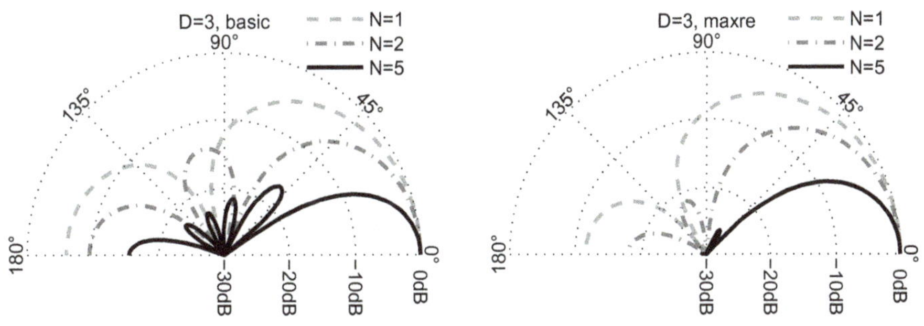

(a) basic $a_n = 1$ (b) max-$\boldsymbol{r}_\mathrm{E}$

Fig. 4.13 3D unweighted $a_n = 1$ basic and weighted max-$\boldsymbol{r}_\mathrm{E}$ Ambisonic panning functions for the orders $\mathrm{N} = 1, 2, 5$

relative to any axis $\boldsymbol{\theta}_\mathrm{s}$ the point set is projected onto. In 3D, the Legendre polynomials $P_n(\zeta)$ are orthogonal polynomials, therefore an Nth-order panning function composed thereof is a polynomial of Nth order. The loudness measure E is calculated by the integral over g_N^2, therefore over a polynomial of the order 2N. The integral to calculate $\boldsymbol{r}_\mathrm{E}$ runs over $g_\mathrm{N}^2\,\zeta$, therefore over a polynomial of the order $2\mathrm{N}+1$. In playback, to get a perfectly panning-invariant loudness measure E of the continuous panning function and also the perfectly oriented $\boldsymbol{r}_\mathrm{E}$ vector of constant spread $\arccos\|\boldsymbol{r}_\mathrm{E}\|$, the parameter t must be $t \geq 2\mathrm{N}+1$. In 3D there are only 5 geometrically regular layouts

- the tetrahedron, $\mathrm{L} = 4$ corners, is a 2-design,
- the octahedron, $\mathrm{L} = 6$ corners, is a 3-design,
- the hexahedron (cube), $\mathrm{L} = 8$ corners, is a 3-design,
- the icosahedron, $\mathrm{L} = 12$ corners, is a 5-design,
- the dodecahedron, $\mathrm{L} = 20$ corners, is a 5-design.

For instance, for $\mathrm{N} = 1$, the octahedron is a suitable spherical design, for $\mathrm{N} = 2$, the icosahedral or dodecahedral layouts are suitable.

Exceeding the geometrically regular layouts, there are designs found by optimization to be regular under the mathematical rule to approximate $\int_{\mathbb{S}^2} Y_n^m(\boldsymbol{\theta})\,\mathrm{d}\boldsymbol{\theta} = \sqrt{4\pi}\,\delta_n$ accurately by $\frac{4\pi}{\mathrm{L}}\sum_l Y_n^m(\boldsymbol{\theta}_l)$ for all $n \leq t$ and $|m| \leq n$. A large collection can be found by Hardin and Sloane [32], Gräf and Potts [33], and Womersley [34] available on the following websites

http://neilsloane.com/sphdesigns/dim3/

http://homepage.univie.ac.at/manuel.graef/quadrature.php

(Chebyshev-type Quadratures on $\mathbb{S}^2$), and

https://web.maths.unsw.edu.au/~rsw/Sphere/EffSphDes/ss.html.

Figure 4.14 gives some graphical examples.

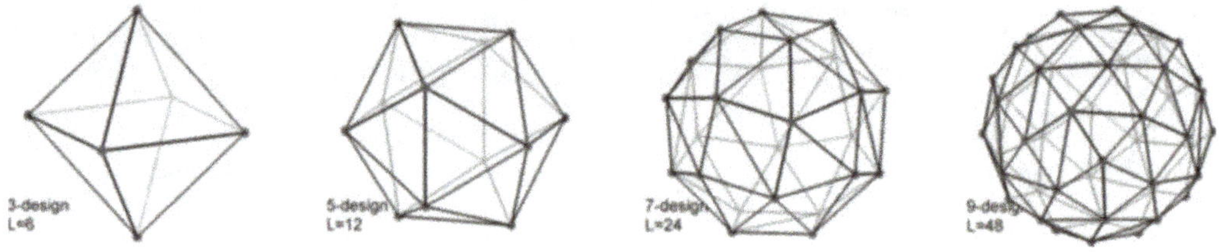

Fig. 4.14 t-designs from Gräf's website (*Chebyshev-type quadrature*)

4.8 Ambisonic Encoding and Optimal Decoding in 3D

To encode a signal s into Ambisonic signals χ_{nm}, we multiply the signal with the encoder representing the direction $\boldsymbol{\theta}_\mathrm{s}$ of the signal by the weights $Y_n^m(\boldsymbol{\theta}_\mathrm{s})$

$$\chi_{nm}(t) = Y_n^m(\boldsymbol{\theta}_\mathrm{s})\, s(t), \tag{4.34}$$

or in vector notation

$$\boldsymbol{\chi}_\mathrm{N} = \boldsymbol{y}_\mathrm{N}(\boldsymbol{\theta}_\mathrm{s})\, s, \tag{4.35}$$

using the column vector $\boldsymbol{y}_\mathrm{N} = [Y_0^0(\boldsymbol{\theta}_\mathrm{s}),\ Y_1^{-1}(\boldsymbol{\theta}_\mathrm{s}),\ \dots,\ Y_\mathrm{N}^\mathrm{N}(\boldsymbol{\theta}_\mathrm{s})]^\mathrm{T}$ of $(\mathrm{N}+1)^2$ components. The Ambisonic signals in $\boldsymbol{\chi}_\mathrm{N}$ are weighted by side-lobe suppressing weights $\boldsymbol{a}_\mathrm{N} = [a_0,\ a_1, a_1, a_1,\ a_2, \dots, a_\mathrm{N}]^\mathrm{T}$, expressed by the multiplication with a diagonal matrix $\mathrm{diag}\{\boldsymbol{a}_\mathrm{N}\}$, and then decoded to the L loudspeaker signals $\boldsymbol{x}$ by a sampling decoder

$$\boldsymbol{D} = \sqrt{\tfrac{4\pi}{\mathrm{L}}}\,\left[\boldsymbol{y}_\mathrm{N}(\boldsymbol{\theta}_1),\ \dots,\ \boldsymbol{y}_\mathrm{N}(\boldsymbol{\theta}_\mathrm{L})\right]^\mathrm{T} = \sqrt{\tfrac{4\pi}{\mathrm{L}}}\,\boldsymbol{Y}_\mathrm{N}^\mathrm{T}, \tag{4.36}$$

using

$$\boldsymbol{x} = \boldsymbol{D}\,\mathrm{diag}\{\boldsymbol{a}_\mathrm{N}\}\,\boldsymbol{\chi}_\mathrm{N}. \tag{4.37}$$

In total, the system for encoding Eq. (4.35) and decoding Eq. (4.36) can also be written to yield loudspeaker gains for one signal

$$\boldsymbol{g} = \boldsymbol{D}\,\mathrm{diag}\{\boldsymbol{a}_\mathrm{N}\}\,\boldsymbol{y}_\mathrm{N}(\boldsymbol{\theta}_\mathrm{s}), \tag{4.38}$$

or in particular for the 3D sampling decoding $\boldsymbol{g} = \sqrt{\tfrac{4\pi}{\mathrm{L}}}\,\boldsymbol{Y}_\mathrm{N}^\mathrm{T}\,\mathrm{diag}\{\boldsymbol{a}_\mathrm{N}\}\,\boldsymbol{y}_\mathrm{N}(\boldsymbol{\theta}_\mathrm{s})$.

4.9 Ambisonic Decoding to Loudspeakers

Ambisonic decoding to loudspeakers has been dealt with by numerous researchers, in the past, particularly because result are not very stable for first-order Ambisonics, and later because they strongly depend on how uniform the loudspeaker layout is for higher-order Ambisonics. Moreover, Solvang found that even the use of too many loudspeakers has a degrading effect [35].

For first-order decoding, the Vienna decoders by Michael Gerzon [36] are often cited, and for higher-order Ambisonic decoding, one can, e.g. find works by Daniel with max-$\boldsymbol{r}_{\mathrm{E}}$ [37] and pseudo-inverse decoding [10], also by Poletti [14, 38, 39].

What turned out to be the most practical solution, is the All-Round Ambisonic Decoding approach (AllRAD) due to its feature of allowing imaginary loudspeaker insertion and downmix as described in the sections above, cf. [40]. It moreover does not have restrictions on the Ambisonics order, which for other decoders often yields poor controllability of panning-dependent fluctuations in loudness and directional mapping errors.

The playable set of directions $\boldsymbol{\theta}_l$ or φ_l is usually finite and discrete, and it is represented by the surrounding loudspeakers' directions. The directional distribution of the surrounding loudspeakers is typically neither a t-design (with $t \geq 2\mathrm{N}+1$ in general, sometimes not even regular polygons with $\mathrm{L} \geq 2\mathrm{N}+2$ loudspeakers for 2D, in particular). In such cases, it is extremely helpful to be aware of the properties of the various decoder design methods.

4.9.1 Sampling Ambisonic Decoder (SAD)

The sampling decoder as introduced above is the simplest decoding method. For dimensions ($\mathrm{D}=2$) and three ($\mathrm{D}=3$), it uses the matrix $\boldsymbol{Y}_{\mathrm{N}} = [\boldsymbol{y}_{\mathrm{N}}(\boldsymbol{\theta}_1), \dots, \boldsymbol{y}_{\mathrm{N}}(\boldsymbol{\theta}_{\mathrm{L}})]$ containing the respective circular or spherical harmonics $\boldsymbol{y}_{\mathrm{N}}(\boldsymbol{\theta})$ sampled at the loudspeaker directions $\{\boldsymbol{\theta}_l\}$,

$$\boldsymbol{D} = \sqrt{\tfrac{S_{\mathrm{D}-1}}{\mathrm{L}}}\,\boldsymbol{Y}_{\mathrm{N}}^{\mathrm{T}}, \tag{4.39}$$

with the circumference of the unit circle denoted as $S_1 = 2\pi$ or the surface of the unit sphere written as $S_2 = 4\pi$. The factor $\sqrt{\frac{S_{\mathrm{D}-1}}{\mathrm{L}}}$ expresses that each loudspeaker synthesizes a fraction of the E measure on the circle or sphere of the surrounding directions. However, the sampling decoder would neither yield perfectly constant loudness and width measures, E, $\|\boldsymbol{r}_{\mathrm{E}}\|$, nor a correct aiming of the localization measure $\boldsymbol{r}_{\mathrm{E}}$ if the loudspeaker layout wasn't optimal. For instance concerning loudness, for panning towards directional regions of poor loudspeaker coverage, sampling misses out the main lobe of the panning function, yielding a noticeably reduced loudness.

4.9.2 Mode Matching Decoder (MAD)

The mode-matching method is used in [10, 39] and yields a fundamentally different decoder design. Its concept is to re-encode the gain vector $\boldsymbol{g}$ of the loudspeakers for any panning direction $\boldsymbol{\theta}_s$ by the encoding matrix $\boldsymbol{Y}_\mathrm{N} = [\boldsymbol{y}_\mathrm{N}(\boldsymbol{\theta}_1), \ldots, \boldsymbol{y}_\mathrm{N}(\boldsymbol{\theta}_\mathrm{L})]$ for all loudspeaker directions $\{\boldsymbol{\theta}_l\}$. Ideally, the re-encoded result should match the encoding of the panning direction with sidelobes suppressed

$$\boldsymbol{Y}_\mathrm{N}\,\boldsymbol{g} = \mathrm{diag}\{\boldsymbol{a}_\mathrm{N}\}\,\boldsymbol{y}_\mathrm{N}(\boldsymbol{\theta}_s).$$

Using the definition $\boldsymbol{g} = \boldsymbol{D}\,\mathrm{diag}\{\boldsymbol{a}_\mathrm{N}\}\,\boldsymbol{y}_\mathrm{N}(\boldsymbol{\theta}_s)$ of the panning gains, we obtain

$$\begin{aligned} \boldsymbol{Y}_\mathrm{N}\boldsymbol{D}\,\mathrm{diag}\{\boldsymbol{a}_\mathrm{N}\}\,\boldsymbol{y}_\mathrm{N}(\boldsymbol{\theta}_s) &= \mathrm{diag}\{\boldsymbol{a}_\mathrm{N}\}\,\boldsymbol{y}_\mathrm{N}(\boldsymbol{\theta}_s), \\ \Rightarrow \boldsymbol{D} &= \sqrt{\tfrac{\mathrm{L}}{S_{\mathrm{D}-1}}}\,\boldsymbol{Y}_\mathrm{N}^\mathrm{T}(\boldsymbol{Y}_\mathrm{N}\boldsymbol{Y}_\mathrm{N}^\mathrm{T})^{-1} \end{aligned} \tag{4.40}$$

so that the decoder $\boldsymbol{D}$ is required to be right-inverse to the matrix $\boldsymbol{Y}_\mathrm{N}$, i.e. $\boldsymbol{Y}_\mathrm{N}\,\boldsymbol{D} = \boldsymbol{Y}_\mathrm{N}\boldsymbol{Y}_\mathrm{N}^\mathrm{T}(\boldsymbol{Y}_\mathrm{N}\boldsymbol{Y}_\mathrm{N}^\mathrm{T})^{-1} = \boldsymbol{I}$, see Eq. (A.63) in Appendix A.4. For the inverse of $\boldsymbol{Y}_\mathrm{N}\boldsymbol{Y}_\mathrm{N}^\mathrm{T}$ to exist, it is necessary to have at least as many loudspeakers as harmonics, i.e. $\mathrm{L} \geq (\mathrm{N}+1)^2$ with $\mathrm{D} = 3$ or $\mathrm{L} \geq 2\mathrm{N}+1$ for $\mathrm{D} = 2$. However, this is not a sufficient criterion yet: In directions poorly covered with loudspeakers, the inversion will boost the loudness, so that the result is often numerically ill conditioned for $(\boldsymbol{Y}_\mathrm{N}\boldsymbol{Y}_\mathrm{N}^\mathrm{T})^{-1}$ unless the loudspeaker layout is uniformly designed, at least. Mode matching decoding is ill-conditioned on hemispherical or semicircular loudspeaker layouts. The solution is equivalently described by the more general pseudo inverse $\boldsymbol{Y}_\mathrm{N}^\dagger$, which is right-inverse for fat matrices.

4.9.3 Energy Preservation on Optimal Layouts

For instance, for an order of $\mathrm{N} = 2$, 2D Ambisonics should work optimally with a ring of 45° spaced loudspeakers on the horizon, a circular $(2\mathrm{N}+1)$-design, or for 3D, a spherical $(2\mathrm{N}+1)$-design. On a t-design selected by $t \geq 2\mathrm{N}$, the loudness measure E is panning-invariant, in general,

$$E = \|\boldsymbol{g}\|^2 = \boldsymbol{y}_\mathrm{N}^\mathrm{T}(\boldsymbol{\theta}_s)\,\mathrm{diag}\{\boldsymbol{a}_\mathrm{N}\}\,\underbrace{\boldsymbol{D}^\mathrm{T}\boldsymbol{D}}_{=\boldsymbol{I}}\,\mathrm{diag}\{\boldsymbol{a}_\mathrm{N}\}\,\boldsymbol{y}_\mathrm{N}(\boldsymbol{\theta}_s) = \|\mathrm{diag}\{\boldsymbol{a}_\mathrm{N}\}\,\boldsymbol{y}_\mathrm{N}(\boldsymbol{\theta}_s)\|^2 = \mathrm{const.}$$

This is because a $t \geq 2\mathrm{N}$-design discretization preserves orthonormality

$$\int \boldsymbol{y}_\mathrm{N}(\boldsymbol{\theta})\,\boldsymbol{y}_\mathrm{N}^\mathrm{T}(\boldsymbol{\theta})\,\mathrm{d}\boldsymbol{\theta} = \tfrac{S_{\mathrm{D}-1}}{\mathrm{L}}\sum_{l=1}^{\mathrm{L}}\boldsymbol{y}_\mathrm{N}(\boldsymbol{\theta}_l)\,\boldsymbol{y}_\mathrm{N}^\mathrm{T}(\boldsymbol{\theta}_l) = \tfrac{S_{\mathrm{D}-1}}{\mathrm{L}}\boldsymbol{Y}_\mathrm{N}\boldsymbol{Y}_\mathrm{N}^\mathrm{T} = \boldsymbol{I}, \tag{4.41}$$

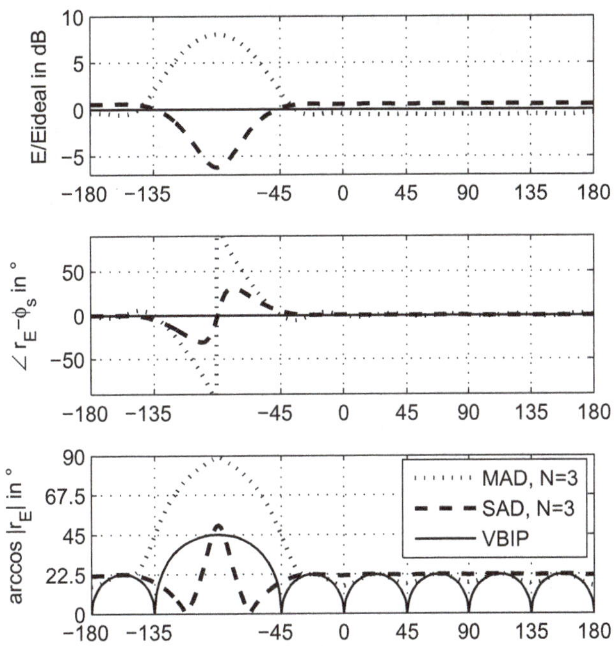

Fig. 4.15 Analysis of loudness, localization error, and width for 3rd-order sampling (SAD) and mode-matching (MAD) Ambisonic decoding for all panning angles on a sub-optimal 45° spaced loudspeaker ring with gap at −90°, compared to VBIP

which implies for the sampling decoder $\boldsymbol{D}^{\mathrm{T}}\boldsymbol{D} = \frac{S_{\mathrm{D}-1}}{\mathrm{L}}\,\boldsymbol{Y}_{\mathrm{N}}\boldsymbol{Y}_{\mathrm{N}}^{\mathrm{T}} = \boldsymbol{I}$, and we notice the panning invariant norm of $g(\boldsymbol{\theta})$ within its coefficients $\boldsymbol{\gamma}_{\mathrm{N}} = \mathrm{diag}\{\boldsymbol{a}_{\mathrm{N}}\}\boldsymbol{y}_{\mathrm{N}}(\boldsymbol{\theta}_{\mathrm{s}})$ by the Parseval theorem $\int g^2(\boldsymbol{\theta})\,\mathrm{d}\boldsymbol{\theta} = \|\boldsymbol{\gamma}_{\mathrm{N}}\|^2$. The panning invariant E measure also holds for the mode-matching decoder using a $t \geq 2\mathrm{N}$-design, as it becomes equivalent to a sampling decoder $\boldsymbol{D} = \sqrt{\frac{\mathrm{L}}{S_{\mathrm{D}-1}}}\boldsymbol{Y}_{\mathrm{N}}^{\mathrm{T}}(\boldsymbol{Y}_{\mathrm{N}}\boldsymbol{Y}_{\mathrm{N}}^{\mathrm{T}})^{-1} = \sqrt{\frac{\mathrm{L}}{S_{\mathrm{D}-1}}}\boldsymbol{Y}_{\mathrm{N}}^{\mathrm{T}}\frac{S_{\mathrm{D}-1}}{\mathrm{L}} = \sqrt{\frac{S_{\mathrm{D}-1}}{\mathrm{L}}}\boldsymbol{Y}_{\mathrm{N}}^{\mathrm{T}}$. Under these ideal conditions, both decoders are energy-preserving.

4.9.4 Loudness Deficiencies on Sub-optimal Layouts

For 2D layouts, Fig. 4.15 shows what happens if a decoder is calculated for a $t \geq 2\mathrm{N}+1$-design with one loudspeaker removed: While, for panning across the gap, the sampling Ambisonic decoder (SAD) yields a quieter signal, moderate localization errors and width fluctuation, the mode-matching decoder (MAD) yields a strong loudness increase and severe jumps in the localization/width. MAD is therefore not very practical with sub-optimal layouts, SAD only slightly more so.

4.9.5 Energy-Preserving Ambisonic Decoder (EPAD)

To establish panning-invariant loudness for decoding to non-uniform surround loudspeaker layouts one can ensure a constant loudness measure E by enforcing $\boldsymbol{D}^{\mathrm{T}}\boldsymbol{D} = \boldsymbol{I}$, which is otherwise only achieved on $t \geq 2\mathrm{N}$-designs. We may search for a decoding matrix $\boldsymbol{D}$ whose entries are closest to the sampling decoder under the constraint to be column-orthogonal:

$$\|\boldsymbol{D} - \sqrt{\tfrac{S_{\mathrm{D}-1}}{\mathrm{L}}}\,\boldsymbol{Y}_{\mathrm{N}}^{\mathrm{T}}\|_{\mathrm{Fro}}^{2} \rightarrow \min \qquad (4.42)$$
$$\text{subject to } \boldsymbol{D}^{\mathrm{T}}\boldsymbol{D} = \boldsymbol{I}.$$

The singular value decomposition of

$$\boldsymbol{Y}_{\mathrm{N}}^{\mathrm{T}} = \boldsymbol{U}\,[\mathrm{diag}\{\boldsymbol{s}\}, \boldsymbol{0}]^{\mathrm{T}}\,\boldsymbol{V}^{\mathrm{T}} \qquad (4.43)$$

can be used to create

$$\boldsymbol{D} = \boldsymbol{U}\,\left[\boldsymbol{I}, \boldsymbol{0}\right]^{\mathrm{T}}\,\boldsymbol{V}^{\mathrm{T}}. \qquad (4.44)$$

by replacing the singular values $\boldsymbol{s}$ with ones. Such a decoder is column-orthogonal, as the singular-value decomposition delivers $\boldsymbol{U}^{\mathrm{T}}\boldsymbol{U} = \boldsymbol{I}$ and $\boldsymbol{V}\boldsymbol{V}^{\mathrm{T}} = \boldsymbol{I}$, and as a consequence[1] $\boldsymbol{D}^{\mathrm{T}}\boldsymbol{D} = \boldsymbol{I}$. The energy-preserving decoder in this basic version requires $\mathrm{L} \geq 2\mathrm{N} + 1$ loudspeakers in 2D or $\mathrm{L} \geq (\mathrm{N} + 1)^2$ in 3D to work.

Note that if the loudspeaker setup directions are already a $t \geq 2\mathrm{N}$ design, the sampling, mode-matching, and energy-preserving decoders are equivalent.

4.9.6 All-Round Ambisonic Decoding (AllRAD)

In Chap. 3 on vector-base amplitude panning methods, a well-balanced panning result in terms of loudness, width, and localization was achieved by MDAP that distributes a signal to an arrangement of several superimposed VBAP virtual sources. Hereby $E = \text{const.}$, $\boldsymbol{r}_{\mathrm{E}} \approx r_{\mathrm{E}}\,\boldsymbol{\theta}_{\mathrm{s}}$, and $r_{\mathrm{E}} \approx \text{const.}$ This works for nearly any loudspeaker layout.

While, to calculate loudspeaker gains, MDAP superimposes an arrangement of discrete virtual sources within a range of $\pm\alpha$ around the panning direction $\boldsymbol{\theta}_{\mathrm{s}}$, one could also think of superimposing a quasi-continuous distribution of virtual sources that are weighted by a continuous panning function $g(\boldsymbol{\theta})$.

The ideal continuous panning function $g(\boldsymbol{\theta})$ of axisymmetric directional spread around the panning direction $\boldsymbol{\theta}_{\mathrm{s}}$ is described by $g(\boldsymbol{\theta}) = \boldsymbol{y}_{\mathrm{N}}^{\mathrm{T}}(\boldsymbol{\theta})\,\mathrm{diag}\{\boldsymbol{a}_{\mathrm{N}}\}\,\boldsymbol{y}_{\mathrm{N}}(\boldsymbol{\theta}_{\mathrm{s}})$, the Ambisonic panning function. This rotation-invariant continuous function is optimal in terms of loudness, width, and localization measures, which are all evaluated by continuous integrals: $E = \int g^2(\boldsymbol{\theta})\,\mathrm{d}\boldsymbol{\theta} = \text{const.}$ expresses panning-invariant loudness, $\boldsymbol{r}_{\mathrm{E}} = \frac{1}{E}\int g^2(\boldsymbol{\theta})\,\boldsymbol{\theta}\,\mathrm{d}\boldsymbol{\theta} = r_{\mathrm{E}}\,\boldsymbol{\theta}_{\mathrm{s}}$ indicates a perfect alignment $\boldsymbol{r}_{\mathrm{E}} \parallel \boldsymbol{\theta}_{\mathrm{s}}$ with the panning direction and a panning-invariant width $r_{\mathrm{E}} = \text{const.}$ However, the optimal values of

[1]In detail, this follows from $\boldsymbol{D}^{\mathrm{T}}\boldsymbol{D} = \boldsymbol{V}[\boldsymbol{I}, \boldsymbol{0}]\boldsymbol{U}^{\mathrm{T}}\boldsymbol{U}[\boldsymbol{I}, \boldsymbol{0}]^{\mathrm{T}}\boldsymbol{V}^{\mathrm{T}} = \boldsymbol{V}[\boldsymbol{I}, \boldsymbol{0}][\boldsymbol{I}, \boldsymbol{0}]^{\mathrm{T}}\boldsymbol{V}^{\mathrm{T}} = \boldsymbol{V}\boldsymbol{V}^{\mathrm{T}} = \boldsymbol{I}$.

these integrals are only preserved by discretization with optimal $t \geq 2\mathrm{N}+1$-design loudspeaker layouts.

All-round Ambisonic decoding (AllRAD) is preceded by the work of Batke and Keiler [16]. They describe Ambisonic panning $\boldsymbol{g}_{\mathrm{AllRAD}}(\boldsymbol{\theta}) = \boldsymbol{D}\,\boldsymbol{y}_{\mathrm{N}}(\boldsymbol{\theta})$ by a decoder $\boldsymbol{D}$, whose result matches best with VBAP $\boldsymbol{g}_{\mathrm{VBAP}}(\boldsymbol{\theta})$. Without max-$\boldsymbol{r}_{\mathrm{E}}$ weights yet, we use this here to define AllRAD by the integral expressing a minimum-mean-square-error problem using the integral over all panning directions $\boldsymbol{\theta}$

$$\min_{\boldsymbol{D}} \int_{\mathbb{S}^2} \left\| \boldsymbol{g}_{\mathrm{VBAP}}(\boldsymbol{\theta}) - \boldsymbol{D}\,\boldsymbol{y}_{\mathrm{N}}(\boldsymbol{\theta}) \right\|^2 \mathrm{d}\boldsymbol{\theta}. \tag{4.45}$$

Equivalently, as described by Zotter and Frank [40] who coined the name, we may define AllRAD as VBAP synthesis on the physical loudspeakers when using as multiple-virtual-source inputs the Ambisonic panning function $g_{\mathrm{AMBI}}(\boldsymbol{\theta}) = \boldsymbol{y}_{\mathrm{N}}^{\mathrm{T}}(\boldsymbol{\theta})$ $\mathrm{diag}\{\boldsymbol{a}_{\mathrm{N}}\}\,\boldsymbol{y}_{\mathrm{N}}(\boldsymbol{\theta}_{\mathrm{s}})$ sampled at an optimal layout of virtual loudspeakers. Here, we write the synthesis as the integral over infinitely many virtual loudspeakers $\boldsymbol{\theta}$,

$$\begin{aligned}
\boldsymbol{g} &= \int \boldsymbol{g}_{\mathrm{VBAP}}(\boldsymbol{\theta})\, g_{\mathrm{AMBI}}(\boldsymbol{\theta})\,\mathrm{d}\boldsymbol{\theta} = \int \boldsymbol{g}_{\mathrm{VBAP}}(\boldsymbol{\theta})\,\boldsymbol{y}_{\mathrm{N}}^{\mathrm{T}}(\boldsymbol{\theta})\,\mathrm{diag}\{\boldsymbol{a}_{\mathrm{N}}\}\boldsymbol{y}_{\mathrm{N}}(\boldsymbol{\theta}_{\mathrm{s}})\,\mathrm{d}\boldsymbol{\theta} \\
&= \underbrace{\int \boldsymbol{g}_{\mathrm{VBAP}}(\boldsymbol{\theta})\,\boldsymbol{y}_{\mathrm{N}}^{\mathrm{T}}(\boldsymbol{\theta})\,\mathrm{d}\boldsymbol{\theta}}_{:=\boldsymbol{D}}\;\mathrm{diag}\{\boldsymbol{a}_{\mathrm{N}}\}\boldsymbol{y}_{\mathrm{N}}(\boldsymbol{\theta}_{\mathrm{s}}).
\end{aligned}$$

We can obviously pull the term $\mathrm{diag}\{\boldsymbol{a}_{\mathrm{N}}\}\boldsymbol{y}_{\mathrm{N}}(\boldsymbol{\theta}_{\mathrm{s}})$ out of the integral. The remaining integral defines the AllRAD matrix $\boldsymbol{D}$. We may interpret it as a transformation of the VBAP loudspeaker gain functions $\boldsymbol{g}_{\mathrm{VBAP}}(\boldsymbol{\theta})$ into spherical harmonic coefficients. In the original paper [40], AllRAD is evaluated by an optimal layout of discrete virtual loudspeakers

$$\boldsymbol{D} = \int \boldsymbol{g}_{\mathrm{VBAP}}(\boldsymbol{\theta})\,\boldsymbol{y}_{\mathrm{N}}^{\mathrm{T}}(\boldsymbol{\theta})\,\mathrm{d}\boldsymbol{\theta} = \tfrac{S_{\mathrm{D}-1}}{\hat{\mathrm{L}}} \sum_{l=0}^{\hat{\mathrm{L}}} \boldsymbol{g}_{\mathrm{VBAP}}(\hat{\boldsymbol{\theta}}_l)\,\boldsymbol{y}_{\mathrm{N}}^{\mathrm{T}}(\hat{\boldsymbol{\theta}}_l) = \tfrac{S_{\mathrm{D}-1}}{\hat{\mathrm{L}}}\,\hat{\boldsymbol{G}}\,\hat{\boldsymbol{Y}}_{\mathrm{N}}^{\mathrm{T}}, \tag{4.46}$$

using the directions $\{\hat{\boldsymbol{\theta}}_l\}$ of a t-design. As VBAP's gain functions aren't smooth (derivatives are non-continuous), they are order-unlimited, and a t-design of sufficiently high t should be used. In the 3D practice, the 5200 pts. Chebyshev-type design from [33] is dense enough. Note that the VBAP part permits improvements by insertion and downmix of imaginary loudspeakers to adapt to asymmetric or hemispherical layouts, as suggested in the original paper [40], cf. Sect. 3.3.

Note that the decoder needs to be scaled properly. For instance, the norm of the omnidirectional component (first column) could be equalized to one, as it would typically be with a sampling decoder; there are alternative strategies to circumvent the scaling problem [41].

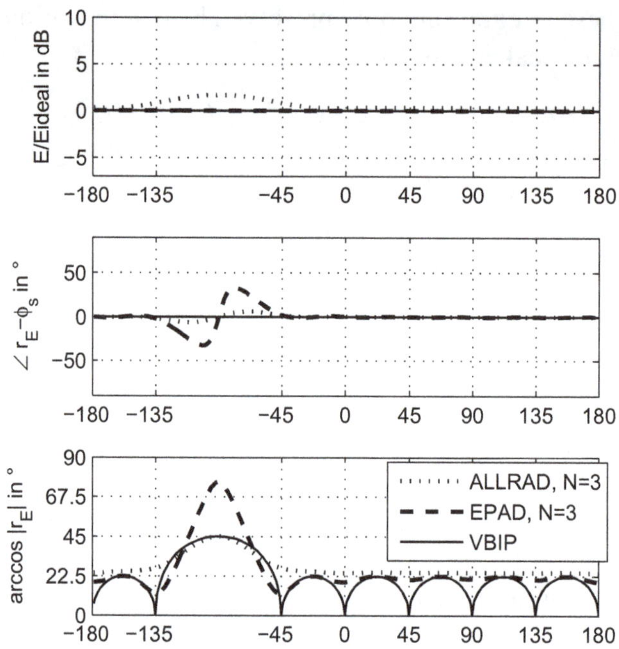

Fig. 4.16 Analysis of loudness, localization error and width measure for energy-preserving Ambisonic decoding (EPAD) and all-round Ambisonic decoding (AllRAD) for all panning angles on a sub-optimal 45° spaced loudspeaker ring with gap at −90°

4.9.7 EPAD and AllRAD on Sub-optimal Layouts

Figude 4.16 shows the improvement achieved with EPAD and AllRAD on an equi-angular arrangement that is suboptimal by the missing loudspeaker at −90°. Both decoders manage to handle either the loudness stabilization perfectly well (EPAD) or keep the directional and spread mapping errors small (AllRAD). We notice that for EPAD, with the constraint that $L \geq (2N + 1)$ just fulfilled for $N = 3$ and $L = 7$ of the simulation, it would not simply be possible to remove any further loudspeakers without degradation.

4.9.8 Decoding to Hemispherical 3D Loudspeaker Layouts

In typical loudspeaker playback situations for large audience, a solid floor and no loudspeakers below ear level are considered practical for several reasons. However, this does not permit decoding by sampling with optimal t-design layouts covering all directions. As shown above, EPAD and AllRAD do not require such arrays. And yet, they still require some care when used with hemispherical loudspeaker layouts, see [15, 40] for further reading.

***EPAD with hemispherical loudspeaker layouts*.** Even for a hemispherical layout, the energy-preserving decoding method requires $L \geq (N + 1)^2$ loudspeakers to achieve a perfectly panning-invariant loudness. However, this is counter-intuitive: *Why should one need at least as many loudspeakers on a hemisphere as are required for same-order playback on a full sphere? Shouldn't the number be half as many?*

Table 4.1 Integration ranges $0 \leq \vartheta \leq \vartheta_{\max}$ to obtain $(\mathrm{N}+1)(\mathrm{N}+2)/2$ Slepian functions with minimum loudness fluctuation $\frac{\max E}{\min E}$ for panning on the hemisphere

N	1	2	3	4	5	6	7	8	9
$\vartheta_{\max}$	115°	114°	113°	111°	108°	106°	104°	102°	101°
$\frac{\max E}{\min E}$	0.5 dB	0.4 dB	0.4 dB	0.3 dB	0.3 dB	0.3 dB	0.2 dB	0.3 dB	0.2 dB

We can show that while the spherical harmonics are orthonormal on the sphere $\mathbb{S}^2$, i.e. $\int_{\mathbb{S}^2} \boldsymbol{y}_\mathrm{N}(\boldsymbol{\theta})\boldsymbol{y}_\mathrm{N}^\mathrm{T}(\boldsymbol{\theta})\,\mathrm{d}\boldsymbol{\theta} = \boldsymbol{I}$, they aren't orthogonal on the hemisphere $S = \mathbb{S}^2 : \vartheta \leq \vartheta_{\max}$

$$\int_S \boldsymbol{y}_\mathrm{N}(\boldsymbol{\theta})\boldsymbol{y}_\mathrm{N}^\mathrm{T}(\boldsymbol{\theta})\,\mathrm{d}\boldsymbol{\theta} = \boldsymbol{G}. \tag{4.47}$$

Here, $\boldsymbol{G}$ is called Gram matrix, and it is evaluated by $\frac{4\pi}{\mathrm{L}}\sum_{l:\theta_{z,l}\geq 0} \boldsymbol{y}_\mathrm{N}(\boldsymbol{\theta}_l)\boldsymbol{y}_\mathrm{N}^\mathrm{T}(\boldsymbol{\theta}_l)$ using a high-enough t-design. By singular-value decomposition of the positive semi-definite matrix $\boldsymbol{G} = \boldsymbol{Q}\,\mathrm{diag}\{\boldsymbol{s}\}\,\boldsymbol{Q}^\mathrm{T}$, with $\boldsymbol{Q}^\mathrm{T}\boldsymbol{Q} = \boldsymbol{Q}\boldsymbol{Q}^\mathrm{T} = \boldsymbol{I}$, we diagonalize $\boldsymbol{G}$ and find new basis functions $\tilde{\boldsymbol{y}}_\mathrm{N}(\boldsymbol{\theta})$, the so-called *Slepian* functions [42], that are orthogonal on S

$$\boldsymbol{Q}^\mathrm{T}\boldsymbol{G}\,\boldsymbol{Q} = \mathrm{diag}\{\boldsymbol{s}\} = \int_S \boldsymbol{Q}^\mathrm{T}\boldsymbol{y}_\mathrm{N}(\boldsymbol{\theta})\boldsymbol{y}_\mathrm{N}^\mathrm{T}(\boldsymbol{\theta})\,\boldsymbol{Q}\,\mathrm{d}\boldsymbol{\theta}, \qquad \Rightarrow \tilde{\boldsymbol{y}}_\mathrm{N}(\boldsymbol{\theta}) = \boldsymbol{Q}^\mathrm{T}\boldsymbol{y}_\mathrm{N}(\boldsymbol{\theta}).$$

Typically, the singular values in $\boldsymbol{s}$ are sorted descendingly $s_1 \geq s_2 \geq \cdots \geq s_{(\mathrm{N}+1)^2}$ so that it is possible to cut out basis functions of significantly large contribution to the upper hemisphere S by

$$\tilde{\boldsymbol{y}}_\mathrm{N}(\boldsymbol{\theta}) = [\boldsymbol{I},\,\boldsymbol{0}]\,\boldsymbol{Q}^\mathrm{T}\boldsymbol{y}_\mathrm{N}(\boldsymbol{\theta}). \tag{4.48}$$

Typically, the numerical integral is extended to slightly below the horizon, see Table 4.1, so that truncation to the $(\mathrm{N}+1)(\mathrm{N}+2)/2$ most significant basis functions, see Fig. 4.17, produces a minimum fluctuation in the loudness measure $\tilde{E} = \|\tilde{\boldsymbol{y}}_\mathrm{N}(\boldsymbol{\theta})\|^2$ for panning on the hemisphere.

With $\tilde{\boldsymbol{y}}_\mathrm{N}(\boldsymbol{\theta})$, EPAD is calculated in the same way as for the ordinary harmonics

$$\tilde{\boldsymbol{Y}}_\mathrm{N}^\mathrm{T} = \tilde{\boldsymbol{U}}[\mathrm{diag}\{\boldsymbol{s}\},\,\boldsymbol{0}]^\mathrm{T}\tilde{\boldsymbol{V}}^\mathrm{T}, \qquad \tilde{\boldsymbol{D}} = \tilde{\boldsymbol{U}}[\boldsymbol{I},\,\boldsymbol{0}]^\mathrm{T}\tilde{\boldsymbol{V}}^\mathrm{T}, \tag{4.49}$$

with the main difference that the lower limit for the number of loudspeakers decreases to $\mathrm{L} \geq (\mathrm{N}+1)(\mathrm{N}+2)/2$. Interfaced to the spherical harmonics by $[\boldsymbol{I},\,\boldsymbol{0}]\,\boldsymbol{Q}^\mathrm{T}$, the hemispherical energy-preserving decoder becomes

$$\boldsymbol{D} = \tilde{\boldsymbol{D}}\,[\boldsymbol{I},\,\boldsymbol{0}]\,\boldsymbol{Q}^\mathrm{T}. \tag{4.50}$$

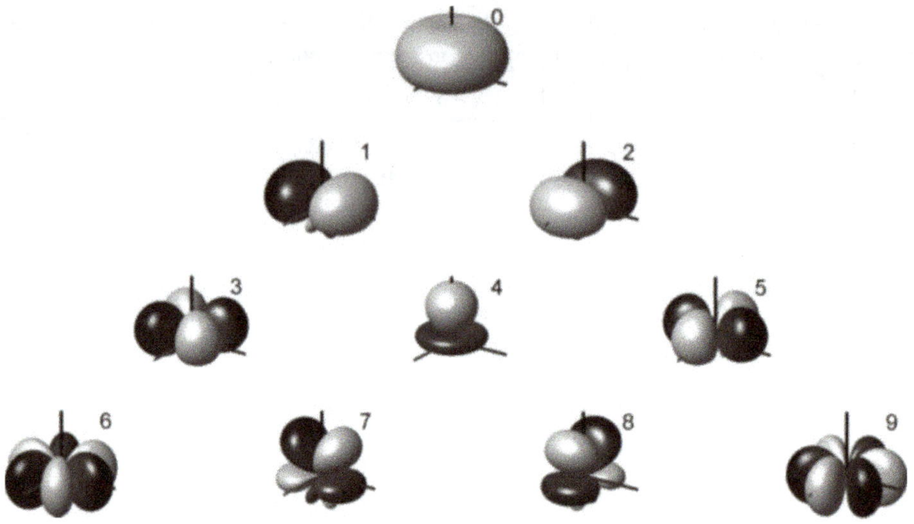

Fig. 4.17 Slepian basis functions for the upper hemisphere, composed of the spherical harmonics up to 3rd order, using $0° \leq \vartheta \leq 113°$ as integration interval and $(\mathrm{N}+1)(\mathrm{N}+2)/2$ functions. To get these nice shapes, the Slepian functions were found separately for every degree m where they were moreover de-mixed by QR decomposition

AllRAD with hemispherical loudspeaker layouts. Because of the vector-base amplitude panning involved, all-round Ambisonic decoding (AllRAD) is comparatively robust to irregular loudspeaker setups. Still, a hemispherical layout does not contain any loudspeaker direction vector pointing to the lower half space, therefore one could just omit information of the lower half space. However, the Ambisonic panning function implies a directional spread, so that panning to exactly the horizon also produces content below, whose omission causes: (i) a loss in loudness, (ii) a slight elevation of the perceived direction, cf. Fig. 4.18.

As discussed in the section on triangulation Sect. 3.3, the insertion of imaginary loudspeakers fixes this behavior. In the case of hemispherical loudspeaker layouts, it is not necessary to downmix the signal of the imaginary loudspeaker at nadir to stabilize both loudness and localization for panning to the horizon.

Signal contributions below but close to the horizon largely contribute to the horizontal loudspeakers, and it is therefore safe to dispose the signal that would feed the imaginary loudspeaker at nadir without loss of loudness. Moreover, this contribution from below also reinforces signals on the horizontal loudspeakers so that localization is pulled back down. Both can be observed in Fig. 4.18 that shows the loudness measure E as well as mislocalization and width by the measure $\boldsymbol{r}_\mathrm{E}$ using max-$\boldsymbol{r}_\mathrm{E}$-weighted AllRAD with 5th-order Ambisonics along a vertical panning circle on the IEM mobile Ambisonics Array (mAmbA). It consists of 25 loudspeakers set up in rings of 8, 8, 4, 4, and 1 loudspeakers at 0, 20, 40, 60, 90 degrees elevation. Rings two and four start at 0 degree, the others are half-way rotated.

Performance comparison on hemispherical layouts. Figure 4.18 shows a comparison of AllRAD and EPAD decoding to the 25-channel mAmbA hemispherical loudspeaker layout.

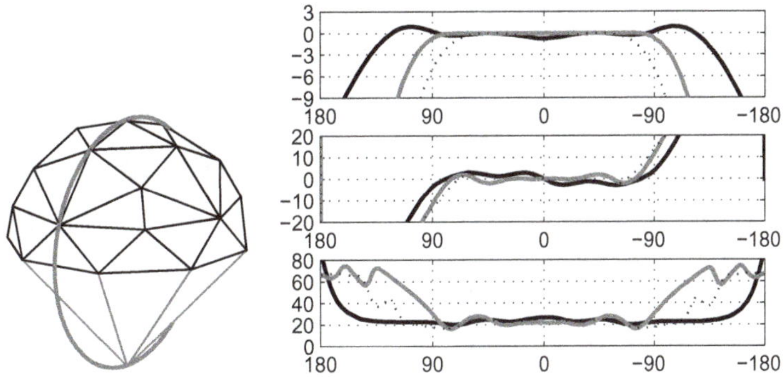

Fig. 4.18 Perceptual measures for 5th max-$\boldsymbol{r}_\mathrm{E}$-weighted AllRAD on the IEM mAmbA layout [43] with (black) and without insertion of the bottom imaginary loudspeaker (black dotted) whose signal is disposed, and max-$\boldsymbol{r}_\mathrm{E}$-weighted EPAD (gray), for panning on a vertical circle: E in dB (top), orientation error of $\boldsymbol{r}_\mathrm{E}$ in degrees (middle), and width expressed by arccos $\|\boldsymbol{r}_\mathrm{E}\|$ in degrees (bottom). The thin dashed line shows AllRAD without imaginary loudspeakers

While (top in Fig. 4.18) AllRAD produces a loudness fluctuation roughly spanning 1 dB for panning on the hemisphere, EPAD only exhibits 0.3 dB, as specified in Table 4.1. While in monophonic playback of noise, loudness differences of less than 0.5 dB can be heard, it is safe to assume that a weak directional loudness fluctuation of less than 1 dB is normally inaudible. In this regard, loudness fluctuation should be no problem with both EPAD and AllRAD.

Concerning the directional mapping, EPAD produces a more strongly pronounced ripple, with $\boldsymbol{r}_\mathrm{E}$ indicating sounds on the horizon $\vartheta_\mathrm{s} = \pm 90^\circ$ to be pulled upwards towards 0° more with EPAD (7°) than with AllRAD (3°). In terms of width, both EPAD and AllRAD exhibit the $\approx 20^\circ$ average associated with max-$\boldsymbol{r}_\mathrm{E}$ weighting. However, EPAD also produces a greater fluctuation, and it widens up to about 30° degree for panning to the horizon $\vartheta_\mathrm{s} = \pm 90^\circ$.

With the 9 loudspeakers of the ITU [44] $4+5+0$ layout (horizontal ring: $\varphi = 0,\ \pm 30^\circ,\ \pm 120^\circ$, upper ring at 40° elevation with $\varphi = \pm 30^\circ,\ \pm 120^\circ$), it is not possible anymore to use EPAD with 5th order, which would be the optimal resolution for the front loudspeaker triplet. EPAD only supports orders up to $\mathrm{N} = 2$, and to lose level towards below-horizon directions, we can use the reduced set of 6 Slepian functions; alternatively all 9 spherical harmonics of $\mathrm{N} = 2$ would also be thinkable. For AllRAD, imaginary loudspeakers are inserted at the sides at azimuth/elevation $\pm 75^\circ/27^\circ$, up at $0^\circ/78^\circ$, back $180^\circ/35^\circ$, and below $0^\circ/-90^\circ$. It is reasonable to downmix the imaginary loudspeakers with a factor one for up, sides, back, and re-normalize the VBAP gain matrix, while disposing the signal of the imaginary loudspeaker below. AllRAD permits to use the order $\mathrm{N} = 5$, which resolves the frontal loudspeaker triplet much better for horizontal panning.

Figure 4.19 shows the result of max-$\boldsymbol{r}_\mathrm{E}$-weighted 2nd-order EPAD and 5th-order AllRAD for the $4+5+0$ layout using a vertical panning curve. While the perfectly constant loudness measure of EPAD might be favored over the almost +3 dB loudness

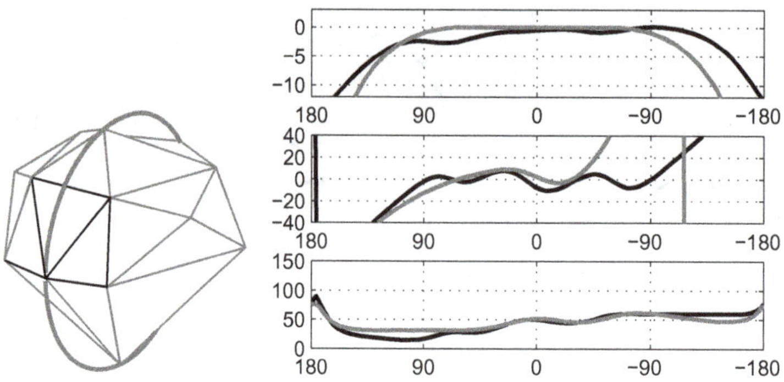

Fig. 4.19 Perceptual measures for Ambisonic panning on the ITU [44] 4 + 5 + 0 layout with insertion and downmix of imaginary loudspeakers at the sides, back, and top, and insertion and disposal of an imaginary loudspeaker below for AllRAD. Measures are evaluated for max-$\boldsymbol{r}_\mathrm{E}$-weighted 5th-order AllRAD (black) and 2nd-order EPAD (gray), for panning on a vertical circle: E in dB (top), orientation error of $\boldsymbol{r}_\mathrm{E}$ in degrees (middle), and width expressed by arccos $\|\boldsymbol{r}_\mathrm{E}\|$ in degrees (bottom)

increase of front and back for AllRAD, AllRAD's lower directional error, narrower width mapping, greater flexibility, and simplicity has often proven to be clearly superior in practice.

4.10 Practical Studio/Sound Reinforcement Application Examples

This section analyzes the application of 3D Ambisonic amplitude panning consisting of encoding and AllRAD to studio (with typical setups of 2 m radius) and sound reinforcement applications (for an audience of, e.g., 250 people). Application scenarios are sketched in [43], and various other examples are given below. Requirements of a constant loudness and width are analyzed below, and as sound reinforcement requires a particularly large sweet area, the $\boldsymbol{r}_\mathrm{E}$ vector model for off-center listening positions from Sect. 2.2.9 is used to depict the sweet area size.

The analysis of decoders above described loudness measures for panning on a circle. To observe them with panning across all directions in Figs. 4.20 and 4.22, world-map-like mappings using a gray-scale representation of the loudness and width measures are more reasonable. For several loudspeaker layouts, its axes are azimuth horizontally and zenith vertically, and the gray-scale map displays the loudness measure E in dB (left column) and the width measure arccos $\|\boldsymbol{r}_\mathrm{E}\|$ in degrees (right column). As 5th-order max-$\boldsymbol{r}_\mathrm{E}$-weighted AllRAD typically produces minor directional mapping errors, they aren't explicitly shown in Figs. 4.20 and 4.22. However, the mappings of the sweet area size of plausible localization in Figs. 4.21 and 4.23 illustrate the usefulness of the systems for the listening areas hosting the number of listeners targeted for either the studio or the sound reinforcement application.

Figure 4.20 illustrates AllRAD's tendency of attenuated signals in too closely spaced loudspeaker ensembles as in the front section of the ITU [44] 4 + 5 + 0. By

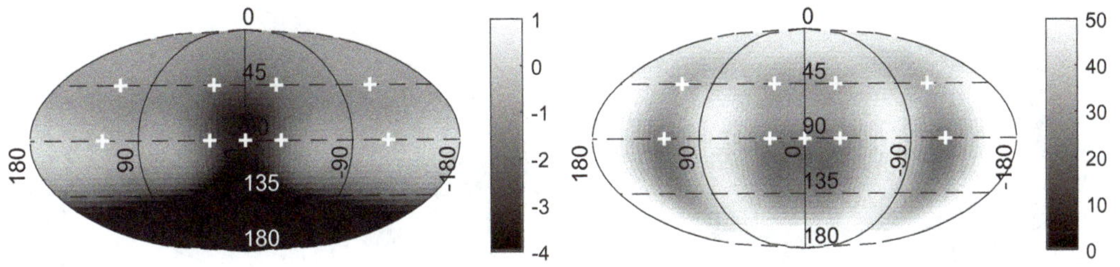

(a) ITU [44] 4+5+0 Studio (1 listener)

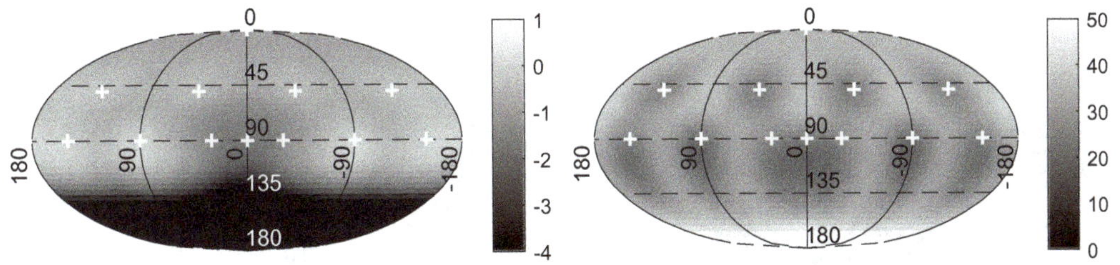

(b) IEM Production Studio (4 listeners)

Fig. 4.20 Comparison of 5th-order max-$\boldsymbol{r}_\mathrm{E}$-weighted AllRAD for panning across all directions on hemispherical loudspeaker layouts in studios. The left column the loudness measure E in dB and the right-most column the width measure $\arccos \|\boldsymbol{r}_\mathrm{E}\|$ in degree, and the loudspeaker position are marked with a white $+$ sign

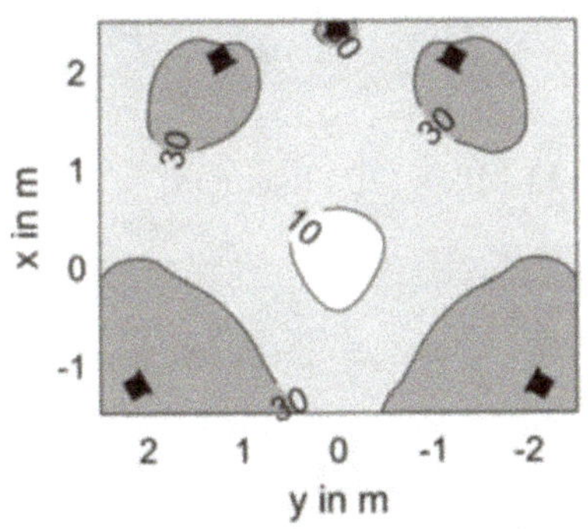

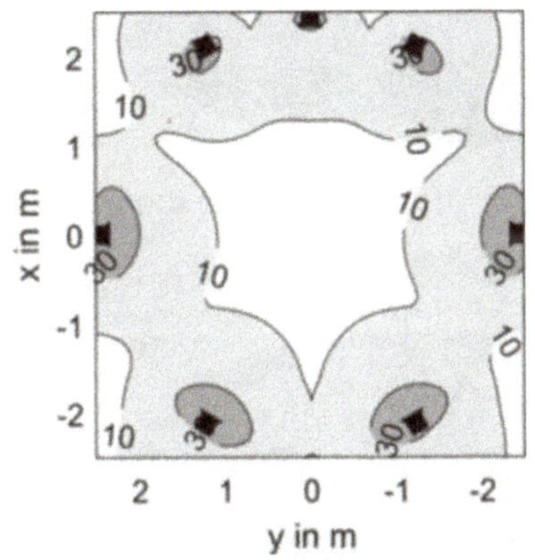

(a) ITU [44] 4+5+0 (1 listener) (b) IEM Production Studio (4 listeners)

Fig. 4.21 Comparison of the calculated sweet are size for 5th-order max-$\boldsymbol{r}_\mathrm{E}$-weighted AllRAD for panning across all directions on hemispherical loudspeaker layouts in studios. As a plausibility definition, the directional mapping errors depending on the listening position should stay within angular bounds (e.g. 10°)

contrast, for instance the mAmbA layout in Fig. 4.22 only has 8 loudspeakers on the horizon, and signals panned to the largely spaced below-horizon triangles tend to get louder. Moreover, it is easier for loudspeaker systems of many channels such as IEM CUBE, mAmbA, Lobby, and Ligeti Hall in Fig. 4.22 to yield smooth loudness and width mappings. Still, also with only a few loudspeakers, slight direction adjustment in the layout can fix some of the behavior, as with the IEM Production Studio, whose ±45° loudspeakers in the elevated layer is superior to a ±30° spacing.

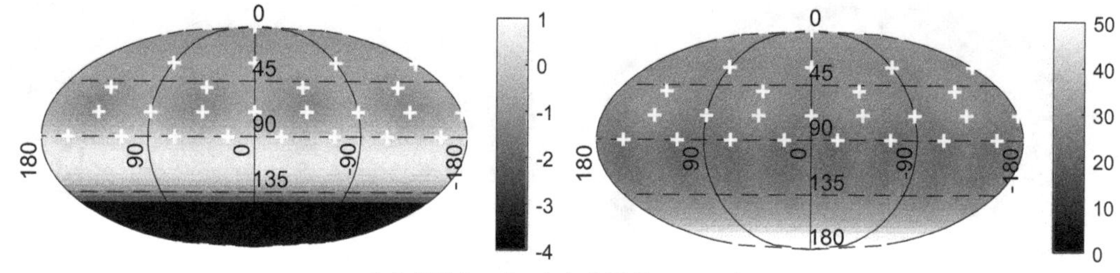

(a) IEM mAmbA (40 listeners)

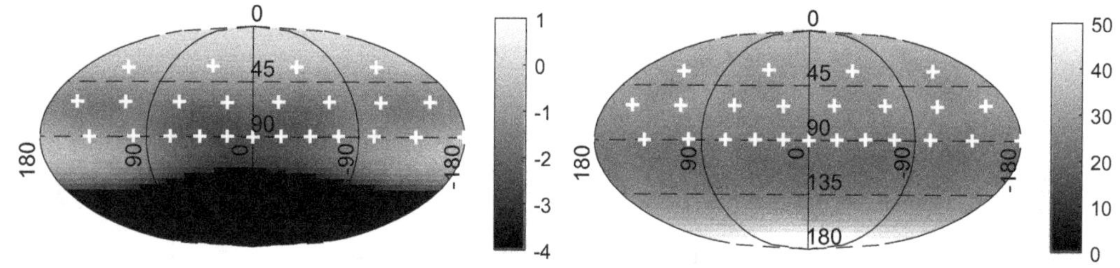

(b) IEM CUBE (80 listeners)

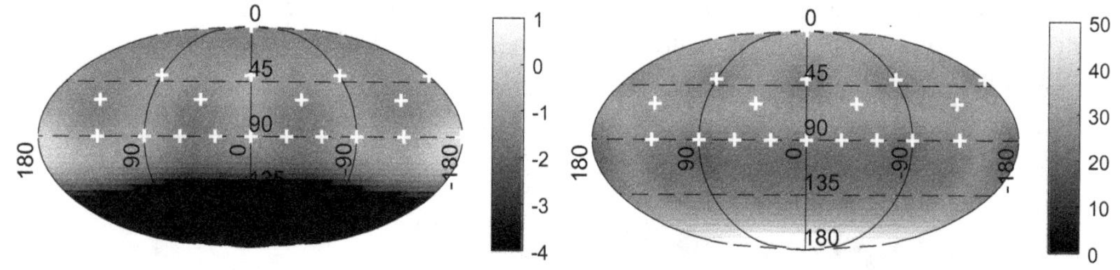

(c) Lobby demo room at AES 144 Milan (200 listeners)

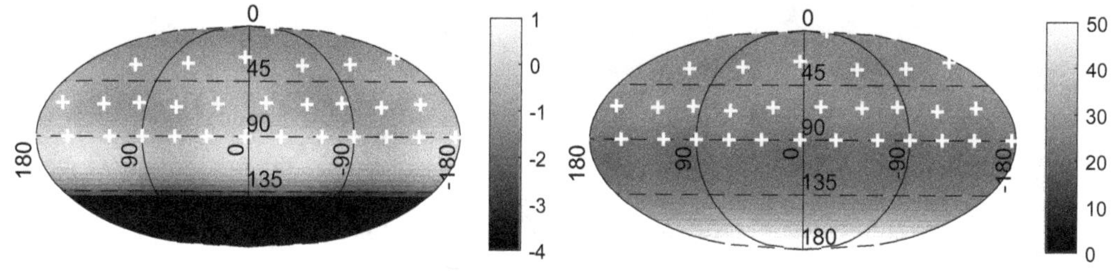

(d) KUG Ligeti Hall (250 listeners)

Fig. 4.22 Comparison of 5th-order max-$\boldsymbol{r}_\mathrm{E}$-weighted AllRAD for panning across all directions on various hemispherical loudspeaker layouts for sound reinforcement. The left column the loudness measure E in dB and the right-most column the width measure $\arccos \|\boldsymbol{r}_\mathrm{E}\|$ in degree, and the loudspeaker position are marked with a white + sign

A hint for designing good decoders sometimes is idealization: often it is better to disregard the true loudspeaker setup locations and feed the decoder design with idealized positions instead. Hereby can one trade slight directional distortions for a more uniform loudness distribution. For instance at the IEM CUBE, loudspeaker

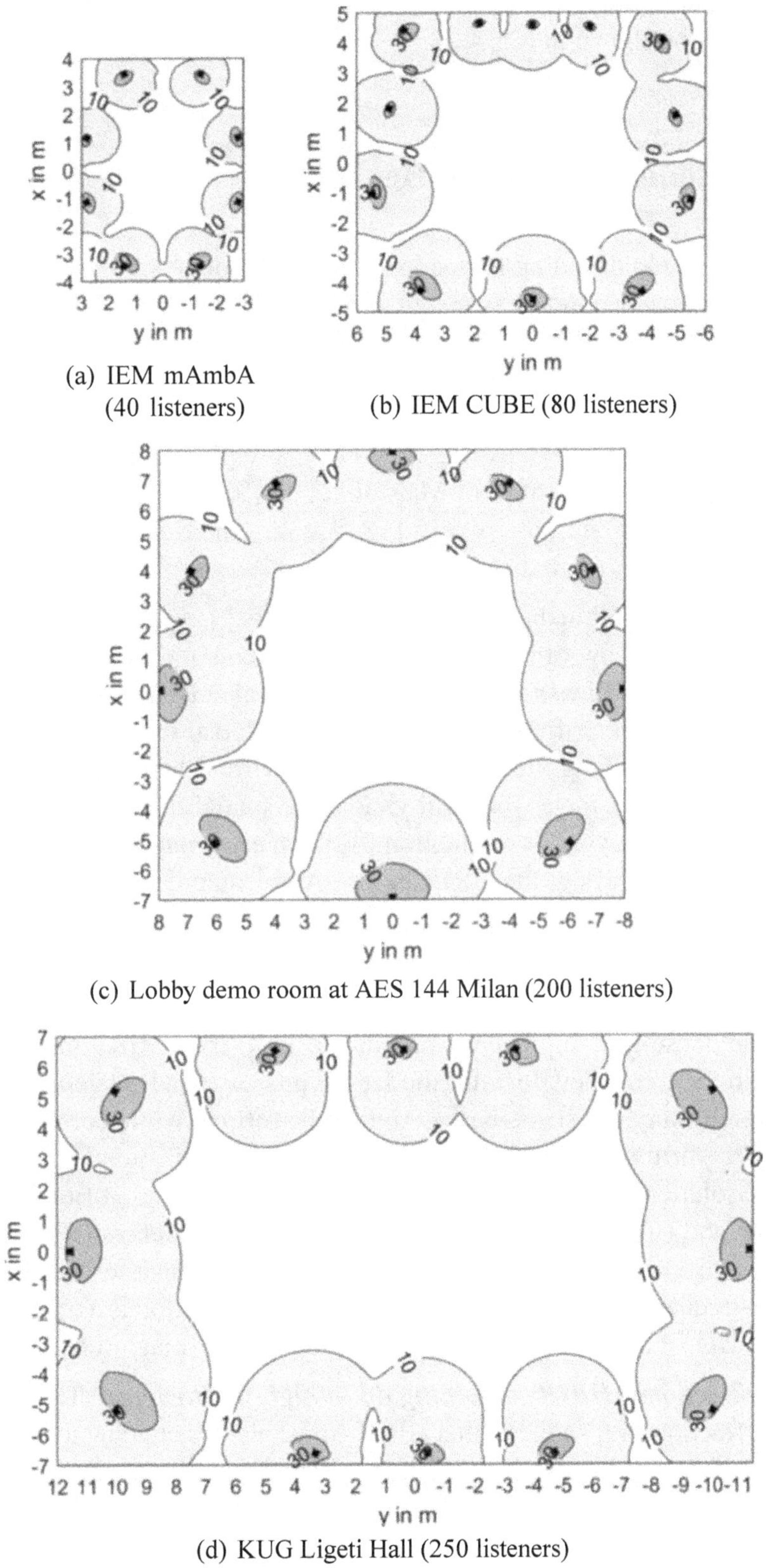

(a) IEM mAmbA (40 listeners)

(b) IEM CUBE (80 listeners)

(c) Lobby demo room at AES 144 Milan (200 listeners)

(d) KUG Ligeti Hall (250 listeners)

Fig. 4.23 Comparison of the calculated sweet are size for 5th-order max-$\boldsymbol{r}_{\mathrm{E}}$-weighted AllRAD for panning across all directions on various hemispherical loudspeaker layouts for sound reinforcement. As a plausibility definition, the directional mapping errors depending on the listening position should stay within angular bounds (e.g. 10°)

locations of the horizontal ring could be idealized to 30° to get a smoother loudness mapping as the one shown in Fig. 4.22.

4.11 Ambisonic Decoding to Headphones

Typically, Ambisonic decoding to headphones can be done similarly as with loudspeakers, except that the loudspeaker signals are rendered to headphones by convolution with the head-related impulse responses (HRIRs) of the corresponding playback directions. Various databases of such HRIRs can be found, e.g., on the website SOFA-conventions.[2] This headphone decoding approach is classically using a small set of so-called *virtual loudspeakers*, as it is found in many places in technical literature, e.g. in the pioneering works of Jean-Marc Jot et al. [9] or Jérôme Daniel [10]. It is relevant in many important other works [18, 45, 46], the SADIE project,[3] and it is employed in Sect. 1.4.2 on first-order Ambisonics.

Coarse. However, as outlined in some research papers [9, 18, 46], these approaches have in common that low-order Ambisonic synthesis is problematic. It can either happen when inserting a *dense grid* of virtual-loudspeaker HRIRs that the Ambisonic smoothing attenuates high-frequency at frontal and dorsal directions. Or, what had been the solution for a long time, a *coarse grid* of virtual-loudspeaker HRIRs does not attenuate high frequencies, but still yields that spatial quality strongly depends on the particular grid layout or orientation [46]. An early paper by Jot [9] proposed to remove the time delays of the HRIR before Ambisonic decomposition, and then to re-insert the otherwise missing interaural time-delay afterwards, for any sound panned in Ambisonics, which unfortunately yields an *object-based* panning system rather than a *scene-based* Ambisonic system.

Dense. Some dense-grid approaches propose to keep the HRIR time delays, or if formulated in the frequency domain: the HRTF phases (head-related transfer function), and hereby stay in a scene-based Ambisonic format, while correcting spectral deficiencies by diffuse-field or interaural-covariance equalization [18, 47]. Finally, most recent solutions proposed by Jin, Sun, and Epain, [17, 48] or Zaunschirm, Schörkhuber, and Höldrich [20, 21] modify the HRIR time delays/HRTF phases but only above, e.g., 3 kHz, without any object-based re-insertion afterwards. The omission of high-frequency interaural time-delay/phase information is a reasonable trade off done in favor of a more important accuracy in spectral magnitude.

What does directional HRIR smoothing do to high frequencies? The geometrical theory of diffraction [49] suggests that HRIRs must always contain at least the delay to the ear of either the shortest direct path or the shortest indirect path via the surface of the head. For a spherical head model with the radius $\mathrm{R} = 0.0875$ m and speed of sound $c = 343\ \frac{\mathrm{m}}{\mathrm{s}}$, the Woodworth-Schlosberg formula [50] is composed of this

[2] https://www.sofaconventions.org.

[3] https://www.york.ac.uk/sadie-project/database.html.

Fig. 4.24 Rigid-sphere model of delay to ear depending on angle ϕ (here: head rotation)

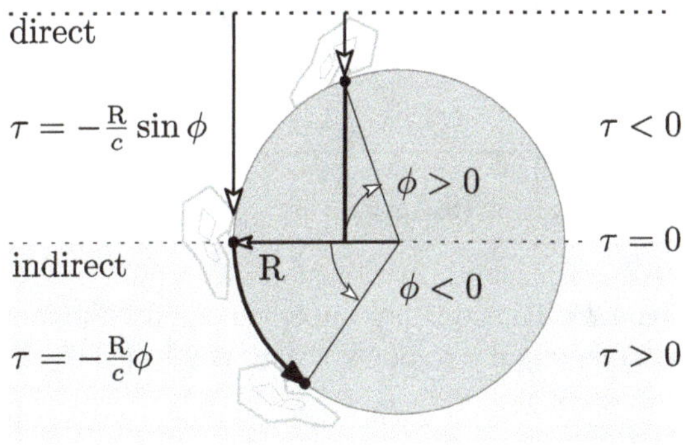

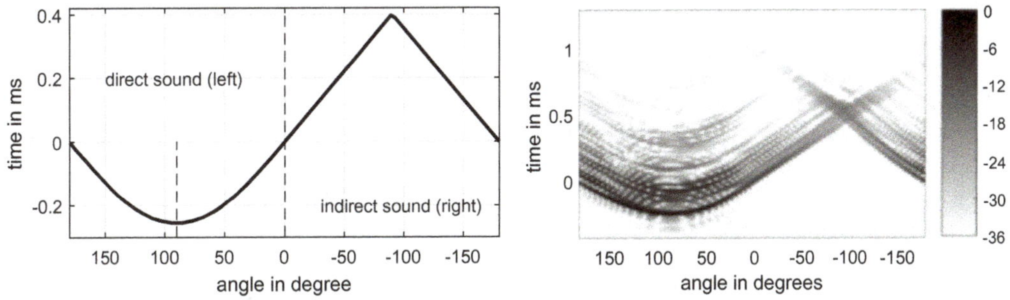

(a) HRIR delay model (b) Measured HRIR

Fig. 4.25 Time delay to the ear depending on azimuth of horizontal sound (a) and (b) 360°-measured KU100 dummy head HRIRs set from TH Köln (color displays dB levels)

consideration, see Fig. 4.24. The left ear receives a distant horizontal sound from the azimuth interval $0 \leq \phi \leq \frac{\pi}{2}$ as direct sound anticipated by $\tau = -\frac{R}{c}\sin\phi$, or for $-\frac{\pi}{2} < \phi \leq 0$ as an indirect sound delayed by $\tau = -\frac{R}{c}\,\phi$,

$$\tau(\phi) = -\tfrac{R}{c}\;\min\{\sin\phi,\ \phi\}, \tag{4.51}$$

as plotted in Fig. 4.25a, and recognizable from dummy-head measurements[4] in Fig. 4.25b.

If the HRIR is smoothed across an angular range, the time-delay curve gets spread across time as well, see Fig. 4.26. In this way, depending on whether the smoothing uses a continuous or discrete set of directions, one either obtains something like a comb filter or a sinc-shaped frequency response. This smoothing is least disturbing around the direct-ear side as shown left in Fig. 4.26, and, as the indirect ear also encounters high-frequency shadowing effects, it is most disturbing mainly for frontal and rear sounds at 0° or 180°, as shown right in Fig. 4.26. The corresponding frequency responses are roughly exemplified with what third-order Ambisonics equivalent smoothing would do to either 45°-spaced HRIRs in Fig. 4.27a or 15°-spaced ones in Fig. 4.27b.

[4]Data HRIR_CIRC360.sofa from http://sofacoustics.org/data/database/thk.

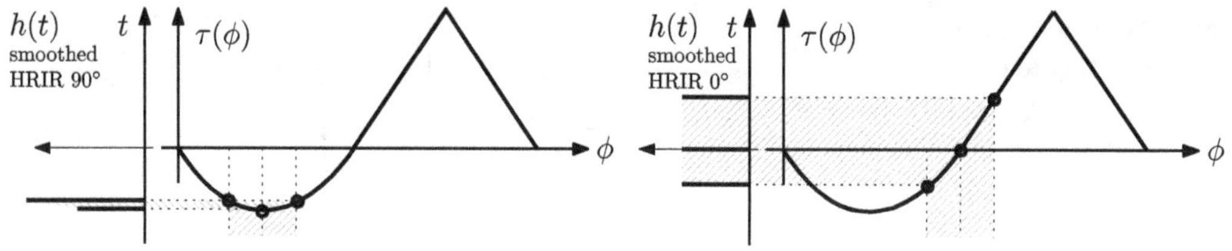

Fig. 4.26 Directionally smoothed playback to multiple HRIRs within a window, e.g. 30°, causes different impulse response shapes at 90° and 0°

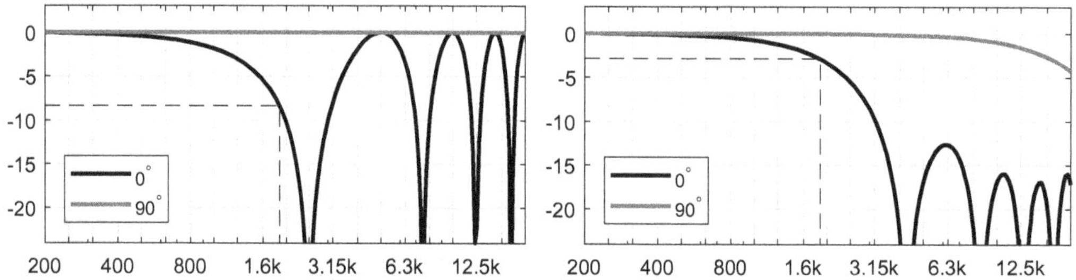

(a) Sparse HRIR set

(b) Dense HRIR set

Fig. 4.27 Differences of directionally smoothed HRTF frequency responses for a horizontal sound from either 0° and 90°, smoothed within a ±22.5° window, which roughly corresponds to the Ambisonics order N = 3; **a** and **b** either use a grid of 45° or 7.5° spaced HRIRs. The dashed line shows the the theoretical frequency limit 1.87 kHz for N = 3

To get an upper frequency limit, it is insightful to work in the frequency domain where the HRIR is denoted head-related transfer function (HRTF). A simplified linearized-phase version around $\phi = 0$ uses $\tau \approx \frac{\mathrm{R}}{c}\,\phi$, and the resulting in the Fourier transform with $\omega = 2\pi\, f$ is

$$H \approx e^{-\mathrm{i}\frac{\omega}{c}\mathrm{R}\,\tau(\phi)} = e^{\mathrm{i}\frac{\omega}{c}\mathrm{R}\,\phi}. \tag{4.52}$$

To represent it by circular or spherical harmonics transformation limited to the order N, a maximum phase change represented by the harmonic $e^{\mathrm{i}\mathrm{N}\phi}$ implies that we can only resolve the phase up to $\frac{\omega}{c}\mathrm{R} \leq \mathrm{N}$, hence the range of accurate operation is limited in frequency

$$f_{\mathrm{N}} \leq \frac{c\,\mathrm{N}}{2\pi\,\mathrm{R}} = \mathrm{N} \cdot 624\,\mathrm{Hz}. \tag{4.53}$$

As high-frequency HRTF phase evolves more rapidly over the angle as what the finite order can represent, this typically yields attenuation of the high frequencies when obtaining circular/spherical harmonics coefficients by transformation integral.

Directional smoothing of the discrete directional HRTFs causes relevant spectral problems, regardless of whether directional smoothing is done by Ambisonics, VBAP, MDAP. Mainly the geometric delay in the HRIRs is responsible for the emerging comb-filter or low-pass behavior. One could pull out the linear phase trend above the frequency limit and re-insert it, but is re-insertion necessary?

4.11.1 High-Frequency Time-Aligned Binaural Decoding (TAC)

As a pre-requisite for their binaural Ambisonic decoders, Schörkhuber et al. [21] tested, above which frequency the removal of the HRTF linear phase trend remains inaudible in direct HRTF-based rendering without panning or smoothing. In fact, most of their listeners could not distinguish the absence of the linear phase trend when removed above 3 kHz for various sound examples (drums, speech, pink noise, rendered at directions 10°, −45°, 80°, −130°). They had their subjects compare the result to a reference with unaltered HRTFs, and the result is analyzed in Fig. 4.28.

By this finding, it is possible to split up each of the 2 × 1 HRIRs $\boldsymbol{h}(t, \boldsymbol{\theta})$ into an unaltered low-pass band and a time-aligned high-pass band to unify the high-frequency HRIR delay

$$\hat{\boldsymbol{h}}(t, \boldsymbol{\theta}) = \boldsymbol{h}_{\leq 3kHz}(t, \boldsymbol{\theta}) + \begin{bmatrix} h_{\mathrm{left}>3kHz}[t - \tau(\arcsin \theta_{\mathrm{y}}), \boldsymbol{\theta}] \\ h_{\mathrm{right}>3kHz}[t + \tau(\arcsin \theta_{\mathrm{y}}), \boldsymbol{\theta}] \end{bmatrix}. \tag{4.54}$$

The time delay model $\tau(\phi)$ uses the angle to the left/right ear on the positive/negative y axis, so arccos $\pm\theta_{\mathrm{y}}$, but shifted by 90°, hence $\phi = \pm \arcsin \theta_{\mathrm{y}}$.

This removal allows use all available HRIRs of dense measurement sets for binaural synthesis of high accuracy, using a suitable linear Ambisonic decoder such as AllRAD. Assuming the resulting modified left and right HRIR for all directions are denoted as 2 × L matrix $\hat{\boldsymbol{H}}(t) = [\hat{\boldsymbol{h}}(t, \boldsymbol{\theta}_1), \ldots, \hat{\boldsymbol{h}}(t, \boldsymbol{\theta}_{\mathrm{L}})]^{\mathrm{T}}$, the $2 \times (\mathrm{N}+1)^2$ filter set for decoding every of the Ambisonic channels to the ears becomes:

$$\hat{\boldsymbol{H}}^{\mathrm{T}}_{\mathrm{SH}}(t) = \hat{\boldsymbol{H}}(t)\, \mathbf{D}\, \mathrm{diag}\{\boldsymbol{a}\}. \tag{4.55}$$

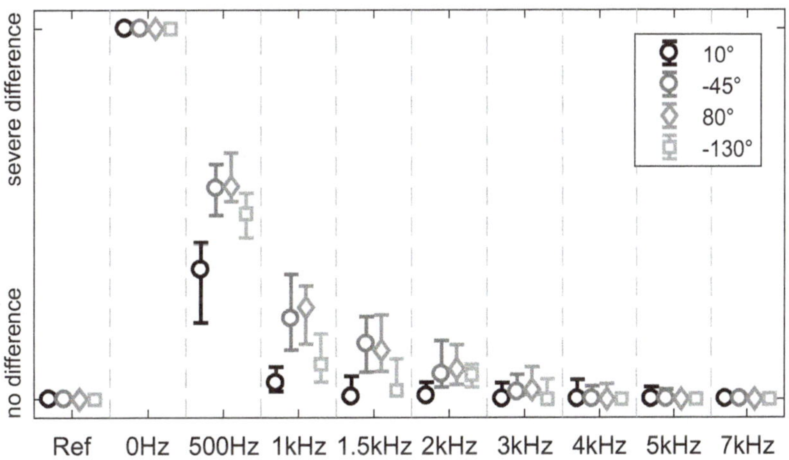

Fig. 4.28 Experiment on audibility of removal of the linear phase trend from HRTF above a varied cutoff frequency from [21] showing medians and 95% confidence intervals

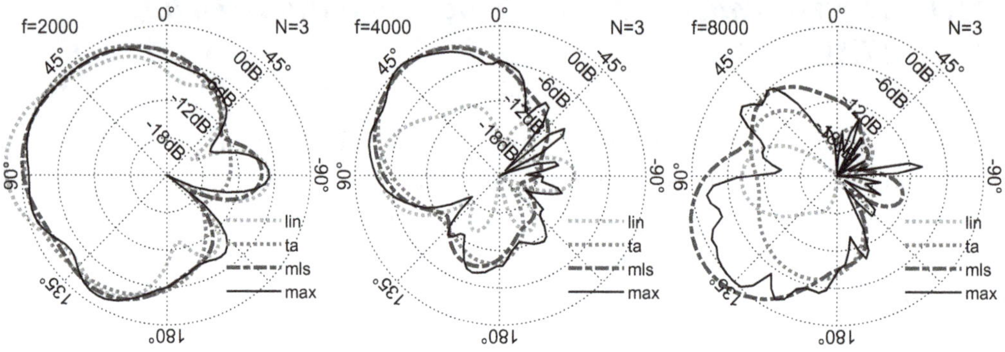

Fig. 4.29 Exemplary horizontal cross-sections of linear (lin), time-aligned (ta), and MagLS/magnitude-least-squares (mls) of third order N = 3, compared to high-order N = 35 (max) Ambisonic left-ear HRTF representations of the TH Cologne HRIR_L2720.sofa set

Results achieved by a pseudo-inverse decoding to hereby time-aligned HRIRs using R = 0.085 cm with N = 3 from the 2702-directions Cologne HRIRs[5] is shown in Fig. 4.29. The resulting polar patterns (ta) clearly outperform the linear decomposition (lin) at frequencies above 2kHz in representing the original HRTFs (max).

4.11.2 Magnitude Least Squares (MagLS)

Alternative to high-frequency time delay disposal, Schörkhuber et al. present an optimum-phase approach [21] that disregards phase match in favor of an improved magnitude match above cutoff. Formulated exemplarily for the left ear, across every HRTF direction $\boldsymbol{\theta}_l$, and for every discrete frequency ω_k, with $h_{l,k} = h(\boldsymbol{\theta}_l, \omega_k)$, this becomes

$$\min_{\hat{\boldsymbol{h}}_{\mathrm{SH},k}} \sum_{l=1}^{\mathrm{L}} \Big[|\boldsymbol{y}_{\mathrm{N}}(\boldsymbol{\theta}_l)^{\mathrm{T}}\, \hat{\boldsymbol{h}}_{\mathrm{SH},k}| - |h_{l,k}| \Big]^2. \tag{4.56}$$

Typically, one would need to solve magnitude least squares or magnitude squares least squares tasks with semidefinite relaxation, see Kassakian [51].

In practice, however, results turn out to be perfect already with an iterative combination of the reconstructed phase $\hat{\phi}_{l,k-1}$ from the previous frequency ω_{k-1} with the HRTF magnitude $|h_{k,l}|$ of the current frequency ω_k, before a linear decomposition thereof into spherical harmonic coefficients $\hat{\boldsymbol{h}}_{\mathrm{SH},k}$.

Every frequency below cutoff $\omega_k < 2\pi f_{\mathrm{N}}$ just uses the linear least-squares spherical harmonics decomposition with the left-inverse of the spherical harmonics $\boldsymbol{Y}_{\mathrm{N}}$ sampled at the HRTF measurement nodes,

[5] Data HRIR_L2702.sofa from http://sofacoustics.org/data/database/thk.

$$\hat{\boldsymbol{h}}_{\mathrm{SH},k} = (\boldsymbol{Y}_{\mathrm{N}}^{\mathrm{T}}\boldsymbol{Y}_{\mathrm{N}})^{-1}\boldsymbol{Y}_{\mathrm{N}}^{\mathrm{T}}\left[h_{l,k}\right]_l \,. \tag{4.57}$$

Continuing with the first frequency above/equal to cutoff $\omega_k \geq 2\pi f_{\mathrm{N}}$, the algorithm proceeds as:

$$\hat{\phi}_{l,k-1} = \angle\left\{\boldsymbol{y}_{\mathrm{N}}(\boldsymbol{\theta}_l)^{\mathrm{T}}\,\hat{\boldsymbol{h}}_{\mathrm{SH},k-1}\right\}\,, \tag{4.58}$$

$$\hat{\boldsymbol{h}}_{\mathrm{SH},k} = (\boldsymbol{Y}_{\mathrm{N}}^{\mathrm{T}}\boldsymbol{Y}_{\mathrm{N}})^{-1}\boldsymbol{Y}_{\mathrm{N}}^{\mathrm{T}}\left[|h_{l,k}|\,e^{\mathrm{i}\hat{\phi}_{l,k-1}}\right]_l\,, \tag{4.59}$$

and then moves to the next frequency $k \leftarrow k+1$. The results are typically transformed back to time domain to get a real-valued impulse response for every spherical harmonic to the regarded ear.

The results of the MagLS approach (mls) outperform the time-alignment approach (ta) in the exemplary results shown for N = 3 in Fig. 4.29, in particular at the highest frequencies, where sphere-model-based delay simplification is not sufficiently helpful, anymore.

4.11.3 *Diffuse-Field Covariance Constraint*

Also for both the above approaches that modify the high-frequency phase, Zaunschirm et al. [20] note that low order rendering degrades envelopment in diffuse fields, so that they introduce an additional covariance constraint as defined by Vilkamo [22]. It can be implemented as a 2×2 filter matrix equalizing the resulting frequency-domain diffuse-field covariance matrix to the one of the original HRTF datasets. On the main diagonal, this covariance matrix shows the diffuse-field ear sensitivities (left and right), and off-diagonal it contains the diffuse-field inter-aural cross correlation.

At every frequency, the 2×2 diffuse-field covariance matrix of the original, very-high-order spherical harmonics HRTF dataset $\boldsymbol{H}_{\mathrm{SH}}^{\mathrm{H}}$ of the dimensions $2 \times (\mathrm{M}+1)^2$ with $(\mathrm{M} \gg \mathrm{N})$ is given by

$$\boldsymbol{R} = \boldsymbol{H}_{\mathrm{SH}}^{\mathrm{H}}\boldsymbol{H}_{\mathrm{SH}}. \tag{4.60}$$

The derivation why this inner product of spherical harmonic coefficients represents the diffuse-field covariance is given in Appendix A.5. The low-order high-frequency modified HRTF coefficient set $\tilde{\boldsymbol{H}}_{\mathrm{SH}}$ of the dimensions $2 \times (\mathrm{N}+1)^2$ also has a 2×2 covariance matrix $\hat{\boldsymbol{R}}$ that will differ from the more accurate $\boldsymbol{R}$,

$$\hat{\boldsymbol{R}} = \hat{\boldsymbol{H}}_{\mathrm{SH}}^{\mathrm{H}}\hat{\boldsymbol{H}}_{\mathrm{SH}}. \tag{4.61}$$

Its diffuse-field reproduction improves after equalizing $\boldsymbol{R} = \hat{\boldsymbol{R}}$ by a 2×2 filter matrix,

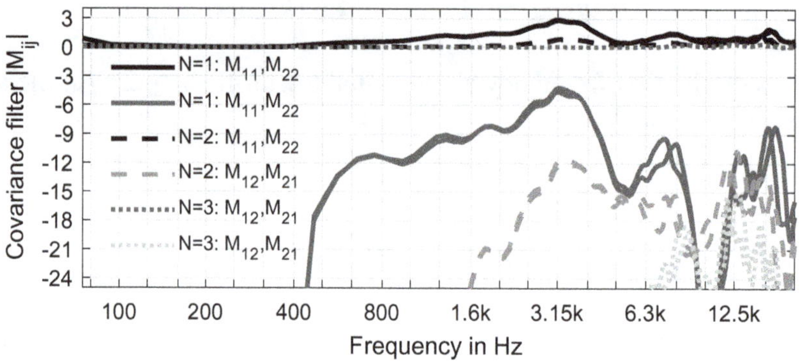

Fig. 4.30 Covariance constraint filters enhance the binaural decorrelation of MagLS by negative crosstalk M_{12} and M_{21}, under corresponding correction of the diffuse-field sensitivities M_{11} and M_{22} at playback orders N < 3

$$\hat{\boldsymbol{H}}_{\mathrm{SH,corr}} = \hat{\boldsymbol{H}}_{\mathrm{SH}}\boldsymbol{M}. \tag{4.62}$$

Appendix A.5 shows the derivation of $\boldsymbol{M}$ based on [20, 22]. In summary, it is composed of factors obtained by Cholesky and SVD matrix decompositions

$$\hat{\boldsymbol{H}}_{\mathrm{corr,SH}} = \hat{\boldsymbol{H}}_{\mathrm{SH}}\,\hat{\boldsymbol{X}}^{-1}\boldsymbol{V}\boldsymbol{U}^{\mathrm{H}}\boldsymbol{X}, \tag{4.63}$$

Cholesky factors:

$$\boldsymbol{H}_{\mathrm{SH}}^{\mathrm{H}}\boldsymbol{H}_{\mathrm{SH}} = \boldsymbol{X}^{\mathrm{H}}\boldsymbol{X},$$

$$\hat{\boldsymbol{H}}_{\mathrm{SH}}^{\mathrm{H}}\hat{\boldsymbol{H}}_{\mathrm{SH}} = \hat{\boldsymbol{X}}^{\mathrm{H}}\hat{\boldsymbol{X}}.$$

SVD:

$$\hat{\boldsymbol{X}}^{\mathrm{H}}\boldsymbol{X} = \boldsymbol{U}\boldsymbol{S}\boldsymbol{V}^{\mathrm{H}},$$

While MagLS binaural decoding with orders higher than 2 or 3 does not require covariance correction, the correction enhances the decorrelation of the ear signals for 1st to 2nd order reproduction, as shown in Fig. 4.30.

4.12 Practical Free-Software Examples

4.12.1 Pd and Circular/Spherical Harmonics

Similar as in the example section on first-order encoding and decoding in pure data (Pd), Fig. 4.31 shows 3rd-order 2D Ambisonic encoding and decoding for an octagon loudspeaker layout. The implementation `[mtx_circular_harmonics]` of the circular harmonics is used from the `iemmatrix` library, and the numbers for $\frac{180}{\pi} = 57.29$ and $a_m = \cos\frac{\pi m}{2\cdot(\mathrm{N}+1)}$ were pre-calculated. Note the similarity to the first-order 2D example of Fig. 1.13, to which the main change is the use of the circular harmonics matrix object.

For decoding to headphones, programming in Pd also looks rather similar as in the first-order example in Fig. 1.14, only more HRIRs matching the respective

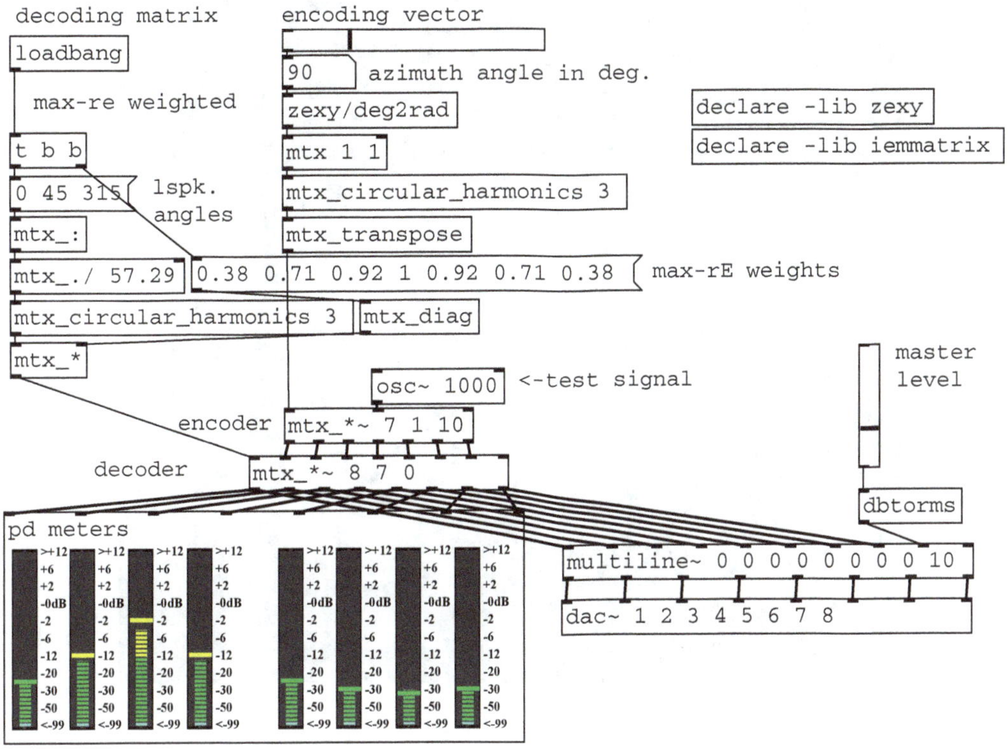

Fig. 4.31 2D encoding and decoding in Pd using `[mtx_circular_harmonics]` with 3rd order, 8 equidistant loudspeakers, and max-$\boldsymbol{r}_{\mathrm{E}}$ weighted decoder

loudspeaker positions need to be employed. To work in 3 dimensions, programming in Pd would also be similar as in the corresponding first-order example of Fig. 1.15, using the matrix object `[mtx_spherical_harmonics]`. Typically, pre-calculated decoders including AllRAD and max-$\boldsymbol{r}_{\mathrm{E}}$ are used and loaded by, e.g., `[mtx D.mtx]` into Pd to keep programming simple.

4.12.2 Ambix Encoder, IEM MultiEncoder, and IEM AllRADecoder

For encoding single- or multi-channel signals into Ambisonics, there are the `ambix_encode_o<N>`, or `ambix_encode_i<L>_o<N>` VST plugins available from Kronlachner's ambix plugin suite or the IEM `MultiEncoder` from the IEM plugin suite. As exemplarily shown in Fig. 4.32, the multi encoder allows to encode channel-based multi-channel audio material, where *channel-based* [52] typically refers to each channel of the multi-channel material meant to be played back on a separate loudspeaker of clearly defined direction, cf. [44]. Elsewhere, the embedding of virtual playback directions can also be found referred to as *beds* or *virtual panning spots*.

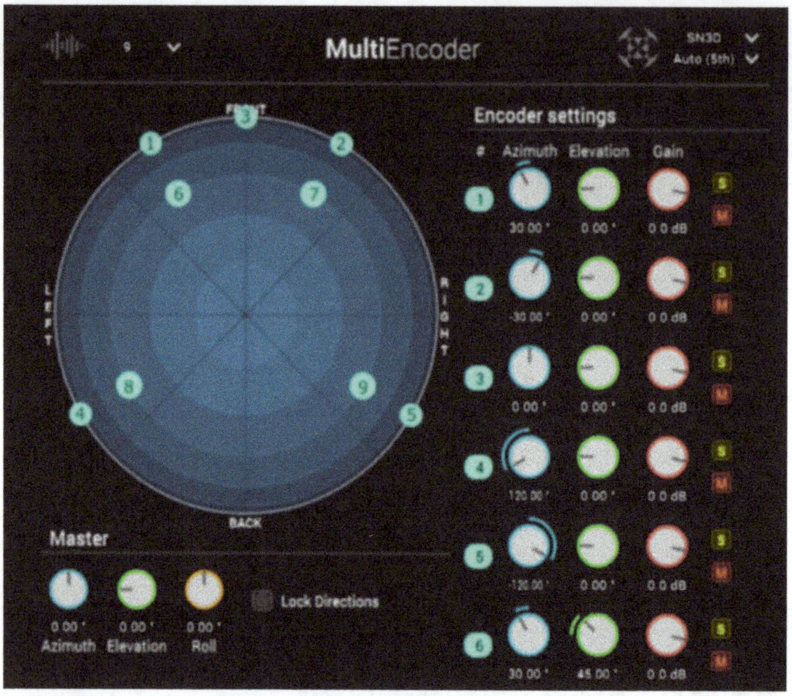

Fig. 4.32 `MultiEncoder` plug-in: encoding of a $4+5+0$ recording

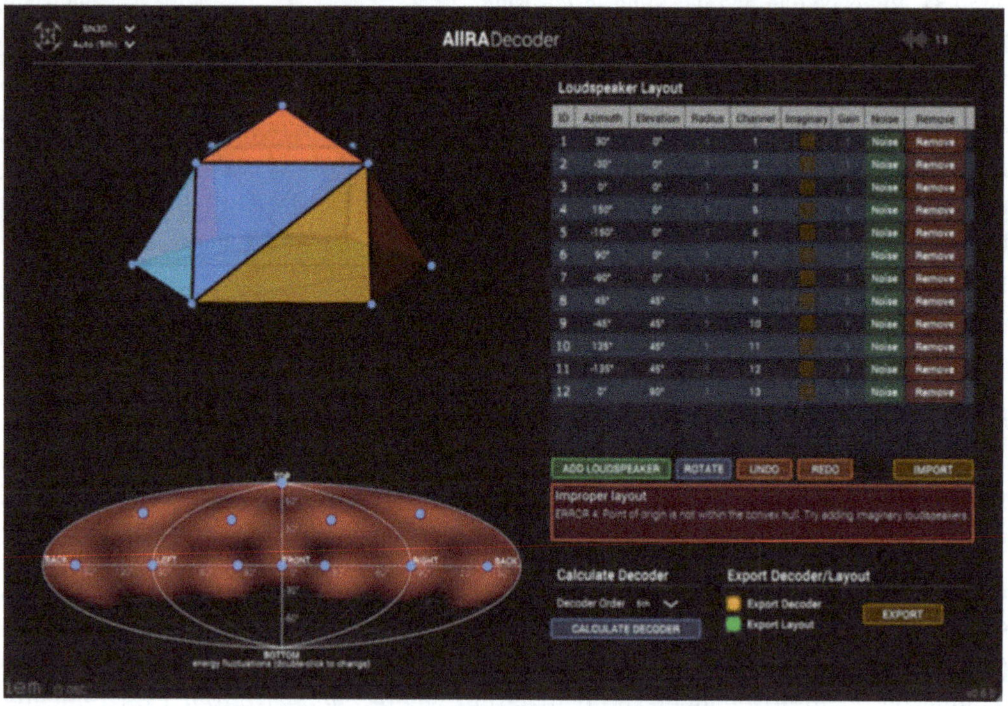

Fig. 4.33 `AllRADecoder` plug-in: $5+7+0$ layout from IEM Production Studio

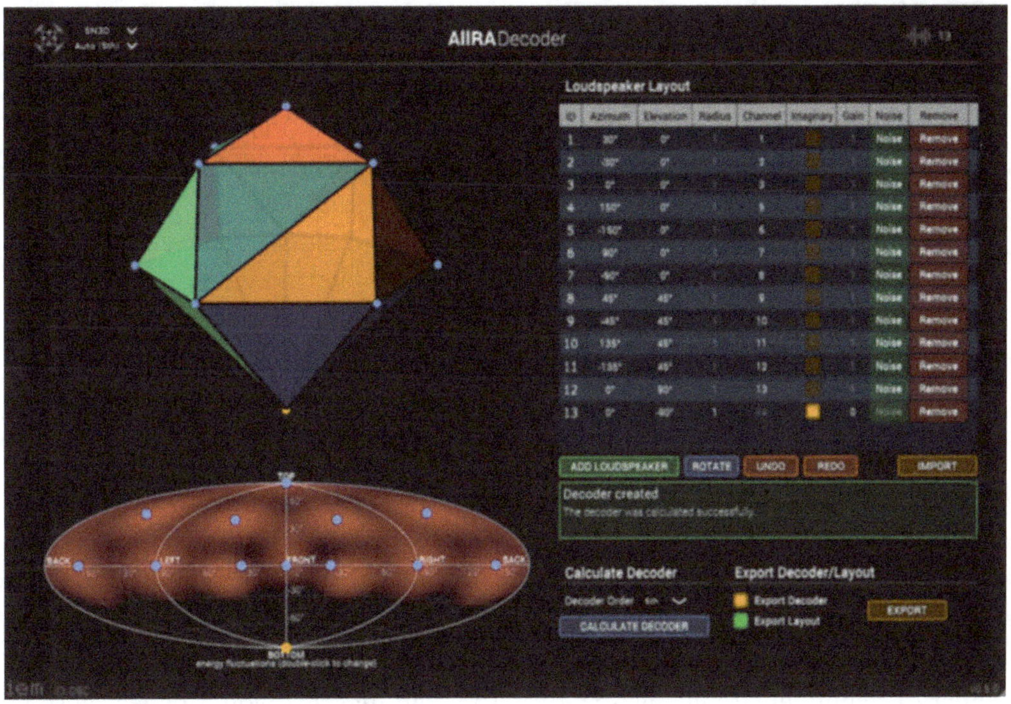

Fig. 4.34 `AllRADecoder` plug-in: 5 + 7 + 0 layout from IEM Production Studio

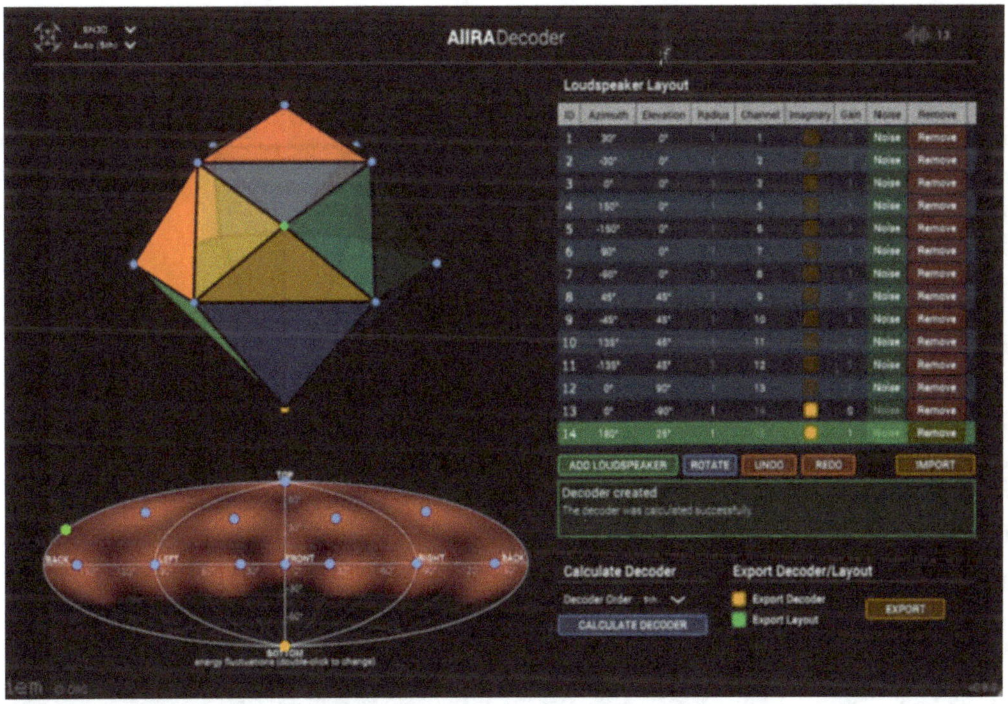

Fig. 4.35 `AllRADecoder` plug-in: 5 + 7 + 0 layout from IEM Production Studio

The IEM `AllRADecoder` permits to manually enter or import the loudspeaker coordinates and channel indices, with the coordinates specified by the azimuth and elevation angle in degrees, as exemplified for the IEM production studio in Fig. 4.33. The figure also shows that just entering the pure $5 + 7 + 0$ layout would produce an error message *Point of origin not within convex hull. Try adding imaginary loudspeakers.*

By adding an imaginary loudspeaker below whose signal is typically omitted, see Fig. 4.34, it becomes geometrically valid to calculate and employ the resulting decoder, however it is better to also insert an imaginary loudspeaker at the rear whose signal is preserved by specifying the gain value 1, as shown in Fig. 4.35.

4.12.3 Reaper, IEM RoomEncoder, and IEM BinauralDecoder

Particularly relevant for head-phone-based listening, rendering of anechoic sounds will typically not *externalize* well, as it does not match the mental expectation of ordinary listening environments [53–56]. To avoid that this would rather cause an in-head localization than the desired external sound image, one can, e.g., use the IEM `RoomEncoder` plugin, see Fig. 4.36. It is based on an image-source room model

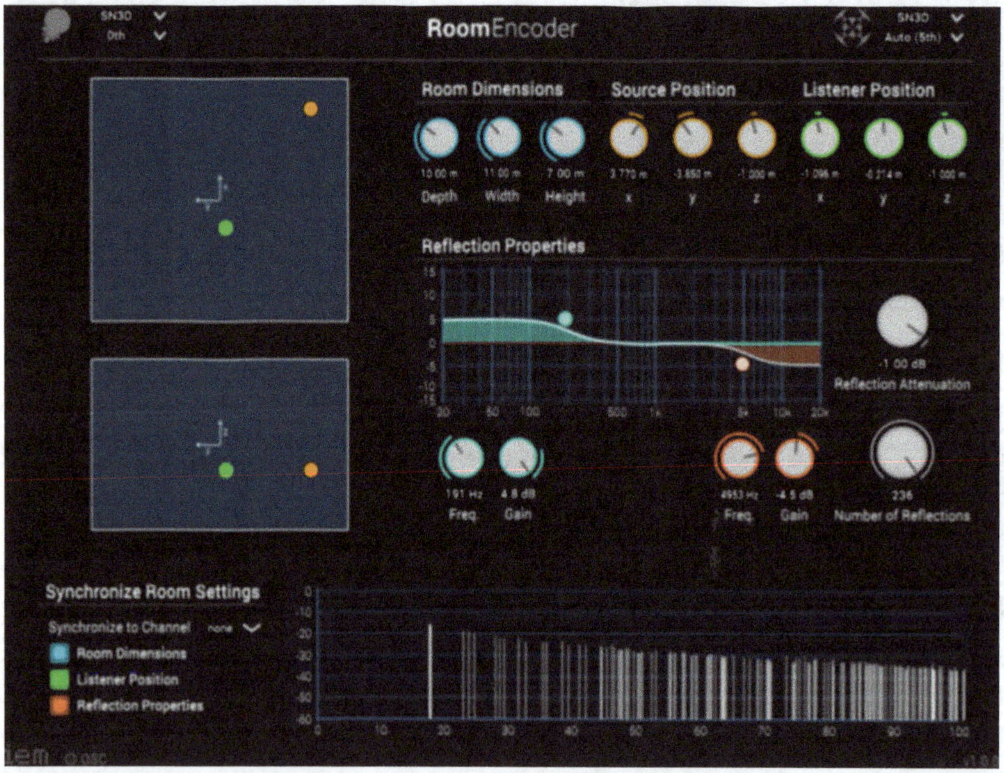

Fig. 4.36 `RoomEncoder` plug-in

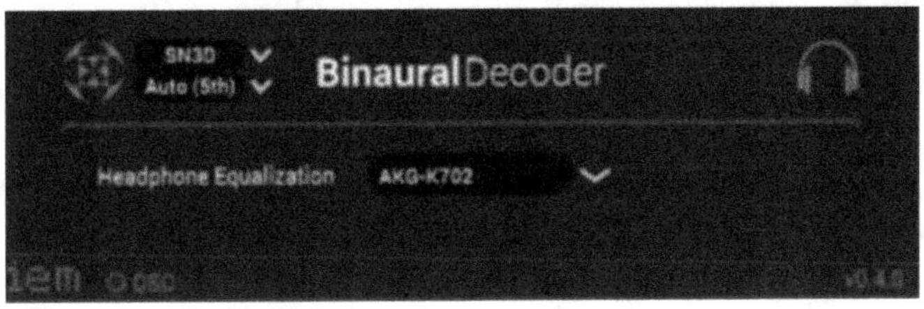

Fig. 4.37 `BinauralDecoder` plug-in

and encodes first-order wall-reflections involving reflection factors and propagation delays together with the desired direct sound.

The MagLS approach for Ambisonic decoding, using the KU100 measurements from Cologne Applied Science University and (optionally) their headphone equalization curves is implemented by the IEM BinauralDecoder, see Fig. 4.37.

In combination of both, IEM `RoomEncoder` and IEM `BinauralDecoder` with an Ambisonics-encoded single-channel sound (e.g. using `ambix_encoder`), one can simply try to place the source and receiver together in the symmetry plane of the room, and then to slightly shift one of both sideways to see how externalization improves by slight asymmetry in the ear signals.

References

1. J.S. Bamford, Ambisonic sound for the masses (1994)
2. D.H. Cooper, T. Shiga, Discrete-matrix multichannel stereo. J. Audio Eng. Soc. **20**(5), 346–360 (1972)
3. P. Felgett, Ambisonic reproduction of directionality in surround-sound systems. Nature **252**, 534–538 (1974)
4. M.A. Gerzon, The design of precisely coincident microphone arrays for stereo and surround sound, in *prepr. L-20 of 50th Audio Engineering Society Convention* (1975)
5. P. Craven, M.A. Gerzon, Coincident microphone simulation covering three dimensional space and yielding various directional outputs, U.S. Patent, no. 4,042,779 (1977)
6. J.S. Bamford, An analysis of ambisonic sound systems of first and second order, Master's thesis, University of Waterloo, Ontario (1995)
7. D.G. Malham, A. Myatt, 3D sound spatialization using ambisonic techniques. Comput. Music J. **19**(4), 58–70 (1995)
8. M.A. Poletti, The design of encoding functions for stereophonic and polyphonic sound systems. J. Audio Eng. Soc. **44**(11), 948–963 (1996)
9. J.-M. Jot, V. Larcher, J.-M. Pernaux, A comparative study of 3-d audio encoding and rendering techniques, in *16th AES Conference* (Rovaniemi, 1999)
10. J. Daniel, Représentation des champs acoustiques, application à la transmission et à la reproduction de scènes sonores complexes dans un contexte multimédia, Ph.D. dissertation, Université Paris 6 (2001)
11. D.B. Ward, T.D. Abhayapala, Reproduction of a plane-wave sound field using an array of loudspeakers. IEEE Trans. Speech Audio Process. **9**(6), 697–707 (2001)
12. G. Dickins, Sound field representation, reconstruction and perception, Master's thesis, Australian National University, Canberra (2003)
13. A. Sontacchi, Neue Ansätze der Schallfeldreproduktion, Ph.D. dissertation, TU Graz (2003)

14. M.A. Poletti, Robust two-dimensional surround sound reproduction for nonuniform loudspeaker layouts. J. Audio Eng. Soc. **55**(7/8), 598–610 (2007)
15. F. Zotter, H. Pomberger, M. Noisternig, Energy-preserving ambisonic decoding. Acta Acust. United Acust. **98**(11), 37–47 (2012)
16. J.-M. Batke, F. Keiler, Using vbap-derived panning functions for 3d ambisonics decoding, in *2nd International Symposium on Ambisonics and Spherical Acoustics* (Paris, 2010)
17. D. Sun, *Generation and perception of three-dimensional sound fields using higher order ambisonics* (University of Sydney, School of Electrical and Information Engineering, 2013). Ph.D thesis
18. Z. Ben-Hur, F. Brinkmann, J. Sheaffer, S. Weinzierl, B. Rafaely, Spectral equalization in binaural signals represented by order-truncated spherical harmonics. J. Acoust. Soc. Am. **141**(6) (2017)
19. F. Brinkmann, S. Weinzierl, Comparison of head-related transfer functions pre-processing techniques for spherical harmonics decomposition, in *AES Conference AVAR* (Redmont, 2018)
20. M. Zaunschirm, C. Schörkhuber, R. Höldrich, Binaural rendering of ambisonic signals by head-related impulse response time alignment and a diffuseness constraint. J. Acoust. Soc. Am. **143**(6), 3616–3627 (2018)
21. C. Schörkhuber, M. Zaunschirm, R. Höldrich, Binaural rendering of ambisonic signals via magnitude least squares, in *Fortschritte der Akustik - DAGA* (Munich, 2018)
22. J. Vilkamo, T. Bäckström, A. Kuntz, Optimized covariance domain framework for time-frequency processing of spatial audio. J. Audio Eng. Soc. **61**(6), 403–411 (2013)
23. F.W.J. Olver, R.F. Boisvert, C.W. Clark (eds.), *NIST Handbook of Mathematical Functions* (Cambridge University Press, Cambridge, 2000), http://dlmf.nist.gov. Accessed June 2012
24. J. Daniel, J.-B. Rault, J.-D. Polack, Acoustic properties and perceptive implications of stereophonic phenomena, in *AES 6th International Conference: Spatial Sound Reproduction* (1999)
25. M. Frank, How to make ambisonics sound good, in *Forum Acusticum, Krakow* (2014)
26. M. Frank, F. Zotter, Extension of the generalized tangent law for multiple loudspeakers, in *Fortschritte der Akustik - DAGA* (Kiel, 2017)
27. M. Frank, A. Sontacchi, F. Zotter, Localization experiments using different 2d ambisonic decoders, in *25th Tonmeistertagung* (2008)
28. P. Stitt, S. Bertet, M.v. Walstijn, Off-centre localisation performance of ambisonics and hoa for large and small loudspeaker array radii. Acta Acust. United Acust. **100**(5), 937–944 (2014)
29. M. Frank, F. Zotter, Exploring the perceptual sweet area in ambisonics, in *AES 142nd Convention* (2017)
30. ISO 31-11:1978, Mathematical signs and symbols for use in physical sciences and technology (1978)
31. ISO 80000-2, quantities and units? Part 2: Mathematical signs and symbols to be used in the natural sciences and technology (2009)
32. R.H. Hardin, N.J.A. Sloane, Mclaren's improved snub cube and other new spherical designs in three dimensions. Discret. Comput. Geom. **15**, 429–441 (1996), http://neilsloane.com/sphdesigns/dim3/
33. M. Gräf, D. Potts, On the computation of spherical designs by a new optimization approach based on fast spherical fourier transforms. Numer. Math. **119** (2011), http://homepage.univie.ac.at/manuel.graef/quadrature.php
34. R.S. Womersley, *Chapter, Efficient spherical designs with good geometric properties in Contemporary Computational Mathematics - A Celebration of the 80th Birthday of Ian Sloan* (Springer, Berlin, 2018), pp. 1243–1285
35. A. Solvang, Spectral impairment of 2d higher-order ambisonics. J. Audio Eng. Soc. **56**(4) (2008)
36. M.A. Gerzon, G.J. Barton, Ambisonic decoders for HDTV, in *prepr. 3345, 92nd AES Convention*, Vienna, 1992
37. J. Daniel, J.-B. Rault, J.-D. Polack, Ambisonics encoding of other audio formats for multiple listening conditions, in *prepr. 4795, 105th AES Convention* (San Francisco, 1998)

38. M.A. Poletti, A unified theory of horizontal holographic sound systems. J. Audio Eng. Soc. **48**(12), 1155–1182 (2000)
39. M.A. Poletti, *Three-dimensional surround sound systems based on spherical harmonics* (J. Audio Eng, Soc, 2005)
40. F. Zotter, M. Frank, *All-round ambisonic panning and decoding* (J. Audio Eng, Soc, 2012)
41. F. Zotter, M. Frank, Ambisonic decoding with panning-invariant loudness on small layouts (allrad2), in *144th AES Convention, prepr. 9943* (Milano, 2018)
42. R. Pail, G. Plank, W.-D. Schuh, Spatially restricted data distributions on the sphere: the method of ortnonormalized functions and applications. J. Geodesy **75**, 44–56 (2001)
43. M. Frank, A. Sontacchi, Case study on ambisonics for multi-venue and multi-target concerts and broadcasts. J. Audio Eng. Soc. **65**(9) (2017)
44. ITU, *Recommendation BS.2051: Advanced sound system for programme production.* ITU (2018)
45. M. Noisternig, A. Sontacchi, T. Musil, R. Höldrich, "A 3d ambisonic based binaural sound reproduction system, in *24th AES Conference* (Banff, 2003)
46. B. Bernschütz, A. V. Giner, C. Pörschmann, J. Arend, Binaural reproduction of plane waves with reduced modal order. Acta Acust. United Acust. **100**(5) (2014)
47. S. Delikaris-Manias, J. Vilkamo, Adaptive mixing of excessively directive and robust beamformers for reproduction of spatial sound, in *Parametric Time-Frequency Domain Spatial Audio*, ed. by V. Pulkki, S. Delikaris-Manias, A. Politis (Wiley, New Jersey, 2017)
48. C.T. Jin, N. Epain, D. Sun, Perceptually motivated binaural rendering of higher order ambisonic sound scenes, in *Fortschritte der Akustik AIA-DAGA* (Merano, 2013). March
49. J.B. Keller, Geometrical theory of diffraction. J. Acoust. Soc. Am. **52**(2), 116–130 (1962)
50. R.S. Woodworth, H. Schlosberg, *Experimental Psychology* (Holt, Rinehart and Winston, 1954)
51. P.W. Kassakian, Convex approximation and optimization with applications in magnitude filter design and radiation pattern synthesis, Ph.D. dissertation, EECS Department, University of California, Berkeley (2006)
52. ITU, *Recommendation BS.2076: Audio Definition Model* (ITU, 2017)
53. S. Werner, G. Götz, F. Klein, Influence of head tracking on the externalization of auditory events at divergence between synthesized andf listening room using a binaural headphone system, in *prepr. 9690, 142nd AES Convention* (Berlin, 2017)
54. F. Klein, S. Werner, T. Mayenfels, Influences of tracking on externalization of binaural synthesis in situations of room divergence. J. Audio Eng. Soc. **65**(3), 178–187 (2017)
55. J. Cubick, Investigating distance perception, externalization and speech intelligibility in complex acoustic environments, Ph.D. dissertation, DTU Copenhagen (2017)
56. G. Plenge, Über das Problem der Im-Kopf-Lokalisation. Acustica **26**(5), 241–252 (1972)

Chapter 5
Signal Flow and Effects in Ambisonic Productions

This system offers significantly enhanced spatialisation technologies [...] with new creative possibilities opening up to anyone with the appropriate number of audio channels available on their computer systems.

Dave G. Malham [1], ICMC, Bejing, 1999.

Abstract This chapter presents the internal working principles of various Ambisonic 3D audio effects. No matter which digital audio workstation or processing software is used in a production, the general Ambisonic signal infrastructure is outlined as an important overview of the signal processing chain. The effects presented are frequency-independent effects such as directional re-mapping (mirror, rotation, warping) and re-weighting (directional level modification), and frequency-dependent effects such as widening/distance/diffuseness, diffuse reverberation, and resolution-enhanced convolution reverberation.

The typical audio processing steps for Ambisonic surround-sound signal manipulation are shown in the block diagram Fig. 5.1 from [2]. The description of the multi-venue application in [3] and one for live effects [4] might be encouraging.

Ambisonic encoding and Ambisonic bus. From the previous section we know that representing single-channel signals $s_c(t)$ together with their direction $\boldsymbol{\theta}_c$ is a matter of encoding, of multiplying the signal by the coefficients $\boldsymbol{y}_{\mathrm{N}}(\boldsymbol{\theta}_c)$ obtained by evaluating the spherical harmonics at the direction from which the signal should appear to come. In productions, there will be multiple signals, either representing *spot microphones*, *virtual playback spots* of embedded channel-based content (*beds*), e.g. stereo or 5.1, material. With all input signals encoded and summed up on an *Ambisonic bus*, we obtain the multi-channel Ambisonic signal representation of an entire audio production

F. Zotter and M. Frank, *Ambisonics*, Springer Topics in Signal Processing 19,
https://doi.org/10.1007/978-3-030-17207-7_5

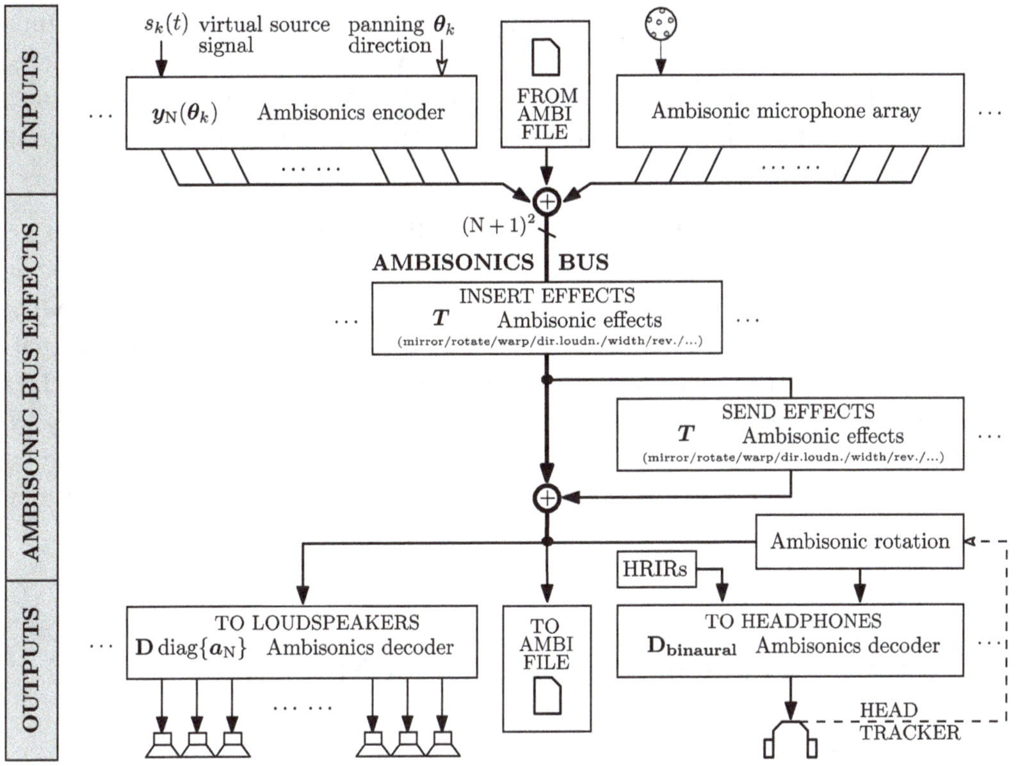

Fig. 5.1 Block diagram as in [2]

$$\chi_N(t) = \sum_{c=1}^{C} y_N(\theta_c)\, s_c(t). \tag{5.1}$$

Ambisonic surround-sound signal. Without decoding to a specific loudspeaker layout, the signal χ_N of the *Ambisonic bus* might appear somewhat virtual. Nevertheless, it allows to be drawn as a surround-sound signal $x(\theta, t)$ whose amplitude can be evaluated and metered at any direction θ, anytime t, using the expansion into spherical harmonics

$$x(\theta, t) = y_N^T(\theta)\, \chi_N(t). \tag{5.2}$$

Upmixing. As first-order recordings are not highly resolved, there are several works on algorithms with resolution enchancement strategies that re-assign time-frequency bins more sharply to directions. A good summary on such *input-specific insert effects* has been given in the book [5, 6]. Available solutions are DirAC, HOA-DirAC, COMPASS, Harpex.

Higher order. Higher-order microphones require more of the acoustic holophonic and holographic basics than presented above, yielding pre-processing filters as *input-specific insert effect*. Higher-order recording is dealt with in the subsequent Chap. 6 after the derivation of the wave equation and the solutions in the spherical coordinate system.

Insert effects: Generic re-mapping and leveling. One can imagine that it should be possible to manipulate the surround-sound signal $x(\theta, t)$ in various ways. For instance, effects based on directional re-mapping can take signals out of their original directional range and place them back into the Ambisonic signal at manipulated directions. Also, directions can be altered in amplitude levels so that, for instance, signals at directions with unwanted content undergo attenuation. Many more useful effects are presented below.

Decoding to loudspeakers/headphones. To map the modified Ambisonic signal $\tilde{\chi}_{\tilde{N}}$ to loudspeakers or headphones, an Ambisonic decoder is needed as discussed in the previous chapter. For decoding to headphones, it should be considered to either take only as few HRIR directions to decode to as possible [7, 8], before signals get convolved and mixed to avoid coloration at frontal directions where delays in the HRIRs change too strongly over the direction to get resolved properly [9, 10]. Alternatively, the approach in [11] proposed removal of the HRIR delay at high frequencies and diffuse-field covariance equalization by a 2×2 filter system, cf. Sect. 4.11.

5.1 Embedding of Channel-Based, Spot-Microphone, and First-Order Recordings

Microphone arrays for near-coincident higher-order Ambisonic recording based on holography will be discussed in the subsequent chapter. Nevertheless it possible to use (i) spot and close microphones and encode their direction into the directional panorama, (ii) first-order microphone arrays to fill the Ambisonic channels only up to the first order, (iii) more classical non-coincident or equivalence-stereophonic microphone arrays whose typical playback directions are encoded in Ambisonics.

The study by Kurz et al. [12] investigated how recordings by first-order encoding of the soundfield microphone ST450 and the Oktava MK4012 tetrahedral microphone arrays compare to the equivalence-stereophonic ORTF, see Fig. 5.2. In addition, ORTF-like mapping of the Oktava MK4012's frontal signals to the $\pm 30°$ directions in 5th order was tested instead of its first-order encoding. Figure 5.3 shows the results of the study in terms of the perceptual attributes localization and spatial depth. It seems that a mixture between ORTF-like 5th-order encoding and first-order encoding of the MK4012 microphone achieves preferred results, while the first-order encoded output of the ST450 Soundfield microphone is rated fair in both attributes, the ORTF microphone only ranked well terms of localization. The results of the ST450 were independent from its orientation, whereas the localization of the first-order-encoded MK4012 was found to be dependent on the orientation. This dependency of the MK4012 is because its microphones are not sufficiently coincident.

As a bottom line of the detailed analysis, one should be encouraged to keep using classical microphone techniques where known to be appropriate and encode their output in higher-order beds or virtual playback directions. However, this should be done with the awareness that stereophonic recording won't necessarily work for a

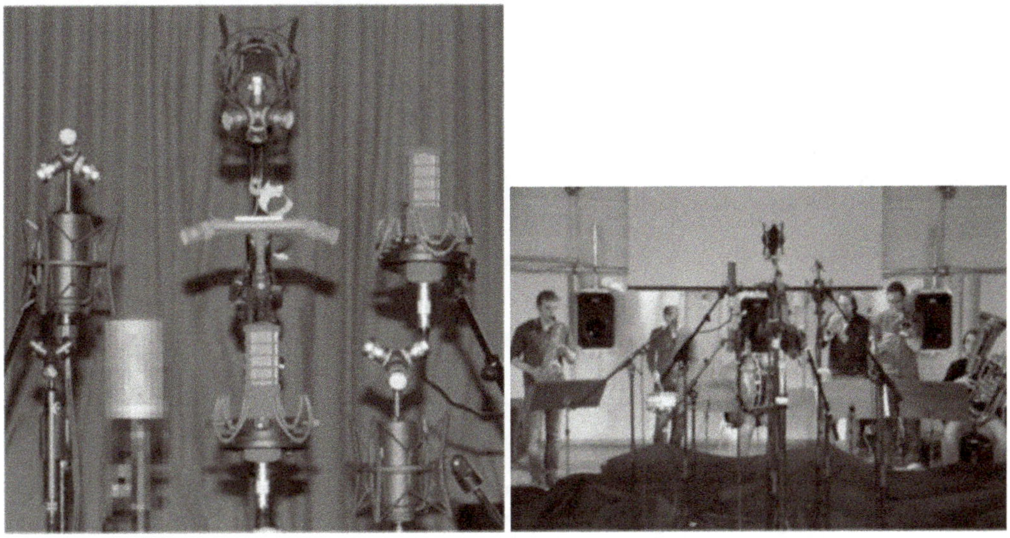

(a) microphone arrays (b) placement

Fig. 5.2 Ensemble of the Ambisonic and reference microphones of the study by Kurz et al. [12]; the pixelized microphone prototype by AKG was excluded from the study

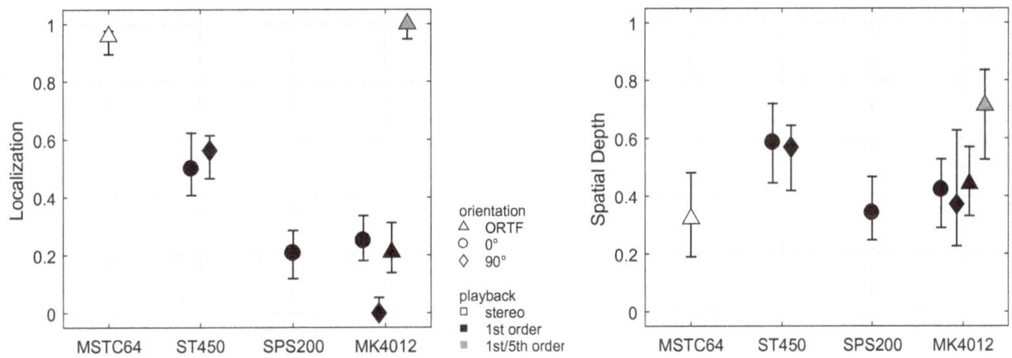

(a) Sound Localization (b) Spatial Depth

Fig. 5.3 Median values and 95% confidence intervals for each attribute from experiments in [12] for different microphones, orientations, and playback processing

large audience area, for which the robustness in directional mapping of equivalence-based techniques seem to be attractive.

An interesting layout is, e.g., specified in Hendrickx et al's work [13], in which they use an equivalence-stereophonic six-channel microphone array. Another interesting idea was used in the ICSA Ambisonics Summer School 2017. A height layer of suitably inclined super-cardioid microphones was added at small vertical distance to the horizontal microphone layer, similarly as the upwards-pointing directional microphones suggested in Lee's and Wallis' work [14, 15] to provide sufficiently attenuated horizontal sounds to the height layer.

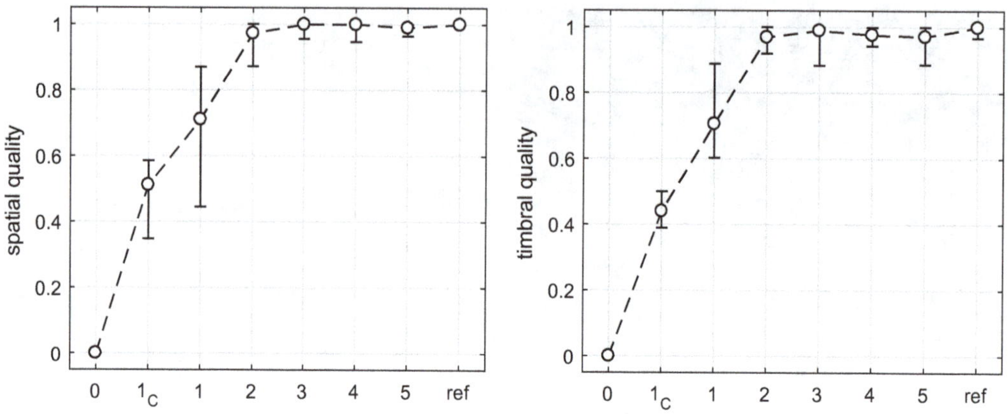

(a) Spatial Quality (b) Timbral Quality

Fig. 5.4 Median values and 95% confidence intervals of listening experiment comparing channel-based orchestra recordings on headphone playback, either directly rendered using the corresponding HRIRs or via binaural Ambisonic decoding of different orders

Binaural rendering study using surround-with-height material. In another study by Lee, Frank, and Zotter [16], static headphone-based rendering of channel-based recordings was compared using direct HRIR-based rendering or Ambisonics-based binaural rendering, cf. Sect. 4.11. The aim was to find whether differently recorded material could be rendered at high quality via binaural Ambisonics renderers, or under which settings this would imply quality degradation when compared to channel-based binaural rendering.

The results from the half of the listening experiment done in Graz is analyzed in Fig. 5.4, and the renderers compared were channel-based "ref", a low-passed mono anchor designed to have poor quality "0", a first-order binaural Ambisonic renderer "1c" based on a cube layout with loudspeakers at $\pm 90°$, $\pm 270°$ azimuth and $\pm 35.3°$ elevation, and MagLS binaural Ambisonic renderers at the orders "1", "2", "3", "4", and "5". Obviously, for orders 2 and above, there is not much quality degradation compared to the reference channel-based binaural rendering. The spatial quality cannot be distinguished from the reference for MagLS with Ambisonic orders 3 and above, and the timbral qualities cannot be distinguished for Ambisonic orders 2 and above.

While this result simplifies the practical requirements for headphone playback remarkably, it can be supposed that due to the limited sweet spot size, loudspeaker playback would still require higher orders, typically.

5.2 Frequency-Independent Ambisonic Effects

Many frequency- and time-independent Ambisonic effects are based on the aforementioned re-mapping of directions and manipulation of directional amplitudes, see e.g. Kronlachner's thesis, [2, 17]; advanced effects can be found in [18]. In general, the surround-sound signal allows to be manipulated by any thinkable transformation that modifies the directional mapping and amplitude of its contents. The formulation

$$\tilde{x}(\tilde{\boldsymbol{\theta}}, t) = g(\boldsymbol{\theta})\; x(\boldsymbol{\theta}, t) \tag{5.3}$$

expresses an operation that is able to pick out every direction $\boldsymbol{\theta}$ of the input signal, weight its signal by a directional gain $g(\boldsymbol{\theta})$, and re-map it to a new direction $\tilde{\boldsymbol{\theta}} = \boldsymbol{\tau}\{\boldsymbol{\theta}\}$ within a transformed signal $\tilde{x}$. To find out how this affects Ambisonic signals, we write both x and $\tilde{x}$ as Ambisonic signals $x(\boldsymbol{\theta}, t) = \boldsymbol{y}_{\mathrm{N}}^{\mathrm{T}}(\boldsymbol{\theta})\; \boldsymbol{\chi}_{\mathrm{N}}(t)$ and $\tilde{x}(\boldsymbol{\theta}, t) = \boldsymbol{y}_{\tilde{\mathrm{N}}}^{\mathrm{T}}(\tilde{\boldsymbol{\theta}})\; \tilde{\boldsymbol{\chi}}_{\mathrm{N}}(t)$ expanded in spherical/circular harmonics,

$$\boldsymbol{y}_{\tilde{\mathrm{N}}}^{\mathrm{T}}(\tilde{\boldsymbol{\theta}})\; \tilde{\boldsymbol{\chi}}_{\tilde{\mathrm{N}}}(t) = g(\boldsymbol{\theta})\; \boldsymbol{y}_{\mathrm{N}}^{\mathrm{T}}(\boldsymbol{\theta})\; \boldsymbol{\chi}_{\mathrm{N}}(t),$$

and use $\int_{\mathbb{S}_{\mathrm{D}}} \boldsymbol{y}_{\tilde{\mathrm{N}}}(\tilde{\boldsymbol{\theta}}) \boldsymbol{y}_{\tilde{\mathrm{N}}}^{\mathrm{T}}(\tilde{\boldsymbol{\theta}})\, \mathrm{d}\tilde{\boldsymbol{\theta}} = \boldsymbol{I}$ by integrating over $\boldsymbol{y}_{\tilde{\mathrm{N}}}(\tilde{\boldsymbol{\theta}}) \mathrm{d}\tilde{\boldsymbol{\theta}} \int_{\mathbb{S}_{\mathrm{D}}}$ to get $\tilde{\boldsymbol{\chi}}_{\tilde{\mathrm{N}}}(t)$ on the left

$$\tilde{\boldsymbol{\chi}}_{\tilde{\mathrm{N}}}(t) = \overbrace{\int_{\mathbb{S}_{\mathrm{D}}} \boldsymbol{y}_{\tilde{\mathrm{N}}}(\tilde{\boldsymbol{\theta}})\; g(\boldsymbol{\theta})\; \boldsymbol{y}_{\mathrm{N}}^{\mathrm{T}}(\boldsymbol{\theta})\; \mathrm{d}\tilde{\boldsymbol{\theta}}}^{:=\boldsymbol{T}}\; \boldsymbol{\chi}_{\mathrm{N}}(t) = \boldsymbol{T}\; \boldsymbol{\chi}_{\mathrm{N}}(t) \tag{5.4}$$

to find the transformed signals being just re-mixed Ambisonic input signals by the matrix $\boldsymbol{T}$ (note that it might require an increased Ambisonic order $\tilde{\mathrm{N}}$). Numerical evaluation of the matrix $\boldsymbol{T} = \int_{\mathbb{S}_{\mathrm{D}}} \boldsymbol{y}_{\tilde{\mathrm{N}}}(\tilde{\boldsymbol{\theta}})\; g(\boldsymbol{\theta})\; \boldsymbol{y}_{\tilde{\mathrm{N}}}(\boldsymbol{\theta})\; \mathrm{d}\tilde{\boldsymbol{\theta}}$ is best done by using a high-enough t-design $\boldsymbol{\Theta} = [\theta_l]$ to discretize the integration variable $\tilde{\boldsymbol{\theta}} = \boldsymbol{\tau}\{\boldsymbol{\theta}\}$. For the discretized input directions $\boldsymbol{\theta}$, an inverse mapping $\boldsymbol{\theta} = \boldsymbol{\tau}^{-1}\{\tilde{\boldsymbol{\theta}}\}$ of the output direction must exist (directional re-mapping must be bijective), so that we can write

$$\boldsymbol{T} = \int_{\mathbb{S}_{\mathrm{D}}} \boldsymbol{y}_{\tilde{\mathrm{N}}}(\tilde{\boldsymbol{\theta}})\; g(\boldsymbol{\tau}^{-1}\{\tilde{\boldsymbol{\theta}}\})\; \boldsymbol{y}_{\mathrm{N}}^{\mathrm{T}}(\boldsymbol{\tau}^{-1}\{\tilde{\boldsymbol{\theta}}\})\; \mathrm{d}\tilde{\boldsymbol{\theta}} = \frac{4\pi}{\hat{\mathrm{L}}}\; \boldsymbol{Y}_{\tilde{\mathrm{N}},\boldsymbol{\Theta}}\; \mathrm{diag}\{\boldsymbol{g}_{\boldsymbol{\tau}^{-1}\{\boldsymbol{\Theta}\}}\}\; \boldsymbol{Y}_{\mathrm{N},\boldsymbol{\tau}^{-1}\{\boldsymbol{\Theta}\}}^{\mathrm{T}}.$$

This formalism is generic and covers simplistic and more complex tasks. It helps understanding that every frequency-independent directional weighting and/or re-mapping is just re-mixing the Ambisonic signals by a matrix, as in Fig. 5.6a.

The ambix VST plugin suite implements several effects, e.g. in the VST plug-ins `ambix_mirror`, `ambix_rotate`, `ambix_directional_loudness`, `ambix_warp`. The sections below explain how these and other effects work inside.

5.2.1 Mirror

Mirroring does not actually require the generic re-mapping and re-weighting formalism from above, yet. The spherical harmonics associated with the Ambisonic channels are shown in Fig. 4.12 and upon closer inspection one recognizes their symmetries, see Fig. 5.5. To mirror the Ambisonic sound scene with regard to planes of symmetry, it is sufficient to sign-invert channels associated with odd-symmetric spherical harmonics as in Fig. 5.6b. Formally, the transform matrix consists of a diagonal matrix $\boldsymbol{T} = \mathrm{diag}\{\boldsymbol{c}\}$ only, with the corresponding sign-change sequence $\boldsymbol{c}$.

Fig. 5.5 Ambisonic singals associated with odd symmetric spherical harmonics are sign-inverted to mirror the sound scene. For every Cartesian axis, illustrations above show spherical harmonics up to the third-order, with the order index n organized in rows and the mode index m in columns. Even harmonics are blurred for visual distinction

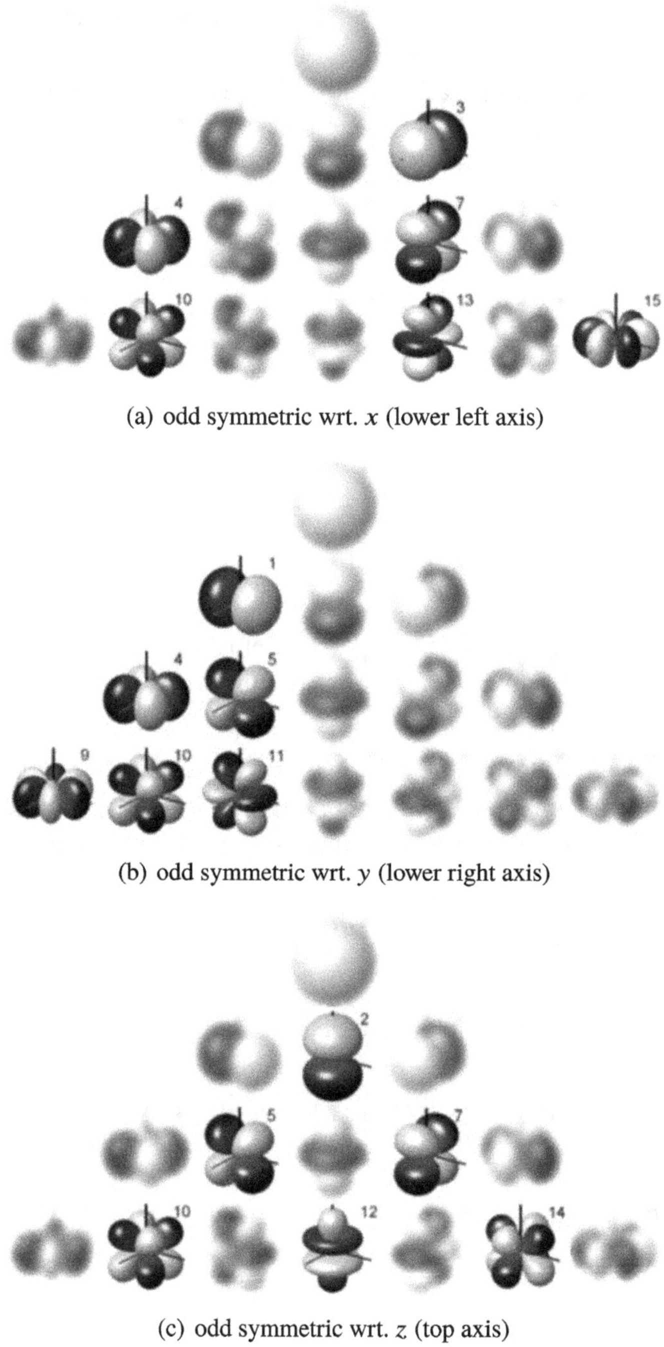

(a) odd symmetric wrt. x (lower left axis)

(b) odd symmetric wrt. y (lower right axis)

(c) odd symmetric wrt. z (top axis)

Up-down: For instance, spherical harmonics with $|m| = n$ are even symmetric with regard to $z = 0$ (up-down), and from this index on, every second harmonic in m is. To flip up and down, it is therefore sufficient to invert the signs of odd-symmetric spherical harmonics with regard to $z = 0$; they are characterized by $n + m$ being an odd number, or $c_{nm} = (-1)^{n+m}$.

Left-right: The $\sin\varphi$-related spherical harmonics with $m < 0$ are odd-symmetric with regard to $y = 0$ (left-right); therefore sign-inverting the signals with the index $m < 0$ exchanges left and and right in the Ambisonic surround signal, i.e. $c_{nm} = (-1)^{m<0}$.

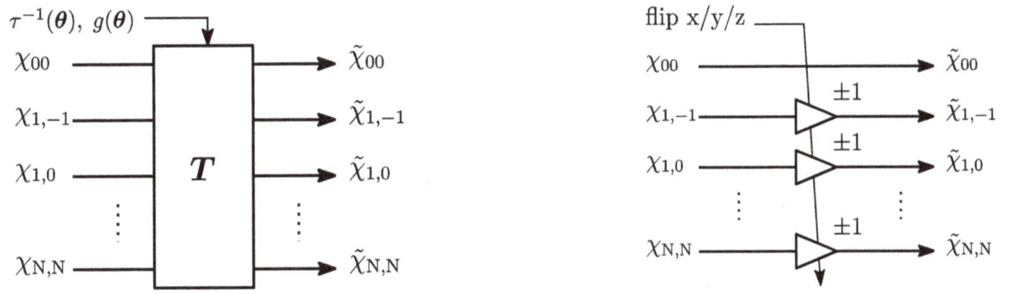

(a) Directional re-mapping, re-weighting effect for Ambisonic signals

(b) Mirroring effect applied on Ambisonic signals

Fig. 5.6 Block diagrams of frequency-independent transformations such as re-mapping and re-weighting (left, matrix operations), or mirroring (right, sign-only operations)

Front-back: Every odd-numbered $m > 0$ is odd-symmetric with regard to $x = 0$ (front-back), and so is every even-numbered harmonic with $m < 0$. Inverting the sign of these harmonics, $c_{nm} = (-1)^{m+(m<0)}$, flips front and back in the Ambisonic surround signal.

5.2.2 3D Rotation

Rotation can be expressed by a general rotation matrix $\boldsymbol{R}$ consisting of a rotation around z by χ, around y by ϑ, and again around z by φ, see Fig. 5.7. This rotation matrix maps every direction $\boldsymbol{\theta}$ to a rotated direction $\tilde{\boldsymbol{\theta}}$:

$$\tilde{\boldsymbol{\theta}} = \boldsymbol{R}(\varphi, \vartheta, \chi)\, \boldsymbol{\theta}, \tag{5.5}$$

$$\boldsymbol{R} = \begin{bmatrix} \cos(\varphi) & -\sin(\varphi) & 0 \\ \sin(\varphi) & \cos(\varphi) & 0 \\ 0 & 0 & 1 \end{bmatrix} \begin{bmatrix} \cos(\vartheta) & 0 & -\sin(\vartheta) \\ 0 & 1 & 0 \\ \sin(\vartheta) & 0 & \cos(\vartheta) \end{bmatrix} \begin{bmatrix} \cos(\chi) & -\sin(\chi) & 0 \\ \sin(\chi) & \cos(\chi) & 0 \\ 0 & 0 & 1 \end{bmatrix}.$$

Using this as a transform rule $\boldsymbol{\tau}(\tilde{\boldsymbol{\theta}}) = \boldsymbol{R}\,\boldsymbol{\theta}$ with neutral gain $g(\boldsymbol{\theta}) = 1$, we find the transform matrix by the inverse mapping $\boldsymbol{\theta} = \boldsymbol{R}^{\mathrm{T}}\tilde{\boldsymbol{\theta}}$ as

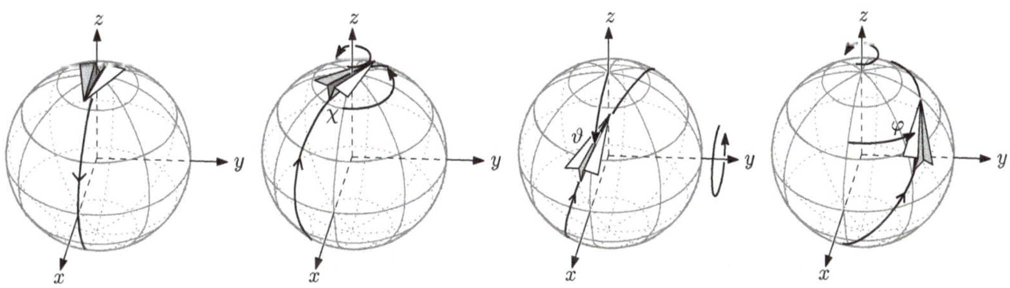

Fig. 5.7 zyz-Rotation on the plain example of great-circle navigation of a paper plane around the earth. With the original location at the zenith, a first rotation around z determines the course, and the subsequent rotations around y and z relocate the plane in zenith and azimuth

$$\boldsymbol{T} = \frac{4\pi}{\hat{L}} \boldsymbol{Y}_{\mathrm{N},\boldsymbol{\Theta}} \boldsymbol{Y}_{\mathrm{N},\boldsymbol{R}^{\mathrm{T}}\boldsymbol{\Theta}}^{\mathrm{T}}. \tag{5.6}$$

Using the $\hat{\mathrm{L}}$ directions of a $t \geq 2\mathrm{N}$-design $\boldsymbol{\Theta}$ is sufficient to sample the harmonics accurately. With the resulting $\boldsymbol{T}$, rotation is implemented as in Fig. 5.6a.

There is plenty of potential for simplification: As only the spherical harmonics of a given order n are required to re-express a rotated spherical harmonic of the same order n, $\boldsymbol{T}$ is actually block diagonal $\boldsymbol{T} = \mathrm{blk\ diag_n}\{\boldsymbol{T}_n\}$, and within each spherical harmonic order, the integral could be more efficiently evaluated using a smaller $t \geq 2n$-design. Moreover, there are various fast and recursive ways to calculate the entries of $\boldsymbol{T}$ as in [19–25] and implemented in most plugins. And yet, in practice a naïve implementation can be fast enough and pragmatic.

Rotation around z. One special case of rotation is important and particularly simple to implement. A directional encoding in azimuth always either is equal to $\Phi_m(\varphi_\mathrm{s})$ in 2D, or contains it in 3D. For $m > 0$, the azimuth encoding $\Phi_m(\varphi_\mathrm{s})$ depends on $\cos m\varphi_\mathrm{s}$, and its negative-sign version $\Phi_{-m}(\varphi_\mathrm{s})$ depends on $\sin(|m|\varphi_\mathrm{s})$. The encoding angle can be offset by the trigonometric addition theorems. They can be written as a matrix:

$$\begin{bmatrix} \sin m(\varphi_\mathrm{s}+\varphi) \\ \cos m(\varphi_\mathrm{s}+\varphi) \end{bmatrix} = \underbrace{\begin{bmatrix} \cos m\varphi & \sin m\varphi \\ -\sin m\varphi & \cos m\varphi \end{bmatrix}}_{\boldsymbol{R}(m\varphi)} \begin{bmatrix} \sin m\varphi_\mathrm{s} \\ \cos m\varphi_\mathrm{s} \end{bmatrix}. \tag{5.7}$$

By this, any Ambisonic signal, be it 2D or 3D, can be rotated around z by the matrices $\boldsymbol{R}(m\varphi)$ for the signal pairs with $\pm m$.

$$\begin{aligned} \begin{bmatrix} \Phi_{-m}(\varphi_\mathrm{s}+\varphi) \\ \Phi_m(\varphi_\mathrm{s}+\varphi) \end{bmatrix} &= \boldsymbol{R}(m\varphi) \begin{bmatrix} \Phi_{-m}(\varphi_\mathrm{s}) \\ \Phi_m(\varphi_\mathrm{s}) \end{bmatrix}, \\ \begin{bmatrix} Y_n^{-m}(\varphi_\mathrm{s}+\varphi, \vartheta_\mathrm{s}) \\ Y_n^m(\varphi_\mathrm{s}+\varphi, \vartheta_\mathrm{s}) \end{bmatrix} &= \boldsymbol{R}(m\varphi) \begin{bmatrix} Y_n^{-m}(\varphi_\mathrm{s}, \vartheta_\mathrm{s}) \\ Y_n^m(\varphi_{\mathrm{s},\vartheta_\mathrm{s}}) \end{bmatrix}. \end{aligned} \tag{5.8}$$

Figure 5.8a shows the processing scheme implementing only the non-zero entries of the associated matrix operation $\boldsymbol{T}$. Combined with a fixed set of 90° rotations around y (read from files), it can be used to access all rotational degrees of freedom in 3D [20].

The rotation effect is one of the most important features when using head-tracked interactive VR playback for headphones. Here, rotation counteracting the head movement has the task to support the impression of a static image of the virtual outside world.

5.2.3 *Directional Level Modification/Windowing*

What might be most important when mixing is the option to treat the gains of different directions differently: it might be necessary to attenuate directions of uninteresting or disturbing content while boosting directions of a soft target signal. For such a

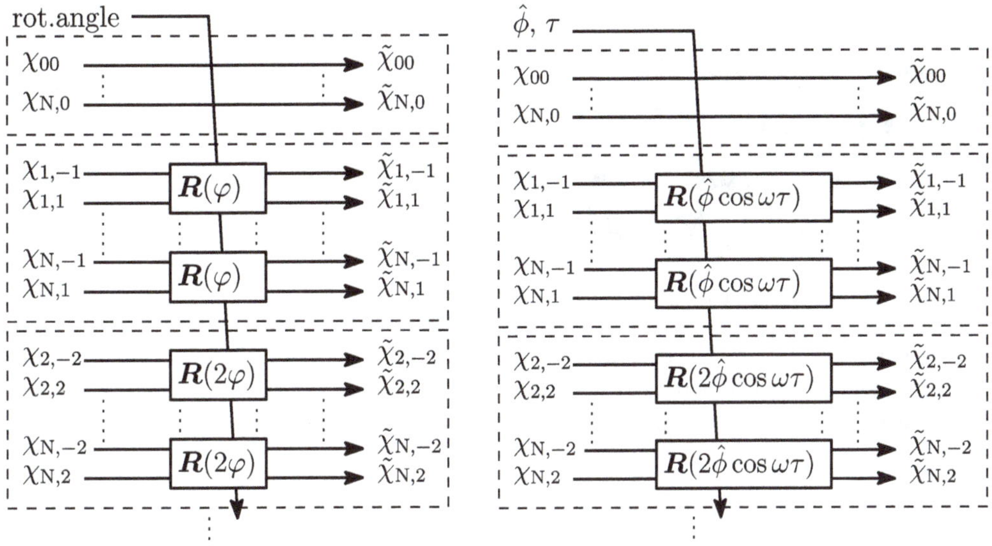

(a) Rotation around z (b) Dispersive rotation around z (widening)

Fig. 5.8 Rotation around z and Ambisonic widening/diffuseness apply simple 2×2 rotation matrices/filter matrices to each Ambisonic signal pair $\chi_{n,m}$ $\chi_{n,-m}$ of the same order n. Note that the order of the input/output channels plotted is not the typical ACN sequence to avoid crossing connections and hereby simplify the diagram

manipulation there is a neutral directional re-mapping $\tilde{\boldsymbol{\theta}} = \boldsymbol{\theta}$ and the transform to define the matrix $\boldsymbol{T}$ that is implemented as in Fig. 5.6a remains

$$\boldsymbol{T} = \frac{4\pi}{\hat{L}} \boldsymbol{Y}_{\tilde{\mathrm{N}},\boldsymbol{\Theta}} \operatorname{diag}\{\boldsymbol{g}_{\boldsymbol{\Theta}}\} \boldsymbol{Y}_{\mathrm{N},\boldsymbol{\Theta}}^{\mathrm{T}}. \tag{5.9}$$

In the simplest version, as implemented in `ambix_directional_loudness`, the gain function just consists of two mutually exclusive regions, e.g. within a region of diameter α around the direction $\boldsymbol{\theta}_{\mathrm{g}}$, and a complementary region outside, with separately controlled gains g_{in} and g_{out}:

$$g(\boldsymbol{\theta}) = g_{\mathrm{in}}\, u(\boldsymbol{\theta}^{\mathrm{T}}\boldsymbol{\theta}_{\mathrm{g}} - \cos\tfrac{\alpha}{2}) + g_{\mathrm{out}}\, u(\cos\tfrac{\alpha}{2} - \boldsymbol{\theta}^{\mathrm{T}}\boldsymbol{\theta}_{\mathrm{g}}), \tag{5.10}$$

where $u(x)$ represents the unit-step function that is 1 for $x \geq 0$ and 0 else. Note that the Ambisonic order of this effect will need to be larger to be lossless. However, with reasonably chosen sizes α and gain ratios $g_{\mathrm{in}}/g_{\mathrm{out}}$, the effect will nevertheless produce reasonable results. Figure 5.9 shows a window at azimuth and elevation at 22.5° with an aperture of 50° using $g_{\mathrm{in}} = 1$ and $g_{\mathrm{out}} = 0$ and the order of $\mathrm{N} = 10$ with a grid of encoded directions to illustrate the influence of the transformation.

For reference: entries of the tensor used to analytically re-expand the product of two spherical functions $x(\boldsymbol{\theta})\, g(\boldsymbol{\theta})$ given by their spherical harmonic coefficients χ_{nm}, γ_{nm} are called Gaunt coefficients or Clebsh-Gordan coefficients [6, 26].

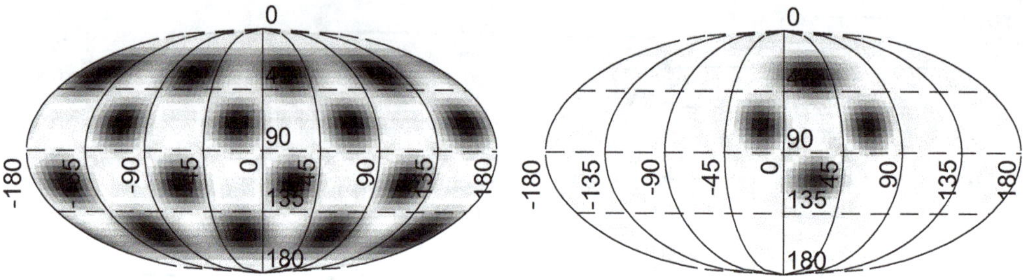

(a) original (b) windowed

Fig. 5.9 Directionally windowed Ambisonic test image at every 90° in azimuth, interleaved in azimuth for the elevations ±60° and ±22.5°, using the order $\mathrm{N} = \tilde{\mathrm{N}} = 10$, a window size of $\frac{\alpha}{2} = 50°$ around azimuth and elevation of 0° and max-$\boldsymbol{r}_\mathrm{E}$ weighting

5.2.4 *Warping*

Gerzon [27, Eq. 4a] described the effect *dominance* that is meant to warp the Ambisonic surround scene to modify how vitally the essential parts in front of the scene are presented.

Warping wrt. a direction. For mathematical simplicity, we describe this bilinear warping with regard to the z direction. To warp with regard to the frontal direction, one first rotates the front upwards, applies the warping operation there, and then rotates back. The bilinear warping modifies the normalized z coordinate $\zeta = \cos\vartheta = \theta_\mathrm{z}$ so that signals from the horizon $\zeta = 0$ are pulled to $\tilde{\zeta} = \alpha$,

$$\tilde{\zeta} = \frac{\alpha + \zeta}{1 + \alpha\zeta}, \tag{5.11}$$

while keeping for the poles $\tilde{\zeta} = \pm 1$ what was originally there $\zeta = \pm 1$. Hereby, the surround signal gets squeezed towards or stretched away from the zenith, or when rotating before and after: towards/from any direction.

The integral can be discretized and solved by a suitable t-design as before, only that for lossless operation, the output order $\tilde{\mathrm{N}}$ must be higher than the input order N. We get a matrix $\boldsymbol{T}$ that is implemented as in Fig. 5.6a is computed by

$$\boldsymbol{T} = \frac{4\pi}{\hat{\mathrm{L}}} \boldsymbol{Y}_{\tilde{\mathrm{N}},\boldsymbol{\Theta}} \,\mathrm{diag}\{\boldsymbol{g}_{\boldsymbol{\tau}^{-1}\{\boldsymbol{\Theta}\}}\}\, \tilde{\boldsymbol{Y}}^\mathrm{T}_{\mathrm{N},\boldsymbol{\tau}^{-1}\{\boldsymbol{\Theta}\}}. \tag{5.12}$$

The inverse mapping yields

$$\zeta = \tau^{-1}(\tilde{\zeta}) = \frac{\hat{\zeta} - \alpha}{1 - \alpha\hat{\zeta}}, \tag{5.13}$$

and it modifies the coordinates of the t-design inserted for $\tilde{\boldsymbol{\theta}}_l = \boldsymbol{\theta}_l = [\theta_{\mathrm{x},l},\ \theta_{\mathrm{y},l},\ \theta_{\mathrm{z},l}]^\mathrm{T}$ with $\tilde{\zeta}_l = \theta_{\mathrm{z},l}$ accordingly

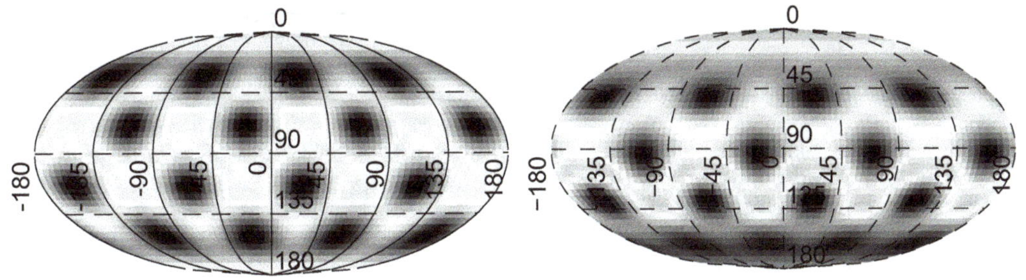

(a) original (b) warped

Fig. 5.10 Warping of the horizontal plane by 22.5° downwards; original Ambisonic test image contains points at every 90° in azimuth, interleaved in azimuth for the elevations ±60° and ±22.5°; orders are $\mathrm{N} = \tilde{\mathrm{N}} = 10$; max-$\boldsymbol{r}_\mathrm{E}$ weighted

$$\tau^{-1}\{\boldsymbol{\Theta}_l\} = \begin{bmatrix} \theta_{\mathrm{x},l}\sqrt{1-(\tau^{-1}\{\tilde{\zeta}_l\})^2} \\ \theta_{\mathrm{y},l}\sqrt{1-(\tau^{-1}\{\tilde{\zeta}_l\})^2} \\ \tau^{-1}\{\tilde{\zeta}_l\} \end{bmatrix} = \begin{bmatrix} \theta_{\mathrm{x},l}\sqrt{1-\left(\frac{\theta_{\mathrm{z},l}-\alpha}{1-\alpha\theta_{\mathrm{z},l}}\right)^2} \\ \theta_{\mathrm{y},l}\sqrt{1-\left(\frac{\theta_{\mathrm{z},l}-\alpha}{1-\alpha\theta_{\mathrm{z},l}}\right)^2} \\ \frac{\theta_{\mathrm{z},l}-\alpha}{1-\alpha\theta_{\mathrm{z},l}} \end{bmatrix}. \tag{5.14}$$

The gain $g(\hat{\zeta})$ of the generic transformation is useful to preserve the loudness of what becomes wider and therefore louder in terms of the E measure after re-mapping. To preserve loudness, the resulting surround signal is divided by the square root of the stretch applied, which is related to the slope of the mapping by $\frac{1}{g} = \sqrt{\frac{\mathrm{d}\zeta}{\mathrm{d}\hat{\zeta}}}$. Expressed as de-emphasis gain, we get

$$g(\hat{\zeta}_l) = \frac{1-\alpha\hat{\zeta}_l}{\sqrt{1-\alpha^2}} = \frac{1-\alpha\theta_{l,\mathrm{z}}}{\sqrt{1-\alpha^2}}. \tag{5.15}$$

Figure 5.10 shows warping of the horizontal plane by 20° downwards, using the test image parameters as with windowing; de-emphasis attenuates widened areas.

In the same fashion, Kronlachner [17] describes another warping curve that warps with regard to fixed horizontal plane and pole, either squeezing or stretching the content towards or away from the horizon, symmetrically for both the upper and lower hemispheres (second option of the `ambix_warp` plugin).

5.3 Parametric Equalization

There are two ways of employing parametric equalizers to Ambisonic channels. Either a single-/multi-channel input of a mono-encoder or a multiple-input encoder is filtered by parametric equalizers. Or each of the Ambisonic signal's channels is filtered by the same parametric equalizer, see Fig. 5.11a.

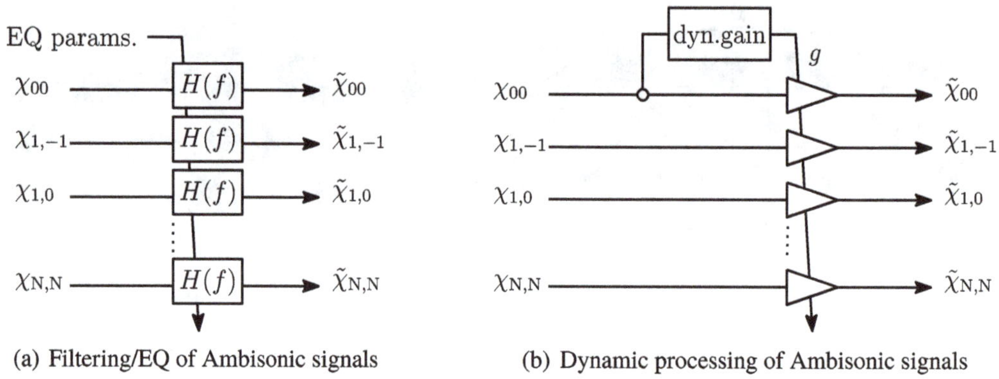

Fig. 5.11 Block diagram of processing that commonly and equally affects all Ambisonic signals, such as parametric equalization and dynamic processing (compression), without recombining the signals

Bass management is often important to not overdrive smaller loudspeaker systems of, e.g., a 5th-order hemispherical playback system with subwoofer signals: All 36 channels from the Ambisonic bus can be sent to a decoder section, in which frequencies below 70–100 Hz are high-cut by a 4th-order filter before running through the Ambisonic decoder, while the first channel from the Ambisonics bus alone, the omnidirectional channel, is being sent to a subwoofer section, in which a 4th-order filter high-cut removes the high frequencies above 70–100 Hz before the signal is sent to the subwoofers. If the playback system is time-aligned between subwoofer and higher frequencies, the 4th-order crossovers should be Linkwitz–Riley filters (either squared Butterworth high-pass or low-pass filters) to preserve phase equality [28].

For more information on parametric equalizers, the reader is referred to Udo Zölzer's book on Digital audio effects [29].

5.4 Dynamic Processing/Compression

Individual compression of different Ambisonic channels would destroy the directional consistency of the Ambisonics signal. Consequently, dynamic processing should rather affect the levels of all Ambisonic channels in the same way. As it typically contains all the audio signals, it is useful to have the first, omnidirectional Ambisonic channel control the dynamic processor as side-chain input, see Fig. 5.11b. For more information on dynamic processing, the reader is referred to Udo Zölzer's book on Digital audio effects [29].

Moreover, it is sometimes useful to compress the vocals of a singer separately. To this end, the directional compression would first extract a part of the Ambisonic signals by a directional window, creating one set of Ambisonic signals without the directional region of the window, and another one exclusively containing it. The compression is applied on the resulting window signal before re-combining it with the rest signals.

5.5 Widening (Distance/Diffuseness/Early Lateral Reflections)

Basic widening and diffuseness effects can be regarded as being inspired by Gerzon [30] and Laitinen [31] who proposed to apply frequency-dependent panning filters, mapping different frequencies to directions dispersed around the panning direction. The resulting effect is fundamentally different from and superior to increasing the spread in frequency-independent MDAP with enlarged spread or Ambisonics with reduced order, which could yield audible comb filtering.

To apply this technique to Ambisonics, Zotter et al. [32] proposed to employ a dispersive, i.e. frequency-dependent, rotation of the Ambisonic scene around the z-axis as in Eq. (5.8) by the matrix $\boldsymbol{R}$ as described above and in Fig. 5.8b, using 2×2 matrices of filters to implement the frequency-dependent argument $m\hat{\phi}\cos\omega\tau$

$$\boldsymbol{R}(m\hat{\phi}\cos\omega\tau) = \begin{bmatrix} \cos(m\hat{\phi}\cos\omega\tau) & \sin(m\hat{\phi}\cos\omega\tau) \\ -\sin(m\hat{\phi}\cos\omega\tau) & \cos(m\hat{\phi}\cos\omega\tau) \end{bmatrix}, \tag{5.16}$$

whose parameters $\hat{\phi}$ and τ allow to control the magnitude and change rate of the rotation with increasing frequency. How this filter matrix is implemented efficiently was described in [33], where a sinusoidally frequency-varying pair of functions

$$g_1(\omega) = \cos\left[\alpha\cos(\omega\,\tau)\right], \qquad g_2(\omega) = \sin\left[\alpha\cos(\omega\,\tau)\right], \tag{5.17}$$

was found to correspond to the sparse impulse responses in the time domain

$$g_1(t) = \sum_{q=-\infty}^{\infty} J_{|q|}(\alpha)\,\cos(\tfrac{\pi}{2}\,|q|)\,\delta(t - q\,\tau) \tag{5.18}$$

$$g_2(t) = \sum_{q=-\infty}^{\infty} J_{|q|}(\alpha)\,\sin(\tfrac{\pi}{2}\,|q|)\,\delta(t - q\,\tau),$$

allowing for truncation to just a few terms in q, typically 11 taps between $-5 \leq q \leq 5$ or fewer, and hereby an efficient implementation. For the implementation of the filter matrix, for each degree m, the value $\alpha = m\hat{\phi}$. (*It might be helpful to be reminded of a phase-modulated cosine and sine from radio communication, whose spectra are the same functions as this impulse response pair.*)

As the algorithm places successive frequencies at slightly displaced directions, the auditory source width increases. Moreover, the frequency-dependent part causes a smearing of the temporal fine-structure in the signal. In [34], it was found that implementations discarding the negative values of q, i.e. keeping $q \geq 0$ sound more natural and still exhibit a sufficiently strong effect. Time constants τ around 1.5 ms yield a *widening* effect, and a *diffuseness* and *distance* impression is obtained with τ around 15 ms. The parameter $\hat{\phi}$ is adjustable between 0 (no effect) and larger values.

Beyond 80° the audio quality starts to degrade. The use as diffusing effect has turned out to be useful as simple simulation of early lateral reflections, because most parts of the spectrum are played back near the reversal points $\pm\hat{\phi}$ of the dispersion contour. For naturally sounding early reflections, additional shelving filters introducing attenuation of high frequencies prove useful.

Figures 5.12 and 5.13 show experimental ratings of the perceived effect strength (width or distance) of the above algorithm in [34], which was implemented as frequency-dependent (dispersive) panning on just a few loudspeakers $\mathrm{L} = 3, 4, 5, 7$ evenly arranged from $-90°$ to $90°$ on the horizon at 2.5 m distance from the central listening position. Loudspeakers were controlled by a sampling decoder of the orders $\mathrm{N} = 1, 2, 3, 5$ with the center of the max-$\boldsymbol{r}_\mathrm{E}$-weighted panning direction at 0° in front. The signal was speech and as a reference it used the frontal loudspeaker with the unprocessed signal "REF". The experiment tested the algorithm with both the symmetric impulse responses suggested by Eq. (5.18), and such truncated to their causal $q \geq 0$-side, for a listening position at the center of the arrangement (bullet marker) and at 1.25 m shifted to the right, off-center (square marker). Figure 5.12 indicates for the widening algorithm with $\tau = 1.5$ ms that the perceived width saturates above $\mathrm{N} > 2$ at both listening positions. Despite the effect of the causal-sided

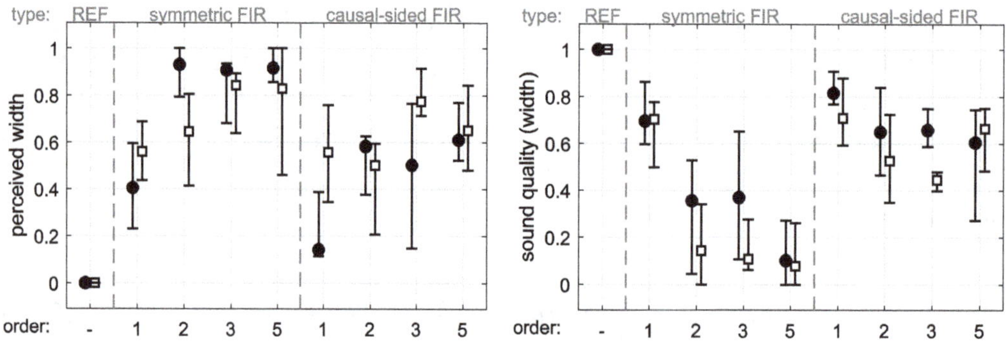

Fig. 5.12 Perceived width (left) and audio quality (right) of frequency-dependent dispersive Ambisonic rotation as widening effect using the setting $\tau = 1.5$ ms, the Ambisonic orders $\mathrm{N} = 1, 2, 3, 5$, and $\mathrm{L} = 3, 4, 5, 7$ loudspeakers on the frontal semi-circle, with listening positions at the center (bullet marker) and half-way right off-center (square marker)

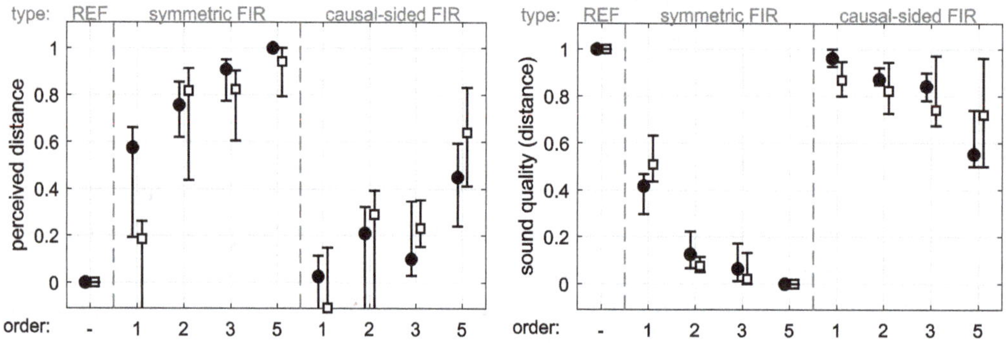

Fig. 5.13 Perceived width (left) and audio quality (right) of frequency-dependent dispersive Ambisonic rotation as distance/diffuseness effect using the setting $\tau = 15$ ms, the Ambisonic orders $\mathrm{N} = 1, 2, 3, 5$, and $\mathrm{L} = 3, 4, 5, 7$ loudspeakers on the frontal semi-circle, with listening positions at the center (bullet marker) and half-way right off-center (square marker)

implementation is weaker in effect strength, it highly outperforms the symmetric FIR implementation in terms audio quality (right diagram), while still producing a clearly noticeable effect when compared to the unprocessed reference (left diagram).

A more pronounced preference of the causal-sided implementation in terms of audio quality is found in Fig. 5.13 for the setting $\tau = 15$ ms, where the algorithm is increasing the diffuseness or perceived distance for orders $\mathrm{N} > 2$ at both listening positions.

5.6 Feedback Delay Networks for Diffuse Reverberation

Feedback delay networks (FDN, cf. [35, 36]) can directly be employed to create diffuse Ambisonic reverberation. A dense response and an individual reverberation for every encoded source can be expected when feeding the Ambisonic signals directly into the inputs of the FDN.

As in Fig. 5.14, FDNs consists of a matrix $\boldsymbol{A}$ that is orthogonal $\boldsymbol{A}^{\mathrm{T}}\boldsymbol{A} = \boldsymbol{I}$ and should mix the signals of the feedback loop well enough to distribute them across all different channels to couple the resonators associated with the different delays τ_i. These delays should not have common divisors to avoid pronounced resonance frequencies, and are therefore typically chosen to be related to prime numbers. Small delays are typically selected to be more closely spaced $\{2, 3, 5, \dots\}$ ms to simulate a diffuse part with densely spaced response at the beginning, and long delays further apart often make the reverberation more interesting. Using unity factors as channel gains $g_{\mathrm{lo}}^{\tau_i}, g_{\mathrm{mi}}^{\tau_i}, g_{\mathrm{hi}}^{\tau_i} = 1$ and any orthogonal matrix $\boldsymbol{A}$, the reverberation time becomes infinite. For smaller channel gains, the FDN produces decaying output.

Reverberation is characterized by the exponentially decaying envelope $10^{-3\frac{t}{T_{60}}}$. For a single delay of the length τ_i, the corresponding gain is g^{τ_i} with $g = 10^{-\frac{3}{T_{60}}}$. This factor with the corresponding exponent provides equal reverberation decay rate in every channel, and hereby exact control of the reverberation time. To make the effect sound natural, it is typical to adjust the gains within a high-mid-low filter set to decrease the reverberation towards higher frequency bands by the gains $g_{\mathrm{hi}}^{\tau_i} \leq g_{\mathrm{mi}}^{\tau_i} \leq g_{\mathrm{lo}}^{\tau_i}$.

The vector gathering the current sample for every feedback path is multiplied by the matrix $\boldsymbol{A}$. For calculation in real-time, Rocchesso proposed to use a scaled Hadamard matrix $A = \frac{1}{\mathrm{M}}\boldsymbol{H}$ of the dimensions $\mathrm{M} = 2^k$ in [37]. It consists of ± 1 entries only and hereby perfectly mixes the signal across the different feedbacks to create a diffuse set of resonances. What is more, this not only replaces the $\mathrm{M} \times \mathrm{M}$ multiply and adds of matrix multiplication multiplies by sums and differences, it is

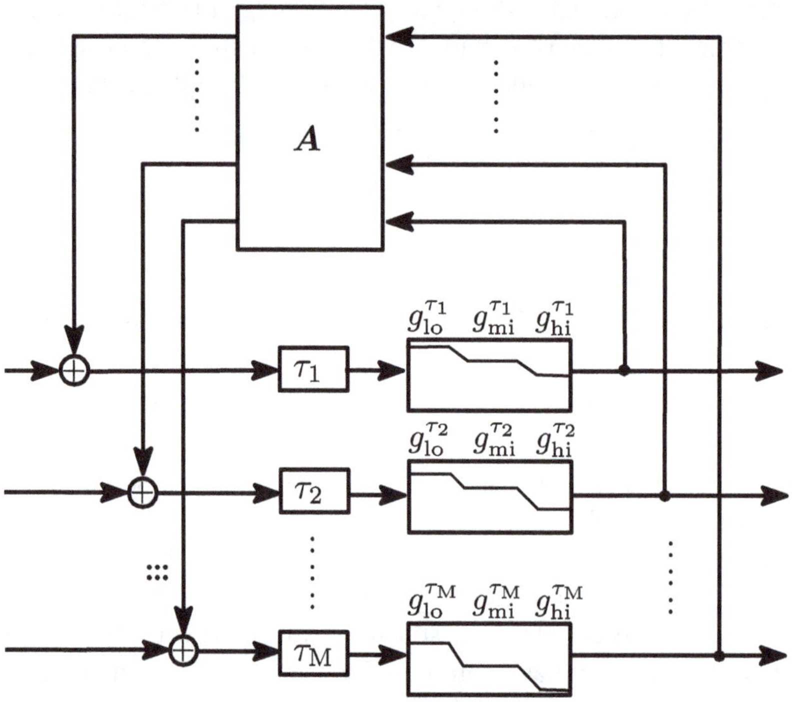

Fig. 5.14 Feedback delay network (FDN) for Ambisonic reverb. The matrix $\boldsymbol{A}$ is unitary and the gain $g = 10^{-\frac{3}{T_{60}}}$ to the power of the delay τ_i allows to adjust a spaitally and temporally diffuse reverberation effect in different bands (lo, mi, hi)

moreover equivalent to the efficient implementation as Fast Walsh-Hadamard Transform (FWHT), a butterfly algorithm. Figure 5.15 shows a graphical implementation example of a 16-channel FWHT in the real-time signal processing environment Pure Data.

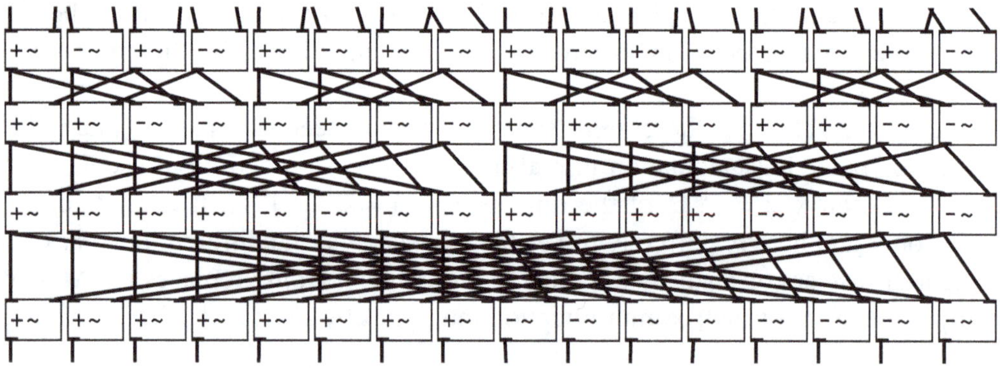

Fig. 5.15 The fast Walsh-Hadamard transform (FWHT) variant implemented in the 16-channel feedback delay network reverberator `[rev3~]` in Pure Data requires only 4×16 sums/differences to replace the 16×16 multiplies of matrix multiplication by $\boldsymbol{A}$

5.7 Reverberation by Measured Room Impulse Responses and Spatial Decomposition Method in Ambisonics

The first-order spatial impulse response of a room at the listener can be improved by resolution enhancements of the spatial decomposition method (SDM) by Tervo [38], which is a broad-band version of spatial impulse response rendering (SIRR) by Merimaa and Pulkki [39, 40]. For reliable measurements, typically loudspeakers are employed, and the typical measurement signals aren't impulses, but swept-sine signals that can are reverted to impulses by deconvolution. A room impulse response is typically sparse in its beginning whenever direct sound and early reflections arrive at the measurement location. Generally, it is likely that those arrival times in the early part do not coincide and are well separated from each other, so that one can assume their *temporal disjointness* at the receiver.

From a room impulse response $h(t)$ that complies with this assumption, for which there consequently is a direction of arrival (DOA) $\boldsymbol{\theta}_{\mathrm{DOA}}(t)$ for every time instant, one could construct an Ambisonic receiver-directional room impulse response as in [41] $h(\boldsymbol{\theta}_{\mathrm{R}}, t) = h(t)\,\delta[1 - \boldsymbol{\theta}_{\mathrm{R}}^{\mathrm{T}}\boldsymbol{\theta}_{\mathrm{DOA}}(t)]$, depending on the direction $\boldsymbol{\theta}_{\mathrm{R}}$ at the receiver. This response can be transformed into the spherical harmonic domain by integrating it over $\boldsymbol{y}_{\mathrm{N}}(\boldsymbol{\theta}_{\mathrm{R}})\,\mathrm{d}\boldsymbol{\theta}_{\mathrm{R}}\int_{\mathbb{S}^2}$, to get the set of Nth-order Ambisonic room impulse responses

$$\tilde{\boldsymbol{h}}_{\mathrm{N}}(t) = h(t)\;\boldsymbol{y}_{\mathrm{N}}[\boldsymbol{\theta}_{\mathrm{DOA}}(t)].$$

A signal $s(t)$ convolved by this vector of impulse responses theoretically generates a 3D Ambisonic image of the mono sound in the room of the measurement. This can be done, e.g., by the plug-in `mcfx_convolver`. Now there are two problems to be solved: (i) how to estimate $\boldsymbol{\theta}_{\mathrm{DOA}}(t)$, (ii) how to deal with the diffuse part of $h(t)$, when there are more sound arrivals at a time than one.

Estimation of the DOA. One could now just detect the temporal peaks of the room impulse response and assign the guessed evolution of the direction of arrival as suggested in [42], and hereby span the envelopment of the room impulse response. Alternatively, if the room impulse response was recorded by a microphone array as in [38], array processing can be used to estimate the direction of arrival $\boldsymbol{\theta}_{\mathrm{DOA}}(t)$. For first-order Ambisonic microphone arrays, when suitably band-limited to the frequency range in which the directional mapping is correct, e.g. between 200 Hz and 4 kHz, the vector $\boldsymbol{r}_{\mathrm{DOA}}$ of Eq. (A.83) in Appendix A.6.2 yields a suitable estimate

$$\tilde{\boldsymbol{r}}_{\mathrm{DOA}}(t) = W(t)\begin{bmatrix}X(t)\\Y(t)\\Z(t)\end{bmatrix} = -\rho c\,h(t)\,\boldsymbol{v}(t), \qquad \boldsymbol{\theta}_{\mathrm{DOA}}(t) = \frac{\tilde{\boldsymbol{r}}_{\mathrm{DOA}}(t)}{\|\tilde{\boldsymbol{r}}_{\mathrm{DOA}}(t)\|}. \tag{5.19}$$

Figure 5.16 shows the directional analysis of the first 100 ms of a first-order directional impulse response taken from the openair lib[1] This response was measured in St. Andrew's Church Lyddington, UK (2600 m^3 volume, 11.5 m source-receiver distance) with a Soundfield SPS422B microphone.

[1] http://www.openairlib.net.

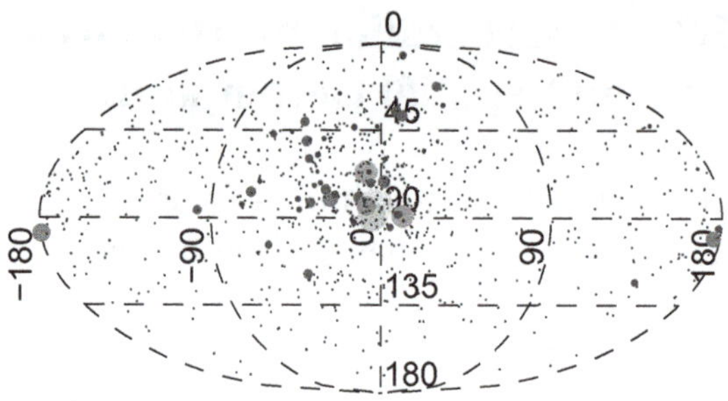

Fig. 5.16 Spatial distribution of the first 100 ms of a first-order directional impulse response measured at St. Andrew's Church Lyddington. Brightness and size of the circles indicate the level

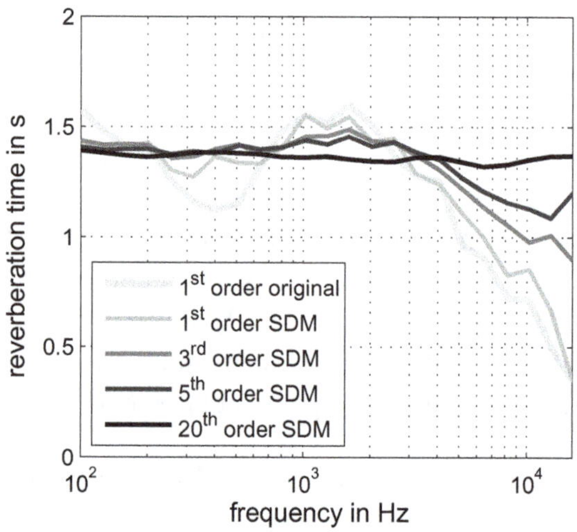

Fig. 5.17 Frequency-dependent reverberation time calculated from original and SDM-enhanced impulse responses without spectral decay recovery

The direct sound from the front is clearly visible, as well as strong early reflections from front and back, and equally distributed weak directions from the diffuse reverb.

Spectral decay recovery for higher-order RIRs. The second task mentioned above is that the multiplication of $h(t)$ by $\boldsymbol{y}_\mathrm{N}[\boldsymbol{\theta}_\mathrm{DOA}(t)]$ to obtain $\tilde{\boldsymbol{h}}(t)$ degrades the spectral decay at higher orders. If there is no further processing, the resulting response typically exhibits a noticeable increased spectral brightness [38, 41, 43]. This unnatural brightness mainly affects the diffuse reverberation tail, where *temporal disjointness* is a poor assumption. There, the corresponding rapid changes of $\boldsymbol{\theta}_\mathrm{DOA}(t)$ cause a strong amplitude modulation in the pre-processing of the late room impulse response at high Ambisonic orders. Typically, long decays of low frequencies leak into high frequencies, and hereby result in an erroneous spectral brightening of the diffuse tail. Figure 5.17 analyses the behavior in terms of an erroneous increase of reverberation time at high frequencies, especially when using high orders.

In order to equalize the spectral decay and hereby the reverberation time of the SDM-enhanced impulse response, there is a helpful pseudo-allpass property of the spherical harmonics for direct and diffuse fields, as described in Eqs. (A.52) and (A.55) of Appendix A.3.7. The signals in the vector $\tilde{\boldsymbol{h}}(t) = [\tilde{h}_n^m(t)]_{nm}$ are first decomposed into frequency bands, yielding the sub-band responses $\tilde{h}_n^m(t, b)$. We can equalize the spectral sub-band decay for every band b and order n by targeting fulfillment of the pseudo-allpass property

$$\sum_{m=-n}^{n} \mathcal{E}\left\{|h_n^m(t,b)|^2\right\} = (2n+1)\mathcal{E}\left\{|h_0^0(t,b)|^2\right\}. \tag{5.20}$$

The formulation above relies on the correct spectral decay of the omnidirectional signal $h_0^0(t,b) = \tilde{h}_0^0(t,b)$, which is unaffected by modulation. Correction is achieved by

$$h_n^m(t,b) = \tilde{h}_n^m(t,b)\sqrt{\frac{(2n+1)\,\mathcal{E}\{|\tilde{h}_0^0(t,b)|^2\}}{\sum_{m=-n}^{n}\mathcal{E}\{|\tilde{h}_n^m(t,b)|^2\}}}; \tag{5.21}$$

here, the expression $\mathcal{E}\{|\cdot|^2\}$ refers to estimation of the squared signal envelope.

Perceptual evaluation. Frank's 2016 experiments [44] measuring the area of the sweet spot also investigated the plausibility of reverberation created by their Ambisonically SDM-processed measurements at different order settings, N = 1, 3, 5. For Fig. 5.18b listeners indicated at which distance from the room's center they heard that envelopment began to collapse to the nearest loudspeakers. One can observe that rendering diffuse reverberation for a large audience benefits from a high Ambisonic order. Moreover, experiments in [43] revealed an improvement of the perceived spatial depth mapping, i.e. a clearer separation between foreground and background sound for the SDM-processed higher-order reverberation, cf. Fig. 1.21b.

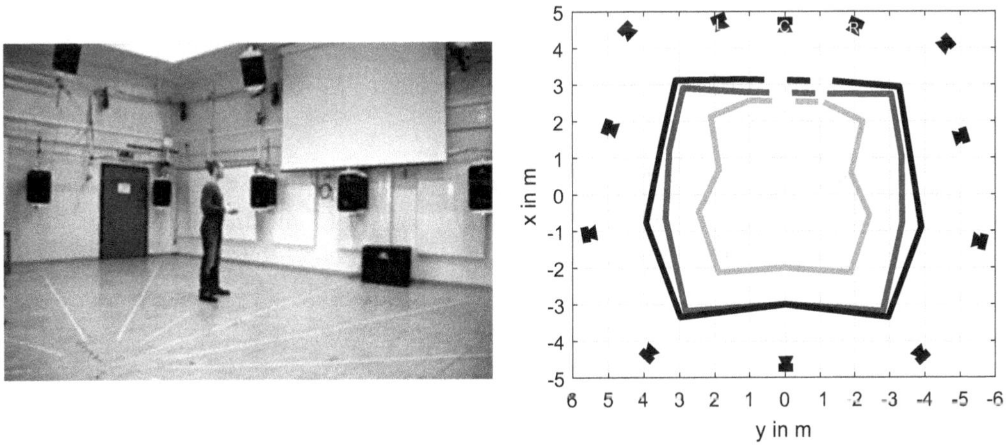

(a) Sweet spot experiment (b) Median size for diuse sound

Fig. 5.18 The perceptual sweet spot size as investigated by Frank [44] for SDM processed RIRs cover an area in IEM CUBE that increases with the SDM order N chosen (black = 5th, gray = 3rd, light gray = 1st order Ambisonics). In comparison to panned direct sound, one should keep some distance to the loudspeakers to avoid breakdown of envelopment

5.8 Resolution Enhancement: DirAC, HARPEX, COMPASS

The concept of parametric audio processing [5] describes ways to obtain resolution-enhanced first-order Ambisonic recordings by parametric decomposition and rendering. One main idea is to decompose short-term stationary signals of a sound scene into a *directional* and a less directional *diffuse* stream.
For synthesis of the *directional* part based on mono signals, it is clear how to obtain the most narrow presentations by amplitude panning or higher-order Ambisonic panning of consistent $\boldsymbol{r}_\mathrm{E}$ vector predictions as in Chap. 2.

The synthesis of *diffuse and enveloping* parts based on a mono signal can require extra processing such as either widening/diffuseness effects or reverberation as in Sects. 5.5 and 5.6, which both also provide a directionally wide distribution of sound. Or more practically, the recording itself could deliver sufficiently many uncorrelated instances of the diffuse sound to be played back by surrounding virtual sources. Envelopment and diffuseness is based on providing a consistently *low interaural covariance or cross correlation* of sufficiently high decorrelation.

DirAC. A main goal of DirAC (Directional Audio Coding [5]) is finding signals and parameters for sound rendering by analyzing first-order Ambisonic recordings. One variant is to use the intensity-vector-based analysis in the short-term Fourier transform (STFT), see also Appendix A.6.2:

$$\boldsymbol{r}_\mathrm{DOA}(t,\omega) = -\frac{\rho c\,\Re\{p(t,\omega)^*\boldsymbol{v}(t,\omega)\}}{|p(t,\omega)|^2} = \frac{\Re\{W(t,\omega)^*[X(t,\omega),Y(t,\omega),Z(t,\omega)]^\mathrm{T}\}}{\sqrt{2}|W(t,\omega)|^2}, \tag{5.22}$$

which can be treated similarly as the $\boldsymbol{r}_\mathrm{E}$ vector, regarding direction and diffuseness $\psi = 1 - \|\boldsymbol{r}_\mathrm{DOA}\|^2$.

Single-channel DirAC is Ville Pulkki's original way to decompose the $W(t,\omega)$ signal in the STFT domain into a directional signal $\sqrt{1-\psi}\,W(t,\omega)$ that is synthesized by amplitude panning and a diffuse signal $\sqrt{\psi}\,W(t,\omega)$ to be synthesized diffusely [45]. *Virtual-microphone DirAC* uses a first-order Ambisonic decoder to the given loudspeaker layout and time-frequency-adaptive sharpening masks increasing the focus of direct sounds, see Vilkamo [46] and [5, Ch. 6], or order e.g. Sect. 5.2.3. Playback of diffuse sounds benefits from an optional diffuseness effect.

HARPEX (high angular-resolution plane-wave expansion [47]) is Svein Berge's patented solution to optimally decode sub-band signals. It is based on the observation he made with Natasha Barrett that decoding to a tetrahedral loudspeaker layout is perceptually outperforming if the tetrahedron nodes are rotationally aligned with the sources of the recording. HARPEX accomplishes convincing diffuse and direct sound reproduction by decoding to a variably adapted virtual loudspeaker layout in every sub band. The layout is adaptively rotation-aligned with sources detected in the band. HARPEX is typically described using an estimator for direction pairs.

COMPASS (COding and Multidirectional Parameterization of Ambisonic Sound Scenes [48]) by Archontis Politis can be seen as an extension of DirAC. In contrast to DirAC, it tries to detect and separate multiple direct sound sources from the ambient or background sound. This is done by applying two different kinds of beamformers: one that contains only the direct sound for each sound source (source signals) and one that contains everything but the direct sound (ambient signal). Similar as before, the source signals are reproduced using amplitude panning and the ambient signal is sent to the decorrelator. In contrast to DirAC, COMPASS is not limited to first-order input but can also enhance the spatial resolution of higher-order inputs.

5.9 Practical Free-Software Examples

5.9.1 IEM, ambix, and mcfx Plug-In Suites

The `ambix_converter` is an important tool when adapting between the different Ambisonic scaling conventions, e.g. the standard SN3D normalization that uses only $\sqrt{\frac{(n-|m|)!}{(n+|m|)!}\,\frac{2-\delta_m}{4\pi}}$ for normalization instead of the full $\sqrt{\frac{(n-|m|)!}{(n+|m|)!}\,\frac{(2-\delta_m)(2n+1)}{4\pi}}$ that is called N3D, see Fig. 5.19. This alternating definition is because of a practical choice of the ambix format [49] to avoid high-order channels becoming louder than the zeroth-order channel. Also it permits to adapt between channel sequences such as ACN's $i = n^2 + n + m$ or SID's $i = n^2 + 2(n - |m|) + (m < 0)$. It is advisable to use test recordings with the main directions, e.g. front, left, top, and to check that the channel separation for decoded material is roughly exceeding 20 dB for 5th-order material. Moreover, it contains inversion of the Condon-Shortley phase that typically causes a 180° rotation around the z axis, and it contains the left-right, front-back, and top-bottom flips discussed in the mirroring operations above.

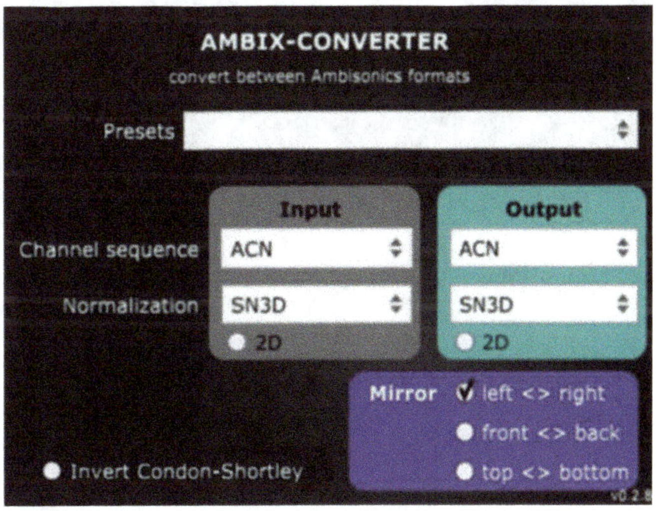

Fig. 5.19 `ambix_converter` plug-in

The `ambix_warping plugin`, see Fig. 5.20, implements the above-mentioned warping operations shifting horizontal sounds towards one of the poles, or into both polar directions. Warping can be applied to any other direction than zenith and nadir when placing it between two mutually inverting `ambix_rotation` or IEM `SceneRotator` objects that intermediately rotate zenith to another direction.

The IEM `SceneRotator` as the `ambix_rotation` plugin can be controlled by head tracking and it essential for an immersive headphone-based experience, see Fig. 5.21. Its processing is done as described above.

The `ambix_directional_loudness` plugin in Fig. 5.22 implements the above-mentioned directional amplitude window in either circular or equi-rectangular spherical shape. Several of these windows can be made, soloed, and remote controlled, each one of which allowing to set a gain for the inside and outside region. This is often useful in practice, when, e.g., reinforcing or attenuating desired or undesired signal parts within an Ambisonic scene.

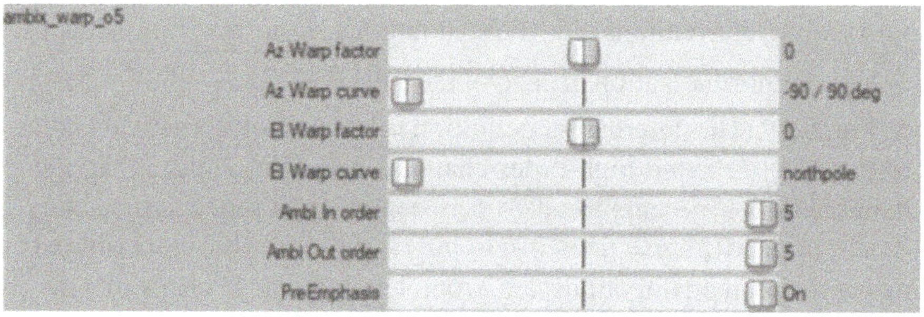

Fig. 5.20 `ambix_warping` plug-in in Reaper

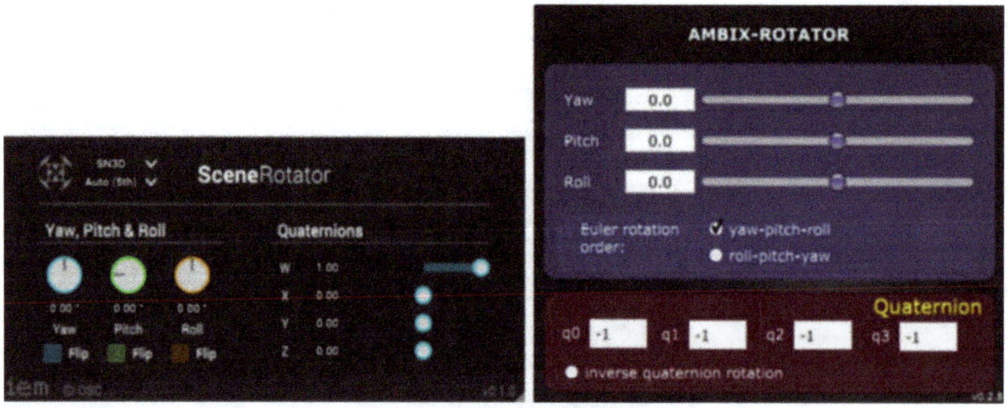

Fig. 5.21 IEM `SceneRotator` and `ambix_rotator` plug-ins

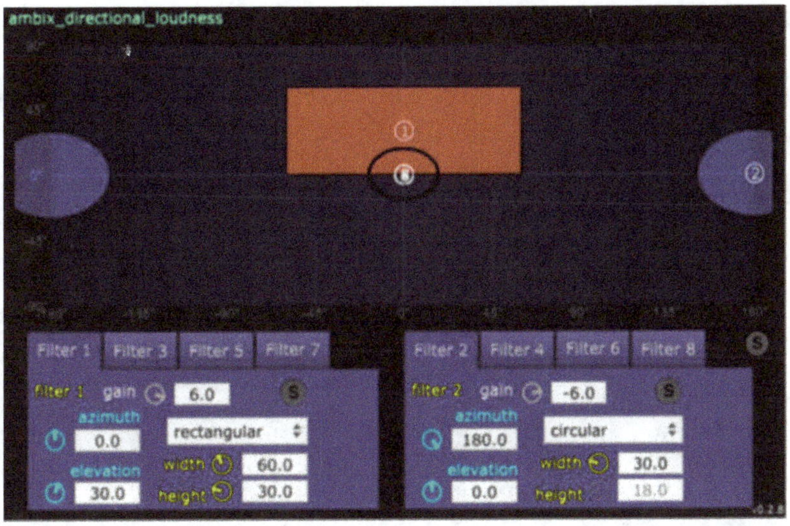

Fig. 5.22 `ambix_directional_loudness` plug-in

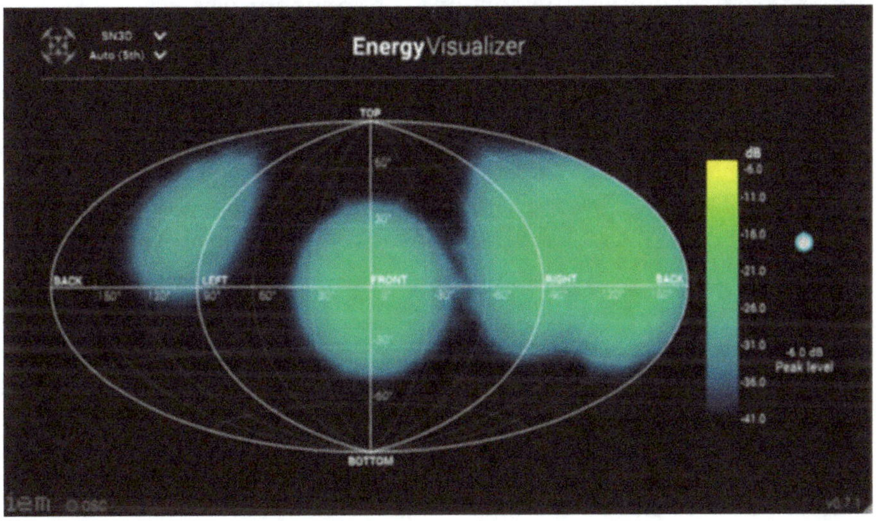

Fig. 5.23 `EnergyVisualizer` plug-in

To observe the changes made to the Ambisonic scene, the IEM `Energy Visualizer` can be helpful, see Fig. 5.23.

If, for instance, the Ambisonic scene requires dynamic compression, as outlined in the section above, the IEM `OmniCompressor` is a helpful tool. It uses the omnidirectional Ambisonic channel to derive the compression gains (as a side-chain for all other Ambisonic channels). Similarly as the `directional_loudness` plug-in, the IEM `DirectionalCompressor` allows to select a window, but this time for setting different dynamic compression within and outside the selected window, see Fig. 5.24.

The multichannel `mcfx_filter` plugin in Fig. 5.25 does not only implement a set of parametric equalizers, a low- and high cut that can be toggled between filter skirts

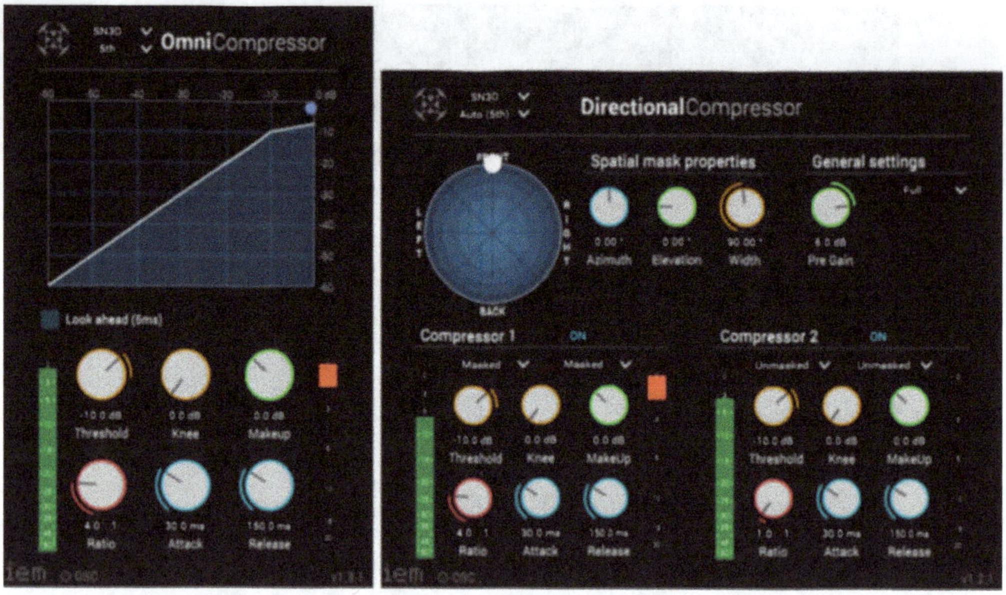

Fig. 5.24 `OmniCompressor` and `DirectionalCompressor` plug-in

Fig. 5.25 `mcfx_filter` plug-in

of either 2nd or 4th order, but it also features a real-time spectrum analyzer to observe the changes done to the signal. It is not only practical for Ambisonic purposes, it's just a set of parametric filters that is equally applied to all channels and controlled from one interface.

The `mcfx_convolver` plug-in in Fig. 5.26 is useful for many purposes, also scientific ones, e.g., when testing binaural filters or driving multi-channel arrays with filters, etc. Its configuration files use the `jconvolver` format that specifies which filter file (typically stored in multi-channel wav files) connects which of its multiple inlets to which of its multiple outlets. It is also used to implement the SDM-based reverberation described in the above sections.

For a cheaper reverberation network, the IEM `FDNReverb` network described above can be used, see Fig. 5.27. It is not in particular an Ambisonic tool, but can

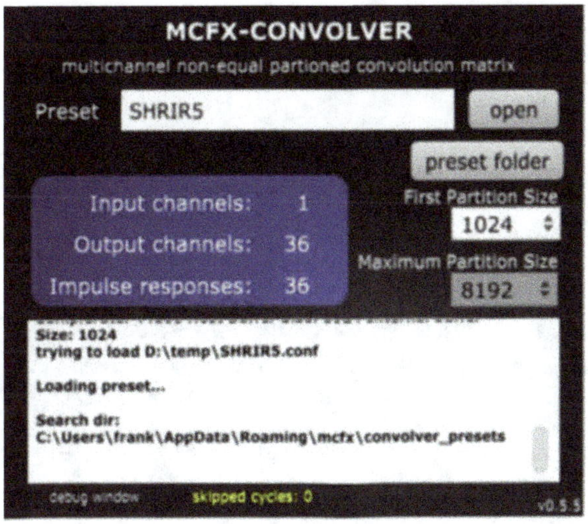

Fig. 5.26 `mcfx_convolver` plug-in

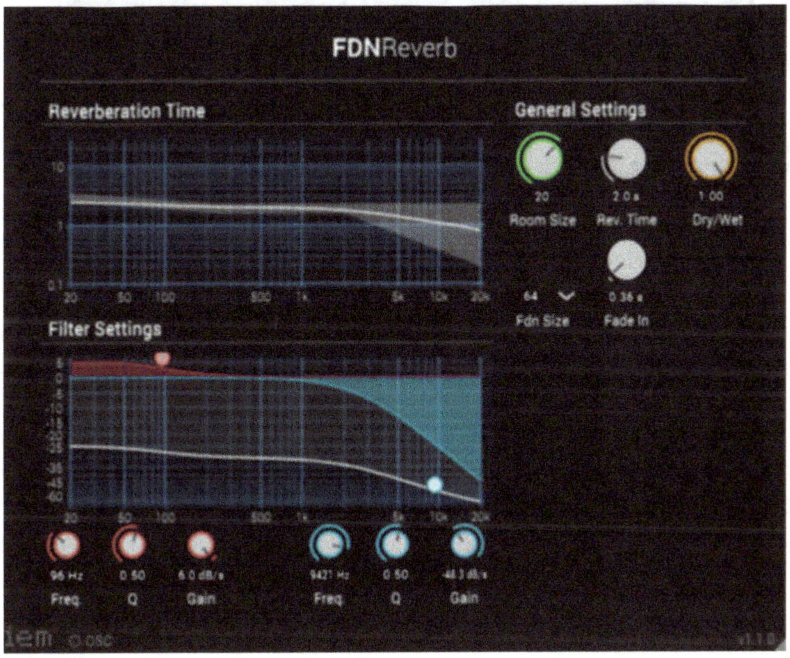

Fig. 5.27 `FDNReverb` plug-in

be used in any multi-channel environment. The particularity of the implementation in the IEM suite is that a slow onset can be adjusted.

The `ambix_widening` plug-in in Fig. 5.28 implements the widening by frequency-dependent, dispersive rotation of the Ambisonic scene around the z axis as described above. It can also be used to cheaply stylize lateral reflections instead of the IEM `RoomEncoder` (Fig. 4.36) with time constant settings exceeding 5 ms, or just as a

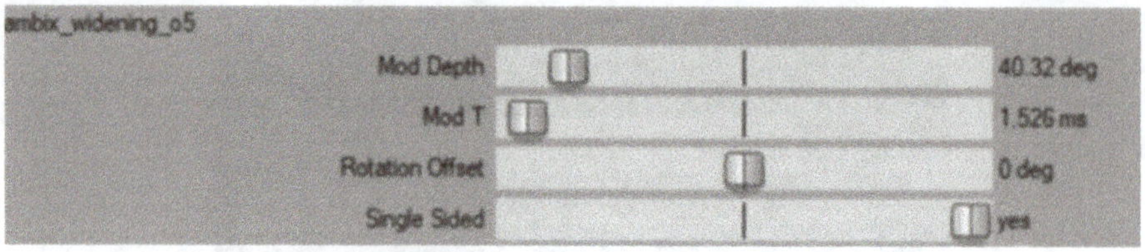

Fig. 5.28 `ambix_widening` plug-in in Reaper

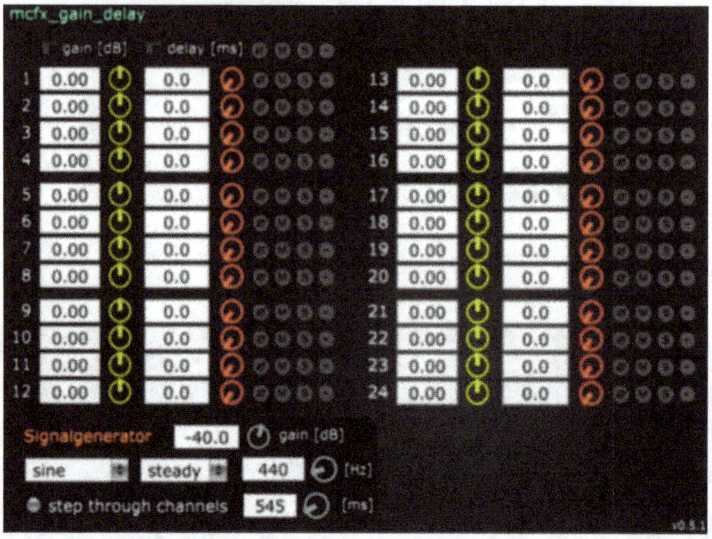

Fig. 5.29 `mcfx_gain_delay` plug-in

widening effect. The setting *single-sided* permits to suppress the slow attack of the Bessel sequence.

Another tool is quite helpful, the `mcfx_gain_delay` plug-in in Fig. 5.29. It permits to to solo or mute individual channels, as well as delay and attenuate them differently. What is more and often even more useful: It is invaluably helpful for testing the signal chain, as one can step through the channels with different signals.

5.9.2 Aalto SPARTA

The SPARTA plug-in suite by Aalto University provides Ambisonic tools for encoding, decoding on loudspeakers and headphones, as well as visualization. A special feature is the `COMPASS decoder` plug-in Fig. 5.30 that can increase the spatial resolution of first-, second-, and third-order recordings. Playback can be done either

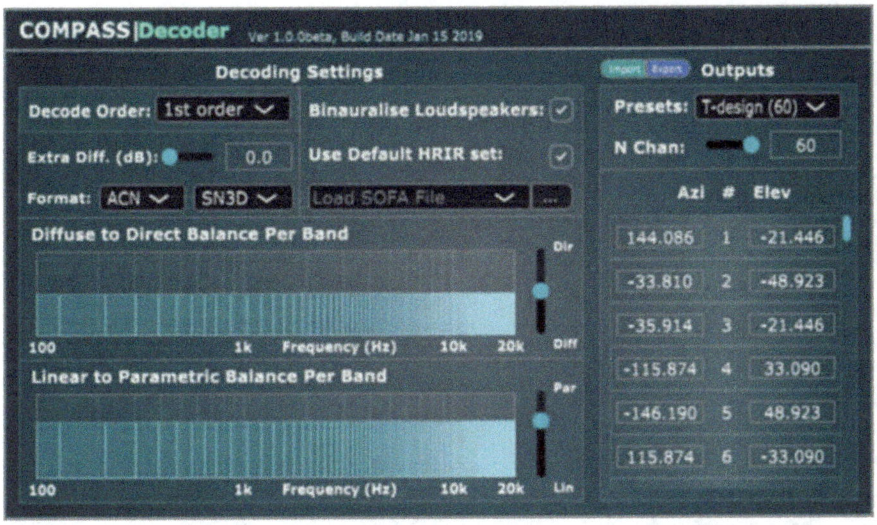

Fig. 5.30 `COMPASS Decoder` plug-in

on arbitrary loudspeaker arrangements or their virtualization on headphones. The signal-dependent parametric processing allows to adjust the balance between direct and diffuse sound in each frequency band. In order to suppress artifacts due to the processing, the parametric playback (`Par`) can be mixed with the static decoding (`Lin`) of the original recording. While it is advisable to keep the parametric contribution below 2/3 for noticable directional improvements and low artifacts, in general, in recordings with cymbals or hihats it is advisable to fade towards `lin` starting at around 4 kHz.

5.9.3 Røde

The `Soundfield` plug-in by Røde in Fig. 5.31 was originally designed to process the signals from the four cardioid microphone capsules of their Soundfield microphone. However, it also supports first-order Ambisonics as input format. It can decode to various loudspeaker arrangements by placing virtual microphones into the directions of the loudspeakers. The directivity of each virtual microphone can be adjusted between first-order cardioid and hyper-cardioid. Moreover, higher-order directivity patterns are possible using a parametric signal-dependent processing, resulting in an increase of the spatial resolution.

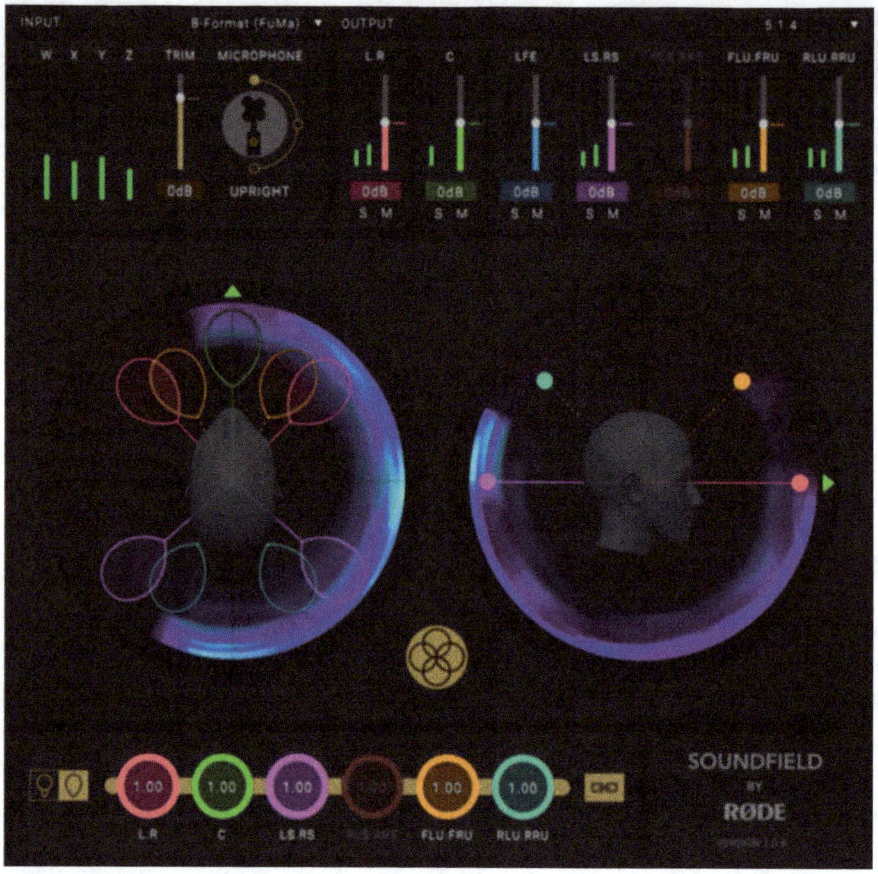

Fig. 5.31 `Soundfield` by Rødé plug-in

References

1. D.G. Malham, Higher-order ambisonic systems for the spatialisation of sound, in *Proceedings of the ICMC* (Bejing, 1999)
2. M. Frank, F. Zotter, A. Sontacchi, Producing 3D audio in ambisonics, in *57th International AES Conference, LA, The Future of Audio Entertainment Technology? Cinema, Television and the Internet* (2015)
3. M. Frank, A. Sontacchi, Case study on ambisonics for multi-venue and multi-target concerts and broadcasts. J. Audio Eng. Soc. **65**(9) (2017)
4. D. Rudrich, F. Zotter, M. Frank, Efficient spatial ambisonic effects for live audio, in *29th TMT*, Cologne, November 2016
5. V. Pulkki, S. Delikaris-Manias, A. Politis (eds.), *Spatial Decomposition by Spherical Array Processing* (Wiley, New Jersey, 2017)
6. A. Politis, Microphone array processing for parametric spatial audio techniques, Ph.D. dissertation, Aalto University, Helsinki (2016)
7. B. Bernschütz, Microphone arrays and sound field decomposition for dynamic binaural recording, Ph.D. Thesis, TU Berlin (TH Köln) (2016), http://dx.doi.org/10.14279/depositonce-5082
8. B. Bernschütz, A.V. Giner, C. Pörschmann, J. Arend, Binaural reproduction of plane waves with reduced modal order. Acta Acustica u. Acustica **100**(5) (2014)

9. Z. Ben-Hur, F. Brinkmann, J. Sheaffer, S. Weinzierl, B. Rafaely, Spectral equalization in binaural signals represented by order-truncated spherical harmonics. J. Acoust. Soc. Am. **141**(6) (2017)
10. C. Andersson, Headphone auralization of acoustic spaces recorded with spherical microphone arrays, M. Thesis, Chalmers TU, Gothenburg (2017)
11. M. Zaunschirm, C. Schörkhuber, R. Höldrich, Binaural rendering of ambisonic signals by head-related impulse response time alignment and a diffuseness constraint. J. Acoust. Soc. Am. **143**(6), 3616–3627 (2018)
12. E. Kurz, F. Pfahler, M. Frank, Comparison of first-order ambisonic microphone arrays, in *International Conference, Spatial Audio (ICSA)* (2015)
13. E. Hendrickx, P. Stitt, J.-C. Messonier, J.-M. Lyzwa, B.F. Katz, C. Boishéraud, Influence of head tracking on the externalization of speech stimuli for non-individualized binaural synthesis. J. Ac. Soc. Am. **141**(3), 2011–2023 (2017)
14. H. Lee, C. Gribben, Effect of vertical microphone layer spacing for a 3d microphone array. J. Audio Eng. Soc. **62**(12), 870–884 (2014)
15. R. Wallis, H. Lee, The reduction of vertical interchannel cross talk: The analysis of localisation thresholds for natural sound sources, MDPI Appl. Sci. **7**(3) (2017)
16. H. Lee, M. Frank, F. Zotter, Spatial and timbral fidelities of binaural ambisonics decoders for main microphone array recordings, in *AES IIA Conference* (York, 2019)
17. M. Kronlachner, F. Zotter, Spatial transformations for the enhancement of ambisonic recordings, in *Proceedings of the ICSA, Erlangen* (2014)
18. M. Hafsati, N. Epain, J. Daniel, Editing ambisonic sound scenes, in *4th International Conference, Spatial Audio (ICSA)* (Graz, 2017)
19. G. Aubert, An alternative to wigner d-matrices for rotating real spherical harmonics. AIP Adv. **3**(6), 062121 (2013). https://doi.org/10.1063/1.4811853
20. F. Zotter, Analysis and synthesis of sound-radiation with spherical arrays, PhD. Thesis, University of Music and Performing Arts, Graz (2009)
21. D. Pinchon, P.E. Hoggan, Rotation matrices for real spherical harmonics: general rotations of atomic orbitals in space-fixed axes. J. Phys. A **40**, 1751–8113 (2007)
22. J. Ivanic, K. Ruedenberg, Rotation matrices for real spherical harmonics. Direct determination by recursion. J. Phys. Chem. **100**, 6342–6347 (1996)
23. J. Ivanic, K. Ruedenberg, Additions and corrections. J. Phys. Chem. A **102**, 9099–9100 (1998)
24. C.H. Choi, J. Ivanic, M.S. Gordon, K. Ruedenberg, Rapid and stable determination of rotation matrices between spherical harmonics by direct recursion. J. Chem. Phys. **111**, 8825–8831 (1999)
25. N.A. Gumerov, R. Duraiswami, Recursions for the computation of multipole translation and rotation coefficients for the 3-d helmholtz equation. SIAM **25**, 1344–1381 (2003)
26. J.R. Driscoll, D.M.J. Healy, Computing fourier transforms and convolutions on the 2-sphere. Adv. Appl. Math. **15**, 202–250 (1994)
27. M.A. Gerzon, G.J. Barton, Ambisonic decoders for HDTV, in *prepr. 3345, 92nd AES Convention* (Vienna, 1992)
28. S.P. Lipshitz, J. Vanderkoy, In-phase crossover network design. J. Audio Eng. Soc. **34**(11), 889–894 (1986)
29. U. Zölzer, *Digital Audio Effects*, 2nd edn. (Wiley, New Jersey, 2011)
30. M.A. Gerzon, Signal processing for simulating realistic stereo images, in *prepr. 3423, Convention Audio Engineering Society* (San Francisco, 1992)
31. M.-V. Laitinen, T. Pihlajamäki, C. Erkut, V. Pulkki, Parametric time-frequency representation of spatial sound in virtual worlds. ACM TAP **9**(2) (2012)
32. F. Zotter, M. Frank, M. Kronlachner, J. Choi, Efficient phantom source widening and diffuseness in ambisonics, in *EAA Symposium on Auralization and Ambisonics* (Berlin, 2014)
33. F. Zotter, M. Frank, Efficient phantom source widening. Arch. Acoust. **38**(1) (2013)
34. F. Zotter, M. Frank, Phantom source widening by filtered sound objects, in *prepr. 9728, Convention, Audio Engineering Society* (Berlin, 2017)

35. J. Stautner, M. Puckette, Designing multi-channel reverberators. Comput. Music J. **6**(1), 52–65 (1982)
36. J.-M. Jot, A. Chaigne, Digital delay networks for designing artificial reverberators, in *90th AES Convention, prepr. 3030* (Paris, 1991)
37. D. Rocchesso, Maximally diffusive yet efficient feedback delay networks for artificial reverberation. IEEE Signal Process. Lett. **4**(9) (1997)
38. S. Tervo, J. Pätynen, A. Kuusinen, T. Lokki, Spatial decomposition method for room impulse responses. J. Audio Eng. Soc. **61**(1/2), 17–28 (2013), http://www.aes.org/e-lib/browse.cfm?elib=16664
39. J. Merimaa, V. Pulkki, Spatial impulse response rendering i: Analysis and synthesis. J. Audio Eng. Soc. **53**(12), 1115–1127 (2005)
40. V. Pulkki, J. Merimaa, Spatial impulse response rendering ii: Reproduction of diffuse sounds and listening tests. J. Audio Eng. Soc. **54**(1/2), 3–20 (2006)
41. M. Zaunschirm, M. Frank, F. Zotter, Brir synthesis using first-order microphone arrays, in *prepr. 9944, 144th AES Convention* (Milano, 2018)
42. C. Pörschmann, P. Stade, J.M. Arend, Binauralization of omnidirectional room impulse responses – algorithm and technical evaluation," in *DAFx-17* (Edinburgh, 2017)
43. M. Frank, F. Zotter, Spatial impression and directional resolution in the reproduction of reverberation, in *Fortschritte der Akustik - DEGA* (Aachen, 2016)
44. M. Frank, F. Zotter, Exploring the perceptual sweet area in ambisonics, in *AES 142nd Convention* (2017)
45. V. Pulkki, Spatial sound reproduction with directional audio conding. J. Audio Eng. Soc. **55**(6), 503–516 (2007)
46. J. Vilkamo, T. Lokki, V. Pulkki, Directional audio coding: virtual microphone-based synthesis and subjective evaluation. J. Audio Eng. Soc. **57**(9) (2009)
47. S. Berge, Device and method converting spatial audio signal, US Patent, no. 8,705,750 B2 (2014)
48. A. Politis, S. Tervo, V. Pulkki, Compass: Coding and multidirectional parameterization of ambisonic sound scenes, in *2018 IEEE International Conference on Acoustics, Speech and Signal Processing (ICASSP)* (IEEE, 2018), pp. 6802–6806
49. C. Nachbar, F. Zotter, A. Sontacchi, E. Deleflie, Ambix - suggesting an ambisonic format, in *Proceedings of the 3rd Ambisonics Symposium* (Lexington, Kentucky, 2011)

Chapter 6
Higher-Order Ambisonic Microphones and the Wave Equation (Linear, Lossless)

... a turning point has been the design of HOA microphones, opening an exciting experimental field in terms of real 3D sound field recording ...

Jérôme Daniel [1] at Ambisonics Symposium 2009.

Abstract Unlike pressure-gradient transducers, single-transducer microphones with higher-order directivity apparently turned out to be difficult to manufacture at reasonable audio quality. Therefore nowadays, higher-order Ambisonic recording with compact devices is based on compact spherical arrays of pressure transducers. To prepare for higher-order Ambisonic recording based on arrays, we first need a model of the sound pressure that the individual transducers of such an array would receive in an arbitrary surrounding sound field. The lossless, linear wave equation is the most suitable model to describe how sound propagates when the sound field is composed of surrounding sound sources. Fundamentally, the wave equation models sound propagation by how small packages of air react (i) when being expanded or compressed by a change of the internal pressure, and to (ii) directional differences in the outside pressure by starting to move. Based there upon, the inhomogeneous solutions of the wave equation describe how an entire free sound field builds up if being excited by an omnidirectional sound source, as a simplified model of an arbitrary physical source, such as a loudspeaker, human talker, or musical instrument. After adressing these basics, the chapter shows a way to get Ambisonic signals of high spatial and timbral quality from the array signals, considering the necessary diffuse-field equalization, side-lobe suppression, and trade off between spatial resolution and low-frequeny noise boost. The chapter concludes with application examples.

Gary Elko and Jens Meyer are the well-known inventors of the first commercially available compact spherical microphone array that is able to record higher-order Ambisonics [2], the Eigenmike. There are several inspiring scientific works with

F. Zotter and M. Frank, *Ambisonics*, Springer Topics in Signal Processing 19,
https://doi.org/10.1007/978-3-030-17207-7_6

valuable contributions that can be recommended for further reading [3–12], above all Boaz Rafaely's excellent introductory book [13].

This mathematical theory might appear extensive, but it cannot be avoided when aiming at an in-depth understanding of higher-order Ambisonic microphones. The theory enables processing of the microphone signals received such that the surrounding sound field excitation is retrieved in terms of an Ambisonic signal. Some readers may want to skip the physical introduction and resume in Sect. 6.5 on spherical scattering or Sect. 6.6 on the processing of the array signals.

6.1 Equation of Compression

Wave propagation involves *reversible* short-term temperature fluctuations becoming effective when air is being compressed by sound, causing the specific stiffness of air in sound propagation. The Appendix A.6.1 shows how to derive this adiabatic compression relation based on the first law of thermodynamics and the ideal gas law. It relates the relative volume change $\frac{V}{V_0}$ to the pressure change $p = -K\,\frac{V}{V_0}$ by the bulk modulus of air. After expressing the bulk modulus by more common constants[1] $K = \rho\,c^2$ and differentially formulating the volume change over time using the change of the sound particle velocity in space, e.g. in one dimension $\dot{p} = -\rho\,c^2\,\frac{\partial v_x}{\partial x}$, cf. Appendix A.6.1, we get the three-dimensional compression equation:

$$\frac{\partial p}{\partial t} = -\rho\,c^2\,\nabla^{\mathrm{T}}\boldsymbol{v}. \tag{6.1}$$

Here the inner product of the Del symbol $\nabla^{\mathrm{T}} = (\frac{\partial}{\partial x}, \frac{\partial}{\partial y}, \frac{\partial}{\partial z})$ with $\boldsymbol{v}$ yields what is called divergence $\mathrm{div}(\boldsymbol{v}) = \nabla^{\mathrm{T}}\boldsymbol{v} = \frac{\partial v_x}{\partial x} + \frac{\partial v_y}{\partial y} + \frac{\partial v_z}{\partial z}$. The equation means: *Independently of whether the outer boundaries of a small package of air are traveling at a common velocity: If there are directions into which their velocity is spatially increasing, the resulting gradual volume expansion over time causes a proportional decrease of interior pressure over time.*

6.2 Equation of Motion

The equation of motion is relatively simple to understand from the Newtonian equation of motion, e.g. for the x direction, $F_x = m\,\frac{\partial v_x}{\partial t}$ equates the external force to mass m times acceleration, i.e. increase in velocity $\frac{\partial v}{\partial t}$. For a small package of air with constant volume $V_0 = \Delta x \Delta y \Delta z$, the mass is obtained by the air density $m = \rho\,V_0$, and the force equals the decrease of in pressure over the three space directions, times the corresponding partial surface, e.g. for the x direction

[1] Typical constants are: density $\rho = 1.2$ kg/m^3, speed of sound $c = 343$ m/s.

$F_\mathrm{x} = -[p(x+\Delta x) - p(x)]\Delta y \Delta z$. For the x direction, this yields after expanding by $\frac{\Delta x}{\Delta x}$

$$-\frac{\Delta p}{\Delta x}\, V_0 = \rho\, V_0\, \frac{\partial v_\mathrm{x}}{\partial t}.$$

Dividing by $-V_0$ and letting $V_0 \to 0$, we obtain the typical shape of the equation of motion for all three space directions

$$\nabla\, p = -\rho\, \frac{\partial \boldsymbol{v}}{\partial t}. \tag{6.2}$$

The equation of motion means: *Independently of the common exterior pressure load on all the outer boundaries of a small air package, an outer pressure decrease into any direction implies a corresponding pushing force on the package causing a proportional acceleration into this direction.*

6.3 Wave Equation

We can combine the compression equation $\frac{\partial p}{\partial t} = -\rho\, c^2\, \nabla^\mathrm{T} \boldsymbol{v}$ with the equation of motion $\nabla\, p = -\rho\, \frac{\partial \boldsymbol{v}}{\partial t}$ by deriving the first one with regard to time $\frac{\partial^2 p}{\partial t^2} = -\rho\, c^2\, \nabla^\mathrm{T} \frac{\partial \boldsymbol{v}}{\partial t}$ and the second one with the gradient ∇^T yielding the Laplacian $\nabla^\mathrm{T}\nabla = \Delta$, hence $\Delta p = -\rho \nabla^\mathrm{T} \frac{\partial \boldsymbol{v}}{\partial t}$. Division of the first result by c^2 and equating both terms yields the lossless wave equation $\Delta p = \frac{1}{c^2}\frac{\partial^2}{\partial t^2} p$ that is typically written as

$$\Big(\Delta - \frac{1}{c^2}\frac{\partial^2}{\partial t^2}\Big) p = 0. \tag{6.3}$$

Obviously, the wave equation relates the curvature in space (expressed by the Laplacian) to curvature in time (expressed by the second-order derivative).

If p is a pure sinusoidal oscillation $\sin(\omega t + \phi_0)$, the second derivative in time corresponds to a factor $-\omega^2$, and by substitution with the wave-number $k = \frac{\omega}{c}$, we can write the frequency-domain wave equation as

$$(\Delta + k^2)\, p = 0, \qquad \text{Helmholtz equation.} \tag{6.4}$$

6.3.1 Elementary Inhomogeneous Solution: Green's Function (Free Field)

The Green's function is an elementary prototype for solutions to inhomogeneous problems $(\Delta + k^2)p = -q$, which is defined as

$$\big(\Delta + k^2\big) G = -\delta.$$

A general excitation q of the equation can be represented by its convolution with the Dirac delta distribution $\int q(\boldsymbol{s})\,\delta(\boldsymbol{r}-\boldsymbol{s})\,\mathrm{d}V(\boldsymbol{s}) = q(\boldsymbol{r})$. Consequently, as the wave equation is linear, the general solution must therefore also equal the convolution of the Green's function with the excitation function $p(\boldsymbol{r}) = \int q(\boldsymbol{s})\,G(\boldsymbol{r}-\boldsymbol{s})\,\mathrm{d}V(\boldsymbol{s})$ over space; if formulated in the time domain: also over time. The integral superimposes acoustical responses of any point in time and space of the source phenomenon, weighted by the corresponding source strength in space and time.

The Green's function in three dimensions is derived in Appendix A.6.3, Eq. (A.91),

$$G = \frac{e^{-\mathrm{i}k\,r}}{4\pi r}, \tag{6.5}$$

with the wave number $k = \frac{\omega}{c}$ and distance between source and receiver $r = \sqrt{\|\boldsymbol{r}-\boldsymbol{r}_\mathrm{s}\|^2}$.

Acoustic source phenomena are characterized by the behavior of the Green's function: far away, the amplitude decays with $\frac{1}{r}$ and the phase $-kr = -\omega\frac{r}{c}$ corresponds to the radially increasing delay $\frac{r}{c}$. Both is expressed in Sommerfeld's radiation condition $\lim_{r\to\infty} r\left(\frac{\partial}{\partial r}p + \mathrm{i}k\,p\right) = 0$.

Plane waves. The radius coordinate of the Green's function is the distance between two Cartesian position vectors $\boldsymbol{r}_\mathrm{s}$ and $\boldsymbol{r}$, the source and receiver location. Letting one of them become large is denoted by re-expressing it in terms of radius and direction vector $\boldsymbol{r}_\mathrm{s} = r_\mathrm{s}\boldsymbol{\theta}_\mathrm{s}$. This permits far-field approximation

$$r_\mathrm{s} = \|\boldsymbol{r}_\mathrm{s}-\boldsymbol{r}\| = \sqrt{(r_\mathrm{s}\boldsymbol{\theta}_\mathrm{s}-\boldsymbol{r})^\mathrm{T}(r_\mathrm{s}\boldsymbol{\theta}_\mathrm{s}-\boldsymbol{r})} = \sqrt{r_\mathrm{s}^2 - 2r_\mathrm{s}\boldsymbol{\theta}_\mathrm{s}^\mathrm{T}\boldsymbol{r} + r^2} \tag{6.6}$$

$$\lim_{r_\mathrm{s}\to\infty} r_\mathrm{s} = \lim_{r_\mathrm{s}\to\infty} r_\mathrm{s}\sqrt{1 - 2\frac{\boldsymbol{\theta}_\mathrm{s}^\mathrm{T}\boldsymbol{r}}{r_\mathrm{s}} + \frac{r^2}{r_\mathrm{s}^2}} = r_\mathrm{s} - \boldsymbol{\theta}_\mathrm{s}^\mathrm{T}\boldsymbol{r}. \qquad (\text{with } \lim_{x\to 0}\sqrt{1-2x} = 1-x).$$

For the *phase approximation*, for instance at a wave-length of 30 cm, we notice even for a relatively small distance difference, e.g. between 15 m and 15 m + 15 cm, we could change the sign of the wave. To approximate the phase of the Green's function, we must therefore at least use $r_\mathrm{s} - \boldsymbol{\theta}_\mathrm{s}^\mathrm{T}\boldsymbol{r}$ as approximation. By contrast, this level of precision is irrelevant for the *magnitude approximation*, e.g., it would be negligible if we used $\frac{1}{15\,\mathrm{m}}$ instead of the magnitude $\frac{1}{15\,\mathrm{m}+15\,\mathrm{cm}}$.

At a large distance r_s assumed to be constant, the Green's function is proportional to a *plane wave* from the source direction $\boldsymbol{\theta}_\mathrm{s}$

$$\lim_{r_\mathrm{s}\to\infty} G = \frac{e^{-\mathrm{i}k\,r_\mathrm{s}}}{4\pi\,r_\mathrm{s}}\,e^{\mathrm{i}k\,\boldsymbol{\theta}_\mathrm{s}^\mathrm{T}\boldsymbol{r}}. \tag{6.7}$$

The plane-wave part is of unit magnitude $|p| = 1$

$$p = e^{\mathrm{i}k\,\boldsymbol{\theta}_\mathrm{s}^\mathrm{T}\boldsymbol{r}} \tag{6.8}$$

and its phase evaluates the projection of the position vector onto the plane-wave arrival direction $\boldsymbol{\theta}_\mathrm{s}$. Towards the direction $\boldsymbol{\theta}_\mathrm{s}$, the phase grows positive, i.e. the signal arrives earlier. Towards the plane-wave propagation direction $-\boldsymbol{\theta}_\mathrm{s}$ the phase grows negatively, implying an increasing time delay, which is constant on any plane perpendicular to $\boldsymbol{\theta}_\mathrm{s}$.

Plane waves are an invaluable tool to locally approximate sound fields from sources that are sufficiently far away, within a small region.[2]

6.4 Basis Solutions in Spherical Coordinates

Figure 4.11 shows spherical coordinates [14, 15] using radius r, azimuth φ, and zenith ϑ. For simplification, zenith is replaced by $\zeta = \cos\vartheta = \frac{z}{r}$, here. We may solve the Helmholtz equation $(\Delta + k^2)p = 0$ in spherical coordinates by the radial and directional parts of the Laplacian $\Delta = \Delta_\mathrm{r} + \Delta_{\varphi,\zeta}$, as identified in Appendix A.3

$$\Delta_\mathrm{r} = \frac{\partial^2}{\partial r^2} + \frac{2}{r}\frac{\partial}{\partial r}, \qquad \Delta_{\varphi,\zeta} = \frac{1-\zeta^2}{r^2}\frac{\partial^2}{\partial \zeta^2} - \frac{2}{r^2}\zeta\frac{\partial}{\partial\zeta} + \frac{1}{r^2(1-\zeta^2)}\frac{\partial^2}{\partial\varphi^2}. \tag{6.9}$$

We already know the spherical harmonics as directional eigensolutions from Sect. 4.7

$$\Delta_{\varphi,\zeta} Y_n^m = -\frac{n(n+1)}{r^2}\, Y_n^m \tag{6.10}$$

and assume them to be a factor of the solution $p_n^m = R\, Y_n^m$ determining the value of $\Delta_{\varphi,\zeta}$ in $(\Delta_\mathrm{r} + k^2 + \Delta_{\varphi,\zeta})p_n^m = 0$. We find a separated radial differential equation after insertion, multiplication by $\frac{r^2}{Y_n^m}$, and re-expressing the differentials $\frac{\partial}{\partial r} = k\frac{\partial}{\partial kr}$ and $\frac{\partial^2}{\partial r^2} = k^2\frac{\partial^2}{\partial (kr)^2}$

$$\left[(kr)^2\frac{\partial^2}{\partial(kr)^2} + 2(kr)\frac{\partial}{\partial(kr)} + (kr)^2 - n(n+1)\right] R = 0. \tag{6.11}$$

Appendix A.6.4 shows how to get physical solutions for R of this, so-called, *spherical Bessel differential equation*: spherical Hankel functions of the second kind $h_n^{(2)}(kr)$ able to represent radiation (radially outgoing into every direction), consistently with Green's function G, diverging with an $(n+1)$-fold pole at $kr = 0$, a physical behavior that would also be observed after spatially differentiating G, see Fig. 6.1; spherical Bessel functions $j_n(kr) = \Re\{h_n^{(2)}(kr)\}$ are real-valued, converge everywhere, exhibit

[2] This is because, strictly speaking, an entire plane-wave sound field is unphysical and of infinite energy: either the exhaustive in-phase vibration of an infinite plane is required, or an infinite-amplitude point-source infinitely far away is required with infinite anticipation $t_\mathrm{s} \to +\infty$ (non-causal).

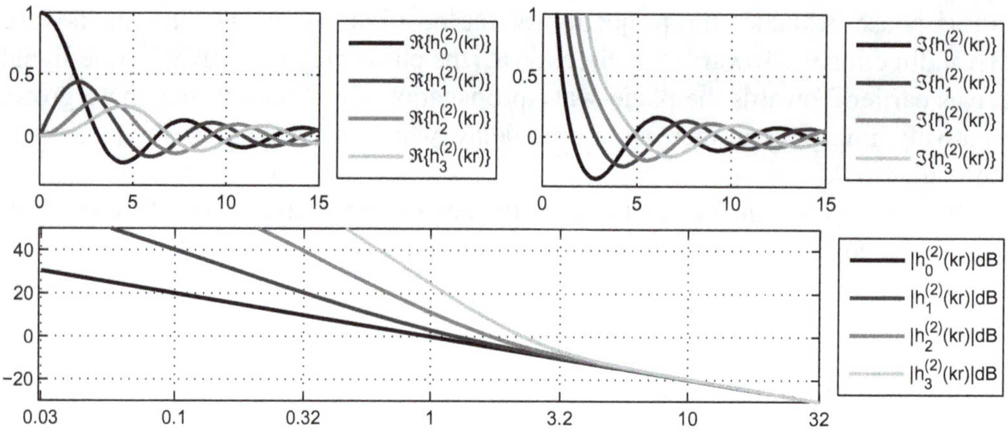

Fig. 6.1 Spherical Bessel functions $j_n(kr) = \Re\{h_n^{(2)}(kr)\}$ (top left), imaginary part of spherical Hankel functions $\Im\{h_n^{(2)}(kr)\}$ (top right), and magnitude/dB of $|h_n^{(2)}(kr)|$ (bottom), over kr

an n-fold zero at $kr = 0$, and can't represent radiation. Implementations typically rely on the accurate standard libraries implementing cylindrical Bessel and Hankel functions:

$$j_n(kr) = \sqrt{\frac{\pi}{2}\frac{1}{kr}}\, J_{n+\frac{1}{2}}(kr), \qquad h_n^{(2)}(kr) = \sqrt{\frac{\pi}{2}\frac{1}{kr}}\, H_{n+\frac{1}{2}}^{(2)}(kr). \tag{6.12}$$

Wave spectra and spherical basis solutions. Any sound field evaluated at a radius r where the air is source-free and homogeneous in any direction can be represented by spherical basis functions for enclosed $j_n(kr)Y_n^m(\boldsymbol{\theta})$ and radiating fields $h_n(kr)Y_n^m(\boldsymbol{\theta})$

$$p = \sum_{n=0}^{\infty}\sum_{m=-n}^{n}\left[b_{nm}\, j_n(kr) + c_{nm}\, h_n(kr)\right]Y_n^m(\boldsymbol{\theta}). \tag{6.13}$$

Here, b_{nm} are the coefficients for *incoming waves* that pass through and emanate from radii larger than r and c_{nm} are the coefficients of *outgoing waves* radiating from sources at radii smaller than r; the coefficients are called *wave spectra* of the incoming and outgoing waves, cf. [16].

Ambisonic plane-wave spectrum, plane wave. Plane waves only use the coefficients b_{nm}, while $c_{nm} = 0$ in Eq. (6.13). The sum of incoming plane waves from all directions, whose amplitudes are given by the spherical harmonics coefficients χ_{nm} as a set of Ambisonic signals are described by the *incoming* wave spectrum, see Appendix A.6.5, Eq. (A.119)

$$b_{nm} = 4\pi\, \mathrm{i}^n\, \chi_{nm}. \tag{6.14}$$

Figure 6.2 shows a single plane wave incoming from the direction $\boldsymbol{\theta}_\mathrm{s}$ represented by

$$b_{nm} = 4\pi\, \mathrm{i}^n\, Y_n^m(\boldsymbol{\theta}_\mathrm{s}) \tag{6.15}$$

Fig. 6.2 Plane wave from y axis $\varphi = \vartheta = \frac{\pi}{2}$ in horizontal cross section; time steps correspond to 0°, 60°, 120°, and 180° phase shifts ϕ in the plot $\Re\{p\, e^{i\phi}\}$ showing p from Eq. (6.13) with $c_{nm} = 0$ and b_{nm} of Eq. (6.15) with $b_{nm} = 4\pi\, i^n Y_n^m(\frac{\pi}{2}, \frac{\pi}{2})$; long wave (top), short wave (bottom); simulation uses N = 25 and area shows $|kx|, |ky| < 2\pi$ and 8π

at four different time steps corresponding to 0°, 60°, 120° and 180° time shifts for the two wave lengths shown.

6.5 Scattering by Rigid Higher-Order Microphone Surface

Higher-order Ambisonic microphone arrays are typically mounted on a rigid sphere of some radius $r = a$, such as the Eigenmike EM32, see Fig. 6.3. The physical boundary of the rigid spherical surface is expressed as a vanishing radial component of the sound particle velocity. The radial sound particle velocity is obtained via the

Fig. 6.3 32-channel higher-order Ambisonic mic. Eigenmike EM32

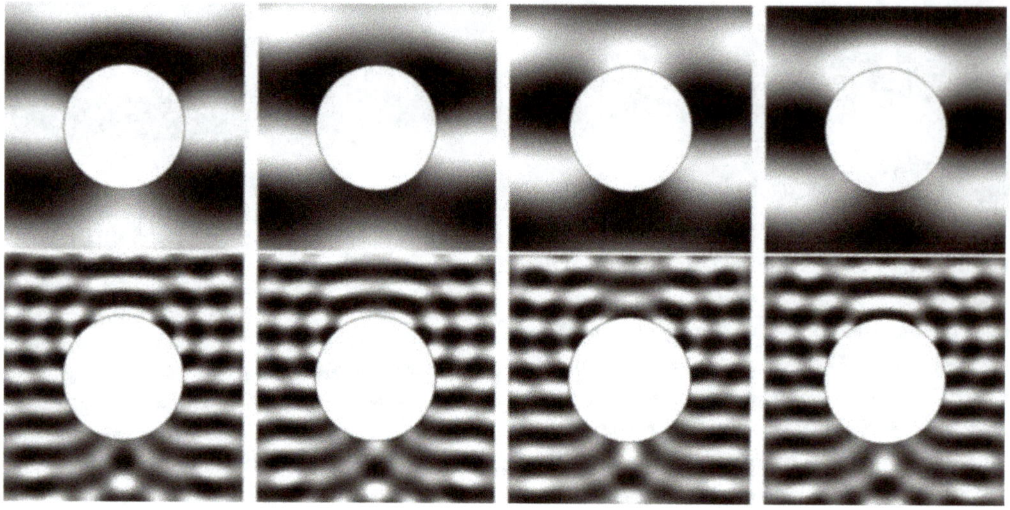

Fig. 6.4 Plane waves scattered by rigid sphere $ka = \pi$ (top) or $ka = 4\pi$ (bottom); time steps correspond to $0°, 60°, 120°$, and $180°$ phase shifts ϕ in the plot $\Re\{p\, e^{\mathrm{i}\phi}\}$ showing p from Eq. (6.13) with b_{nm} and c_{nm} from Eq. (6.15) with $b_{nm} = 4\pi \mathrm{i}^n Y_n^m(\frac{\pi}{2}, \frac{\pi}{2})$ and Eq. (6.16); simulation uses N = 25

equation of motion Eq. (6.2) by deriving Eq. (6.13). This requires to evaluate differentiated spherical radial solutions $j_n'(x)$ as well as $h_n'^{(2)}(x)$, which is implemented by $f_n'(x) = \frac{n}{x} f_n(x) - f_{n+1}(x)$ for either of the functions, cf. e.g. [16]. A sound-hard boundary condition at the radius a requires

$$v_\mathrm{r}\Big|_{r=\mathrm{a}} = \frac{\mathrm{i}}{\rho\, c} \sum_{n=0}^{\infty} \sum_{m=-n}^{n} \left[b_{nm}\, j_n'(kr) + c_{nm}\, h_n'^{(2)}(kr)\right]_{r=\mathrm{a}} Y_n^m(\boldsymbol{\theta}) = 0,$$

which is fulfilled by a vanishing term in square brackets. The rigid boundary responds to incoming surround-sound by velocity-canceling outgoing waves $h_n'^{(2)}(k\mathrm{a})\, c_{nm} = -j_n'(k\mathrm{a})\, b_{nm}$. The coefficients ψ_{nm} yield the sound pressure in Fig. 6.4,

$$p = \sum_{n=0}^{\infty} \sum_{m=-n}^{n} \psi_{nm}\, Y_n^m(\boldsymbol{\theta}), \quad \text{with } \psi_{nm} = \left[j_n(kr) - h_n^{(2)}(kr) \frac{j_n'(k\mathrm{a})}{h_n'^{(2)}(k\mathrm{a})}\right]_{r=\mathrm{a}} b_{nm}. \tag{6.16}$$

The two terms of the bracket are typically further simplified by a common denominator and recognizing the Wronskian Eq. (A.97) in the numerator $\frac{j_n(x) h_n'(x) - j_n'(x) h_n(x)}{h_n'(x)} = \frac{\mathrm{i}}{x^2 h_n'(x)}$

$$\psi_{nm}|_{r=\mathrm{a}} = \frac{\mathrm{i}}{(k\mathrm{a})^2\, h_n'^{(2)}(k\mathrm{a})}\, b_{nm}. \tag{6.17}$$

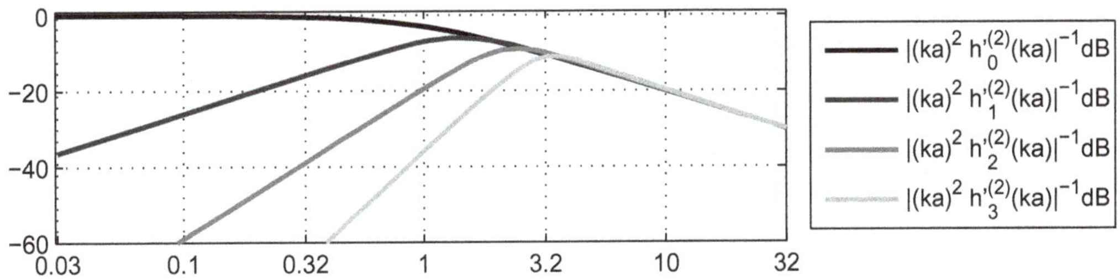

Fig. 6.5 Attenuation/dB of Ambisonic signals of different orders for varying values of ka

Relation of recorded sound pressure to Ambisonic signal. The scattering equation relates the recorded sound pressure expanded in spherical harmonics to the Ambisonic signal of surround sound scene, see frequency responses in Fig. 6.5,

$$\psi_{nm}|_{r=\mathrm{a}} = \frac{4\pi\,\mathrm{i}^{n+1}}{(k\mathrm{a})^2\,h_n^{\prime(2)}(k\mathrm{a})}\,\chi_{nm}. \tag{6.18}$$

It is formally convenient that as soon as the sound pressure is given in terms of its spherical harmonic coefficient signals ψ_{nm}, the Ambisonic signals χ_{nm} of a concentric playback system are obviously just an inversely filtered version thereof, with no need for further unmixing/matrixing.

Recognizable from Fig. 6.6 and following our intuition, waves of lengths larger than the diameter 2a of the sphere will only weakly map to complicated high-order patterns. It is therefore easily understood that the transfer function $\mathrm{i}^{n+1}[(k\mathrm{a})^2\,h_n^{\prime(2)}(k\mathrm{a})]^{-1}$ attenuates the reception of high-order Ambisonic signals at low frequencies, see Fig. 6.5.

6.6 Higher-Order Microphone Array Encoding

The block diagram of Ambisonic encoding of higher-order microphone array signals is shown in Fig. 6.7. The first processing step is about decomposing the pressure samples $\boldsymbol{p}(t)$ from the microphone array into its spherical harmonics coefficients $\boldsymbol{\psi}_\mathrm{N}(t)$: To which amount do the samples contain omnidirectional, figure-of-eight, and other spherical harmonic patterns, up to which the microphone arrangement allows decomposition. The frequency-independent matrix $(\boldsymbol{Y}_\mathrm{N}^\mathrm{T})^\dagger$ does the conversion. It is the left-inverse to the spherical harmonics sampled at the microphone positions, as shown in the upcoming section.

The second step then sharpens the sound pressure image to an Ambisonic signal by filtering the spherical harmonic coefficient signals. The basic relation between sound pressure coefficients and Ambisonic signals is given in Eq. (6.18) and describes a filter for every coefficient signal, differing only in filter characteristics for different spherical harmonic orders. Robustness to noise, microphone matching and position-

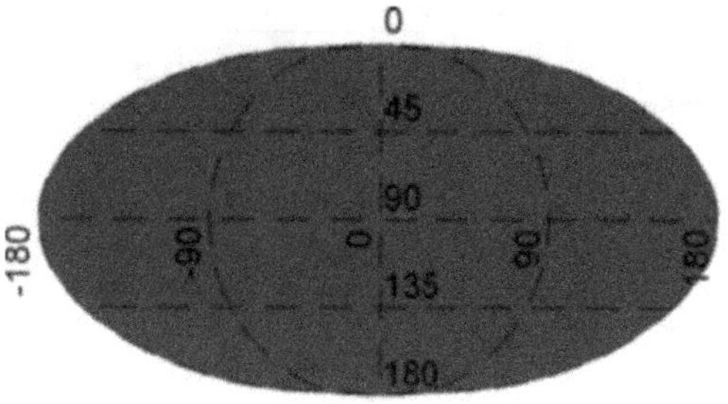

(a) $\frac{1}{32}$-wave-length diameter,i.e. $k\text{a} = \frac{\pi}{32}$

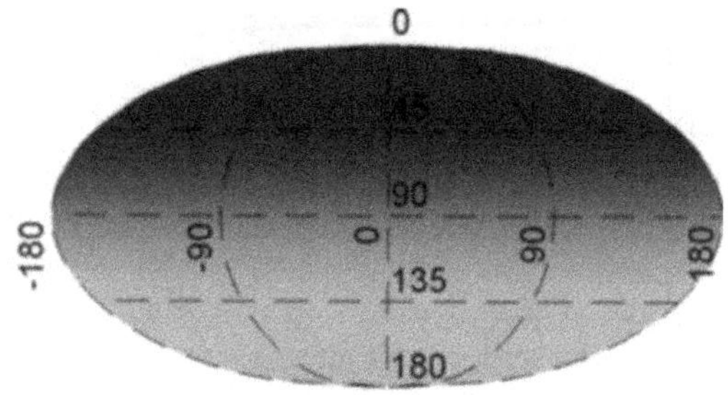

(b) $\frac{1}{2}$-wave-length diameter,i.e. $k\text{a} = \frac{\pi}{2}$

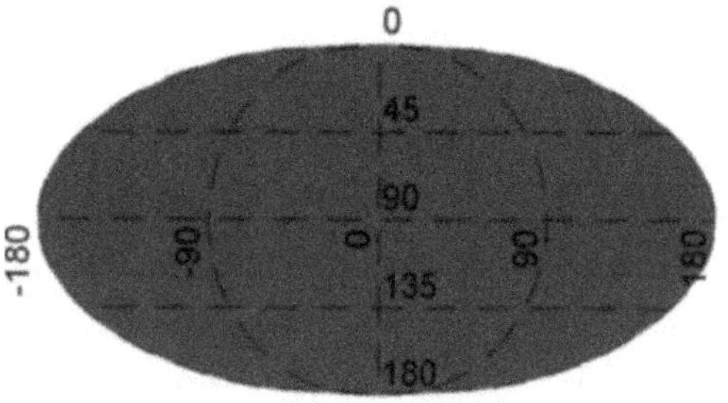

(c) $\frac{1}{16}$-wave-length diameter,i.e. $k\text{a} = \frac{\pi}{16}$

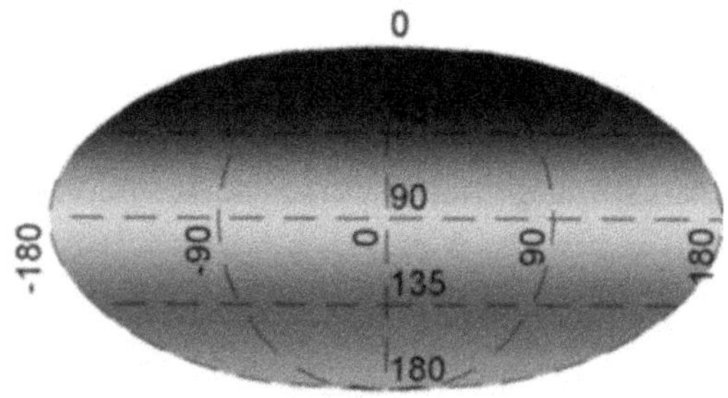

(d) 1-wave-length diameter,i.e. $k\text{a} = \pi$

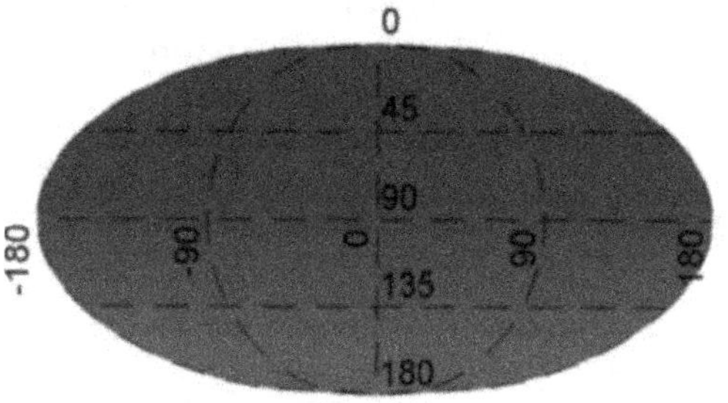

(e) $\frac{1}{8}$-wave-length diameter,i.e. $k\text{a} = \frac{\pi}{8}$

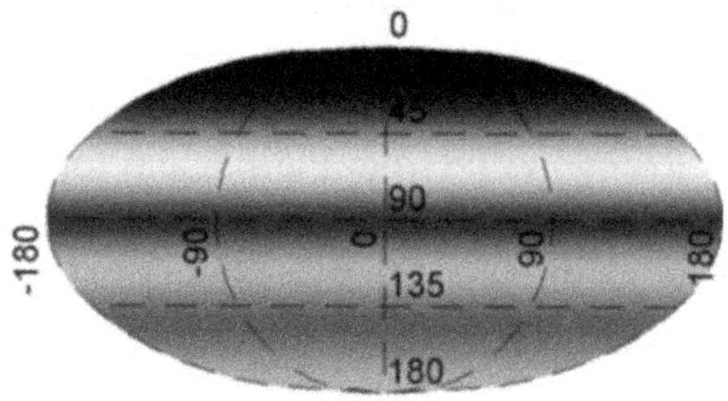

(f) 2-wave-length diameter,i.e. $k\text{a} = 2\pi$

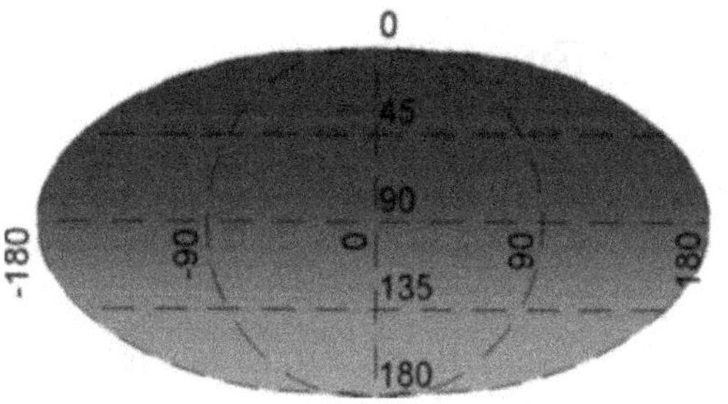

(g) $\frac{1}{4}$-wave-lengths diameter,i.e. $k\text{a} = \frac{\pi}{4}$

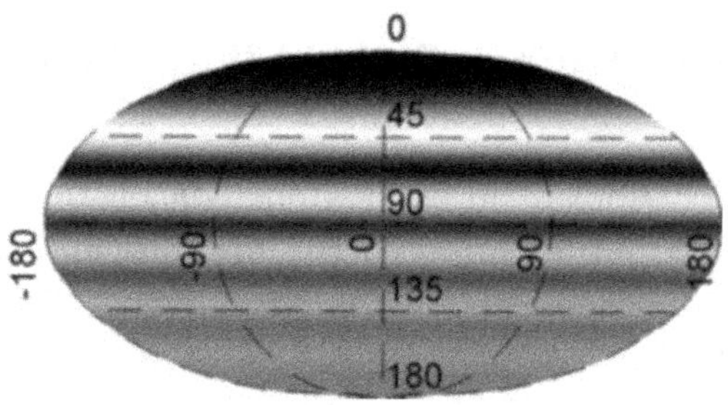

(h) 4-wave-lengths diameter,i.e. $k\text{a} = 4\pi$

Fig. 6.6 Plane-wave sound pressure image $\Re\{p\,e^{-\mathrm{i}k\mathrm{a}}\}$ on rigid sphere with varying ka using ψ_{nm} from Eq. (6.17) expanded over the spherical harmonics $p = \sum \psi_{nm} Y_n^m$ and $\chi_{nm} = Y_n^m(0,0)$ for a plane wave from z. With the wave length $\lambda = \frac{c}{f}$, the value ka is related to a diameter 2a of $\frac{ka}{\pi} = \frac{2\pi\, f\, \mathrm{a}}{\pi\, c} = \frac{2\mathrm{a}}{\lambda}$ in wave lengths to express frequency dependency; simulation uses $\mathrm{N} = 50$; for $a = 4.2$ cm, ka values correspond to $f = 125, 250, 500, 1000, 2000, 4000, 8000, 16000$ Hz

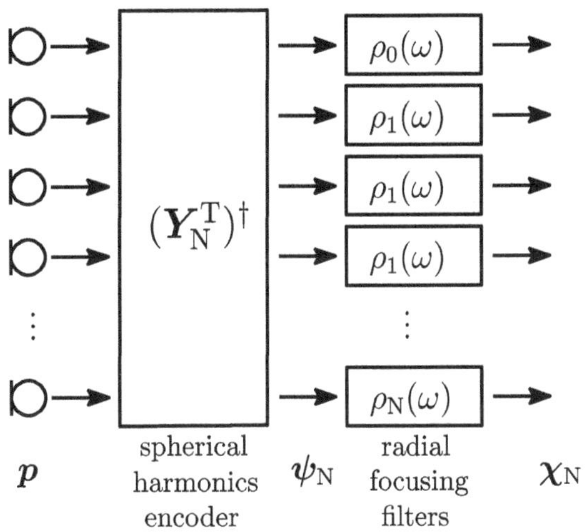

Fig. 6.7 Higher-order Ambisonic microphone encoding: sound pressure samples $\boldsymbol{p}(t)$ are spherical-harmonics decomposed by the matrix $(\boldsymbol{Y}_\mathrm{N}^\mathrm{T})^\dagger$, and the resulting coefficient signals $\boldsymbol{\psi}_\mathrm{N}(t)$ are converted to Ambisonic signals $\boldsymbol{\chi}_\mathrm{N}(t)$ by the sharpening filters $\rho_n(\omega)$

ing is the key here, and only achieved by the careful design of these filters, as shown in a further sections below. The design considers a gradually increasing sharpening over frequency, for which it moreover employs a filter bank with separate, max-$\boldsymbol{r}_\mathrm{E}$ weighted and E normalized bands, in order to provide (i) limitation of noise and errors, (ii) a frequency response perceived as flat, and (iii) optimal suppression of the sidelobes.

6.7 Discrete Sound Pressure Samples in Spherical Harmonics

To determine the Ambisonics signals χ_{nm}, we obviously need to find ψ_{nm} based on all sound pressure samples $p(\boldsymbol{\theta}_i)$ recorded by the microphones distributed on the rigid-sphere array. To accomplish this, we set up a system of model equations equating the pressure samples to the unknown coefficients ψ_{nm} expanded over the spherical harmonics $Y_n^m(\boldsymbol{\theta}_i)$ sampled at every microphone position. A vector and matrix notation $\boldsymbol{p} = [p(\boldsymbol{\theta}_i)]_i$ and $\boldsymbol{Y}_\mathrm{N}^\mathrm{T} = [\boldsymbol{y}(\boldsymbol{\theta}_i)^\mathrm{T}]_{i,nm}$ is helpful

$$\begin{bmatrix} p(\boldsymbol{\theta}_1) \\ \vdots \\ p(\boldsymbol{\theta}_\mathrm{M}) \end{bmatrix} = \begin{bmatrix} Y_0^0(\boldsymbol{\theta}_1) & \dots & Y_\mathrm{N}^\mathrm{N}(\boldsymbol{\theta}_1) \\ \vdots & \vdots & \vdots \\ Y_0^0(\boldsymbol{\theta}_\mathrm{M}) & \dots & Y_\mathrm{N}^\mathrm{N}(\boldsymbol{\theta}_\mathrm{M}) \end{bmatrix} \begin{bmatrix} \psi_{00} \\ \vdots \\ \psi_\mathrm{NN} \end{bmatrix}$$

$$\boldsymbol{p}_\mathrm{N} = \boldsymbol{Y}_\mathrm{N}^\mathrm{T}\,\boldsymbol{\psi}_\mathrm{N}. \qquad (6.19)$$

Left inverse (MMSE). The equation can be (pseudo-)inverted if the matrix $\boldsymbol{Y}_\mathrm{N}$ is well conditioned. Typically more microphones are used than coefficients searched $\mathrm{M} \geq (\mathrm{N}+1)^2$. Inversion is a matter of mean-square error minimization: As the M dimensions may contain more degrees of freedom than $(\mathrm{N}+1)^2$, the coefficient vector $\boldsymbol{\psi}_\mathrm{N}$ giving the closest model $\boldsymbol{p}_\mathrm{N}$ to the measurement $\boldsymbol{p}$ is searched,

$$\min_{\boldsymbol{\psi}_\mathrm{N}} \|\boldsymbol{e}\|^2, \qquad \text{with } \boldsymbol{e} = \boldsymbol{p}_\mathrm{N} - \boldsymbol{p} = \boldsymbol{Y}_\mathrm{N}^\mathrm{T}\boldsymbol{\psi}_\mathrm{N} - \boldsymbol{p}. \tag{6.20}$$

The minimum-mean-square-error (MMSE) solution is, see Appendix A.4, Eq. (A.65),

$$\boldsymbol{\psi}_\mathrm{N} = (\boldsymbol{Y}_\mathrm{N}\boldsymbol{Y}_\mathrm{N}^\mathrm{T})^{-1}\boldsymbol{Y}_\mathrm{N}\,\boldsymbol{p} = (\boldsymbol{Y}_\mathrm{N}^\mathrm{T})^\dagger\,\boldsymbol{p}. \tag{6.21}$$

The resulting left inverse $(\boldsymbol{Y}_\mathrm{N}\boldsymbol{Y}_\mathrm{N}^\mathrm{T})^{-1}\boldsymbol{Y}_\mathrm{N}$ inverts the thin matrix $\boldsymbol{Y}_\mathrm{N}^\mathrm{T}$ from the left. $(\boldsymbol{Y}_\mathrm{N}^\mathrm{T})^\dagger$ symbolizes the pseudo inverse; it is left-inverse for thin matrices.

If the microphones are arranged in a t-design and the order N is chosen suitably, then the transpose matrix times $\frac{4\pi}{\mathrm{L}}$ is equivalent to the left inverse. A more thorough discussion on spherical point sets can be found in [17–19].

The *maximum determinant points* [20] are a particular kind of *critical* directional sampling scheme that allows to use exactly as few microphones $\mathrm{M} = (\mathrm{N}+1)^2$ as spherical harmonic coefficients obtained, yielding a well-conditioned square matrix $\boldsymbol{Y}_\mathrm{N}$, so that it can be inverted directly without left/pseudo-inversion. The 25 maximum-determinant points for $\mathrm{N} = 4$ are used in the simulation example below.[3]

Finite-order assumption and spatial aliasing. An important implication of estimating ψ_{nm} is that we need to assume that the distribution of the sound pressure is of limited spherical harmonic order on the measurement surface. This could be done by restricting the frequency range, as high-order harmonics are attenuated well-enough according above suitable frequency limits, cf. Fig. 6.5. However, low-pass filtered signals are unacceptable in practice. Instead, one has to accept *spatial aliasing* at high frequencies, i.e. directional mapping errors and direction-specific comb filters. Figure 6.8 shows spatial aliasing of $\boldsymbol{\psi}_\mathrm{N} = (\boldsymbol{Y}_\mathrm{N}^\mathrm{T})^{-1}\,\boldsymbol{p}$ in the angular domain $p = \sum \psi_{nm} Y_n^m$.

6.8 Regularizing Filter Bank for Radial Filters

The filters $\mathrm{i}^n\big[(ka)^2\, h_n'^{(2)}(ka)\big]^{-1}$ of Fig. 6.5 exhibit an nth-order zero at 0 Hz, $ka = 0$. To retrieve the Ambisonic signals χ_{nm} from the sound pressure signals ψ_{nm}, their inverse would have a n-fold (unstable) pole at 0 Hz. Considering that microphone self noise and array imperfection cause erroneous signals louder than the acoustically expected nth-order vanishing signals around 0 Hz, filter shapes will moreover cause an excessive boost of erroneous signals unless implemented with precaution. Filters of the different orders n must be stabilized by high-pass slopes of at least the order n, see also [6, 9, 21–25], and with $(n+1)$th-order high-pass slopes, see Fig. 6.9, such errors are being cut off by first-order high-pass slopes at exemplary cut-on frequencies at 90, 680, 1650, 2600 Hz for the Ambisonic orders 1, 2, 3, 4, yielding a

[3]md04.0025 on https://web.maths.unsw.edu.au/~rsw/Sphere/Images/MD/md_data.html.

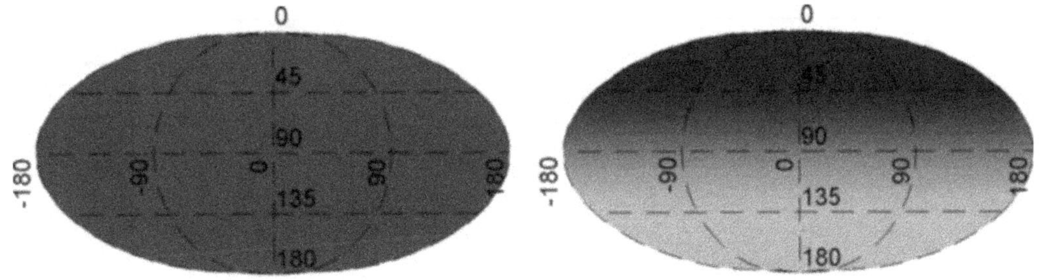

(a) $\frac{1}{32}$-wave-length diameter, i.e. $ka = \frac{\pi}{32}$ (b) $\frac{1}{2}$-wave-length diameter, i.e. $ka = \frac{\pi}{2}$,

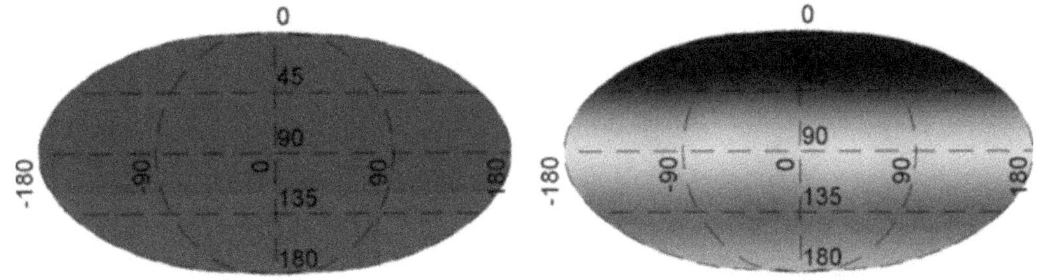

(c) $\frac{1}{16}$-wave-length diameter, i.e. $ka = \frac{\pi}{16}$ (d) 1-wave-length diameter, i.e. $ka = \pi$

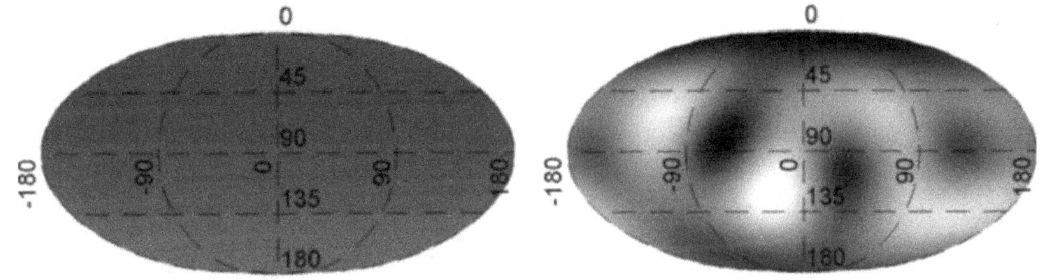

(e) $\frac{1}{8}$-wave-length diameter, i.e. $ka = \frac{\pi}{8}$ (f) 2-wave-length diameter, i.e. $ka = 2\pi$

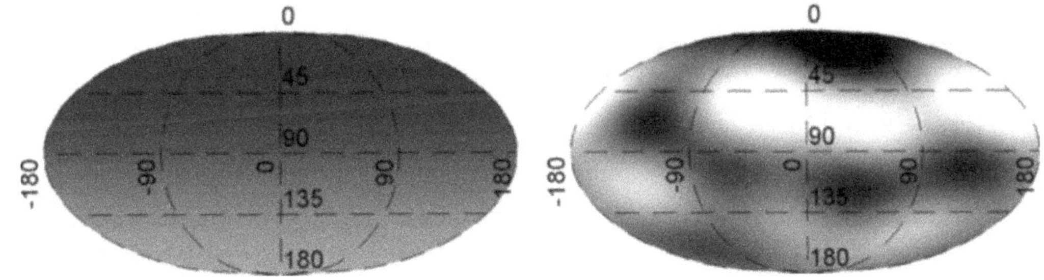

(g) $\frac{1}{4}$-wave-lengths diameter, i.e. $ka = \frac{\pi}{4}$ (h) 4-wave-lengths diameter, i.e. $ka = 4\pi$

Fig. 6.8 Interpolated plane-wave sound pressure image $\Re\{p\, e^{-ika}\}$ on rigid-sphere array with 25 microphones allowing decomposition up to the order $\mathrm{N} = 4$; simulation uses orders up to 25, and the aliasing-free operation can only be expected within $kr < \mathrm{N}$

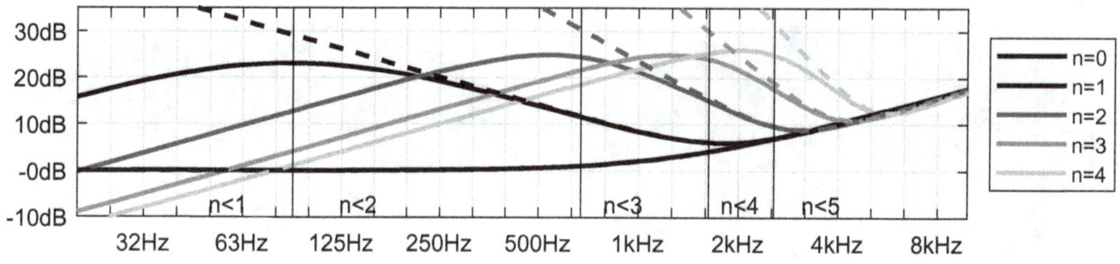

Fig. 6.9 Filters $(ka)^2\, h_n'^{(2)}(ka)$/dB over frequency/Hz, regularized with $(n+1)$th-order high-pass filters

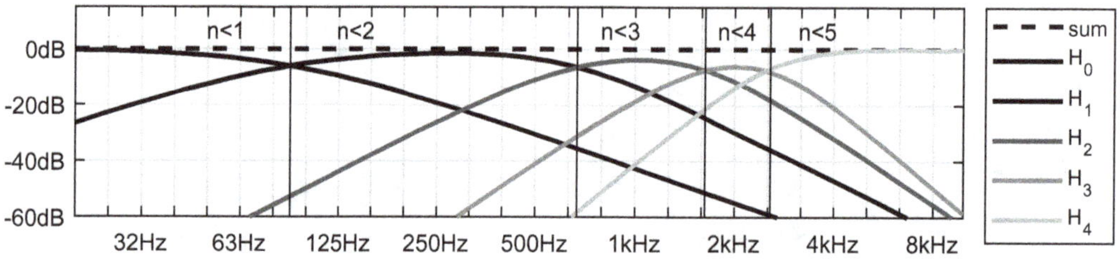

Fig. 6.10 Stabilizing filter bank/dB over frequency/Hz: signal orders $n > b$ are excluded from the band b

noise boost of 20 dB for a 4th-order microphone with a $=$ 4.2 cm, at most. However, just cutting on the frequencies of each order is not enough: every cut-on frequency causes a noticeable loudness drop below due to the discarded signal contributions. It is better to design a filter bank with crossovers instead, which allows compensation for the loudness loss in every band. A zero-phase, nth-order Butterworth high-pass response is defined by $H_{\mathrm{hi}} = \frac{\omega^n}{1+\omega^n}$ and amplitude-complementary to the low pass $H_{\mathrm{lo}} = \frac{1}{1+\omega^n}$, so that $H_{\mathrm{hi}} + H_{\mathrm{lo}} = 1$.

Using this filter type, the filter bank in Fig. 6.10 can be constructed as follows: The band-pass filters $H_b(\omega)$ are composed of a $(b+1)$th-order high- and $(b+2)$th-order low-pass skirt at ω_b, and ω_{b+1}, respectively, except for the band $b = 0$ (low-pass) and $b = \mathrm{N}$ (high-pass)

$$\hat{H}_0(\omega) = \frac{1}{1+\left(\frac{\omega}{\omega_1}\right)^2}, \quad \hat{H}_b(\omega) = \frac{\left(\frac{\omega}{\omega_b}\right)^{b+1}}{1+\left(\frac{\omega}{\omega_b}\right)^{b+1}}\,\frac{1}{1+\left(\frac{\omega}{\omega_{b+1}}\right)^{b+2}}, \quad \hat{H}_{\mathrm{N}}(\omega) = \frac{\left(\frac{\omega}{\omega_{\mathrm{N}}}\right)^{\mathrm{N}+1}}{1+\left(\frac{\omega}{\omega_{\mathrm{N}}}\right)^{\mathrm{N}+1}}. \tag{6.22}$$

To make the bands perfectly reconstructing, filters are normalized by the sum response

$$H_b = \frac{\hat{H}_b}{\sum_{b=0}^{\mathrm{N}} \hat{H}_b(\omega)}. \tag{6.23}$$

By adjusting the cut-on frequencies ω_b of the different orders $b = 1, \ldots, \mathrm{N}$, the noise and mapping behavior of the microphone array is adjusted; only the zeroth order is present in every band down to 0 Hz.

This filter bank design moreover allows to adjust loudness and sidelobe suppression in every frequency band, separately.

6.9 Loudness-Normalized Sub-band Side-Lobe Suppression

The filter bank design shown above would only yield Ambisonic signals whose order increases with the frequency band. Ideally, this variation of the order comes with the necessity of individual max-$\boldsymbol{r}_\mathrm{E}$ sidelobe suppression in every band. Moreover, Ambisonic signals of different orders are differently loud, so also diffuse-field equalization of the E measure is desirable in every band.

To fulfill the above constraints, we propose to use the following set of FIR filter responses as given in [26, 27], that are modified by a filter bank employing diffuse-field normalized max-$\boldsymbol{r}_\mathrm{E}$-weights in separate frequency bands $b = 0, \ldots, \mathrm{N}$, cf. Fig. 6.11, with the nth order discarded for bands below $b < n$:

$$\rho_n(\omega) = \left[\sum_{b=n}^{\mathrm{N}} a_{n,b}\, H_b(\omega)\right] \mathrm{i}^{-n-1}\,(ka)^2\, h_n'^{(2)}(ka)\, e^{\mathrm{i}ka}. \tag{6.24}$$

Here, $e^{\mathrm{i}ka}$ removes the linear phase of $h_n'^{(2)}$, and $a_{n,b}$ is the set of diffuse-field ($\sqrt{E}$) equalized max-$\boldsymbol{r}_\mathrm{E}$ weights for the band b in which the Ambisonic orders retrieved are $0 \leq n \leq b$

$$a_{n,b} = \begin{cases} P_n\!\left(\cos\frac{137.9^\circ}{b+1.51}\right)\sqrt{\dfrac{\sum_{n=0}^{\mathrm{N}}(2n+1)\left[P_n\!\left(\cos\frac{137.9^\circ}{\mathrm{N}+1.51}\right)\right]^2}{\sum_{n=0}^{b}(2n+1)\left[P_n\!\left(\cos\frac{137.9^\circ}{b+1.51}\right)\right]^2}}, & \text{for } n \leq b \\ 0, & \text{otherwise.} \end{cases} \tag{6.25}$$

Figure 6.12 shows the polar patterns of the corresponding direction-spread functions.

For the implementation of $\rho_n(\omega)$ by fast block filtering, $\omega = 2\pi f$ and $k = \omega/c$ are uniformly sampled with frequency, and the inverse discrete Fourier transform yields the associated impulse responses (attention: the value at 0 Hz must be replaced for stable results, and cyclic time-domain shifts and windows are necessary).

The direction-spread function of a plane-wave sound pressure mapped to a directional Ambisonic signal becomes frequency-dependent as shown in Fig. 6.13, and it has minimal side lobes.

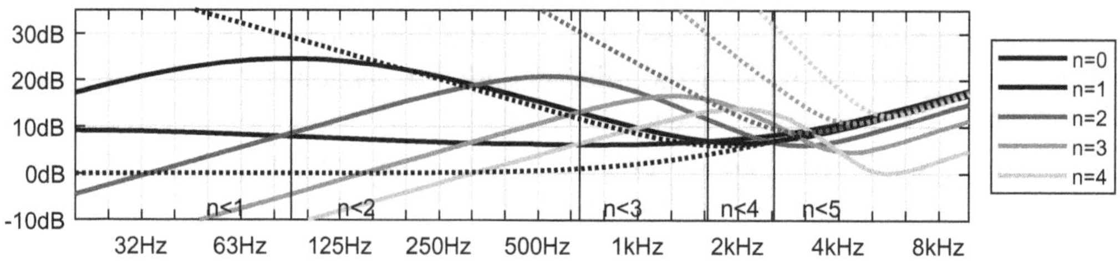

Fig. 6.11 Filter-bank-regularized/dB over frequency/Hz, diffuse-field equalized max-$\boldsymbol{r}_\mathrm{E}$ weighted spherical microphone array responses using $\mathrm{i}^n \rho_n(\omega) = \sum_{b=n}^{\mathrm{N}} a_{n,b}\, H_b(\omega)\, (k\mathrm{a})^2\, h_n'^{(2)}(k\mathrm{a})$

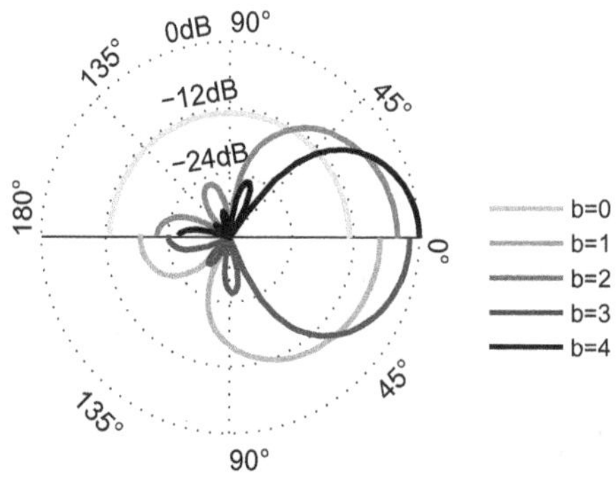

Fig. 6.12 Diffuse-field equalized (to $E = 1$) max-$\boldsymbol{r}_\mathrm{E}$ direction-spread functions; even orders are plotted on upper, odd orders on lower semi-circle

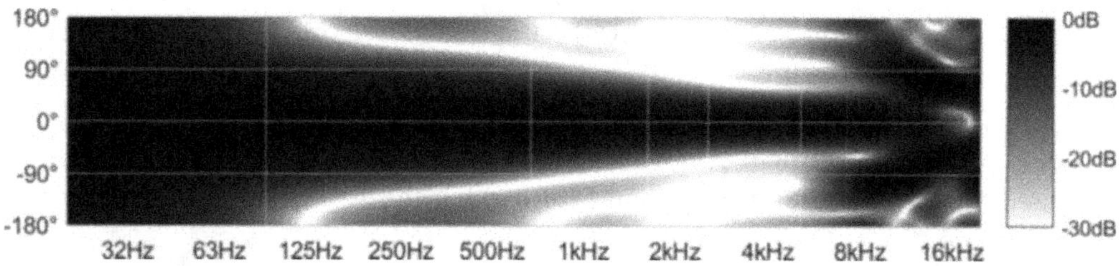

Fig. 6.13 Direction spread/dB over frequency/Hz in zenithal cross section/degrees through Ambisonic signal of simulated microphone processing response to plane wave from zenith and the parameters a = 4.2 cm, M = 25 mics., max-$\boldsymbol{r}_\mathrm{E}$-weighted in bands 90, 680, 1650, 2600 Hz for the cut on of the orders 1, 2, 3, 4. Simulation is done with the order $\mathrm{N}_\mathrm{sim} = 30$ and spatial aliasing will occur above 5.2 kHz. Gain matching was assumed to be up to $< \pm 0.5$ dB accurate; the map shows the direction spread normalized to its value at 0° for every frequency to make its shape easier to read

6.10 Influence of Gain Matching, Noise, Side-Lobe Suppression

Typical gain mismatch between the microphones is not always more accurate than 0.5 dB. The result is that the physically dominant omnidirectional signal will leak into the higher-order signals by directionally random gain variations. However, acoustically, higher-order components are expected to be weak and to require amplification.

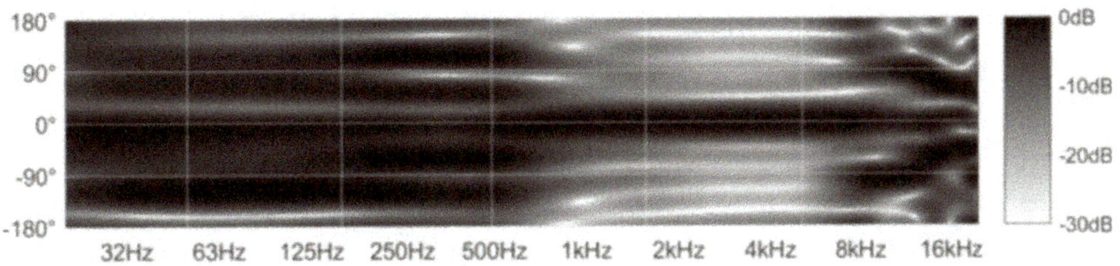

(a) 50, 160, 500, 1600 cut on with < ±0.5 dB gain matching, no sidelobe suppression

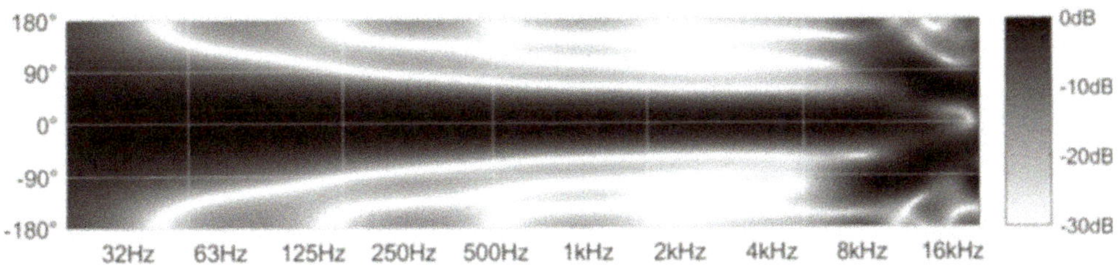

(b) 50, 160, 500, 1600 cut on with only 4th-order sidelobe suppression, assuming perfect gain match

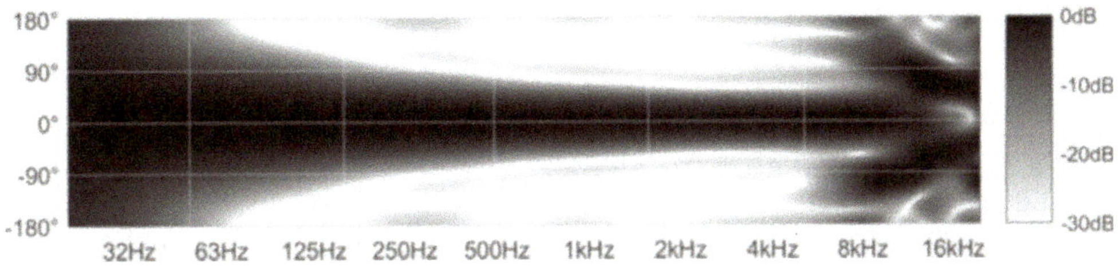

(c) 50, 160, 500, 1600 cut on with individual max-$\boldsymbol{r}_{\mathrm{E}}$ sidelobe suppression per band, assuming perfect gain match

Fig. 6.14 Influence of carelessly selected cut-on frequencies for regularization (top), and of non-individual sidelobe suppression per band (middle), in contrast to ideal results (bottom); the maps show direction spreads normalized to their values at 0° for every frequency to make side lobes easier to read

The effect on mapping is equivalent to one of microphone self noise, however gain mismatch yields a correlated signal exciting the microphones, whereas self-noise yields low-frequency noise.

If regularization filters were set to 50, 160, 500, 1600 and sidelobe suppression turned off for testing, one would get the poor image as in Fig. 6.14a, where high-order signals at low frequencies are highly boosted.

If a noise-free case is assumed, and only the max-$\boldsymbol{r}_{\mathrm{E}}$ side-lobe suppression of the highest band is used for all bands, one gets the image in Fig. 6.14b, which improves with individual max-$\boldsymbol{r}_{\mathrm{E}}$ weights in Fig. 6.14c.

Self-noise behavior. Assuming that self-noise of the microphones is uncorrelated, it will also remain uncorrelated and of equal strength after decomposing the M microphone signals $p_i = \mathcal{N}$ into the $(\mathrm{N}+1)^2$ spherical harmonic coefficient signals $\psi_{nm} = \frac{(\mathrm{N}+1)^2}{\mathrm{M}}\mathcal{N}$, if $\mathrm{M} \approx (\mathrm{N}+1)^2$ and the microphone arrangement permits a well-conditioned pseudo inversion $\boldsymbol{Y}_{\mathrm{N}}^{\dagger}$. The spectral change of the microphone self noise due to the radial filters $\rho_n(\omega)$ can be described by the noise of the $(2n+1)$

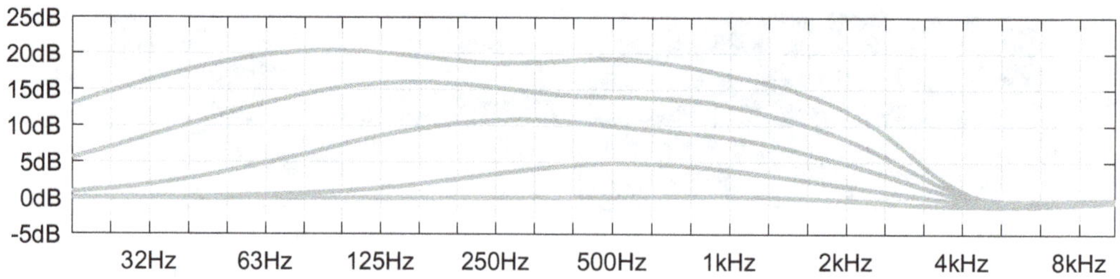

Fig. 6.15 Self-noise modification $|G(\omega)|^2$/dB over frequency/Hz for the filter bank configurations using the cut on frequencies $2k, 3k, 4k, 5k$ (no noise amplification), 600, $2k, 3.5k, 4.2k$ (5 dB noise amplification), 280, $1.3k, 2.6k, 3.6k$ (10 dB noise amplification), 150, 950, $2k, 3.15k$ (15 dB noise amplification), and 90, 680, $1.65k, 2.6k$ (20 dB noise amplification)

signals of the same order, amplified by $|\rho_n(\omega)|^2$, in comparison to the zeroth-order signal:

$$|G(\omega)|^2 = \frac{\sum_{n=0}^{N}(2n+1)|\rho_n(\omega)|^2}{|(ka)^2\, h_0^{\prime(2)}(ka)|^2}. \tag{6.26}$$

Figure 6.15 analyzes the noise amplification for the simulation example (max-$\boldsymbol{r}_\mathrm{E}$ weighting in each sub band, a = 4.2 cm) and shows the dependency on exemplary cut on frequencies configured to tune the filterbank to 0, 5, 10, 15, and 20 dB noise boosts. The trade here is: the more noise boost one can allow, the more directional resolution one gets, see Fig. 6.16.

Open measurement data (SOFA format) characterizing the directivity patterns of the 32 Eigenmike em32 transducers are provided under the link http://phaidra.kug.ac.at/o:69292. They are measured on a 12° × 11.25° azimuth× zenith grid, yielding 480 × 256 pt impulse responses for each of the 32 transducers.

6.11 Practical Free-Software Examples

6.11.1 *Eigenmike Em32 Encoding Using Mcfx and IEM Plug-In Suites*

We give a practical signal processing example for the Eigenmike em32 which is applicable e.g. in digital audio workstations. First the 32 signals are encoded by matrix multiplication (IEM `MatrixMultiplier`), cf. Fig. 6.17a, yielding 25 fourth-order signals. The preset (json file) is provided online http://phaidra.kug.ac.at/o:79231. The radial filtering that sharpens the surround sound image uses `mcfx-convolver`, see Fig. 6.17b, with 25 SISO filters, one for each Ambisonic signal, using the 5 different filter curves for the orders $n = 0, \ldots, 4$ as defined above. The convolver presets (wav

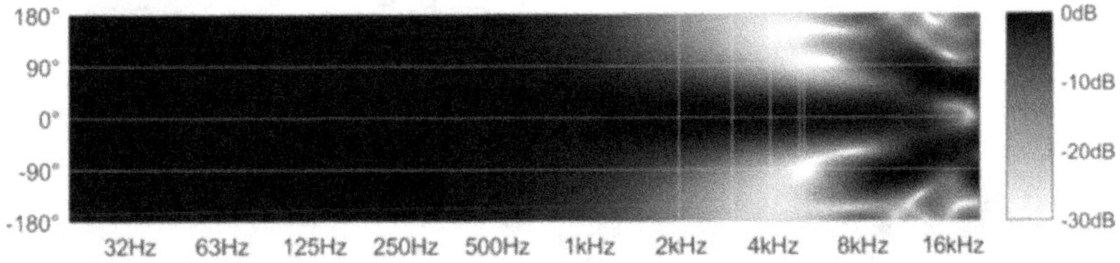

(a) 0 dB noise boost $(2k, 3k, 4k, 5k)$ individual max-$\boldsymbol{r}_{\mathrm{E}}$ weighting

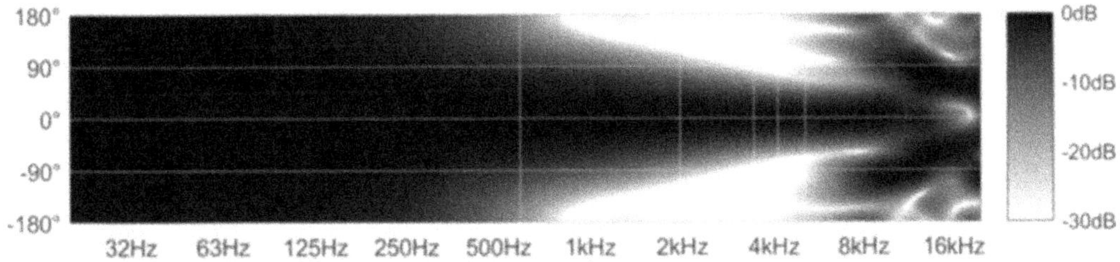

(b) 5 dB noise boost $(600, 2k, 3.5k, 4.2k)$ individual max-$\boldsymbol{r}_{\mathrm{E}}$ weights

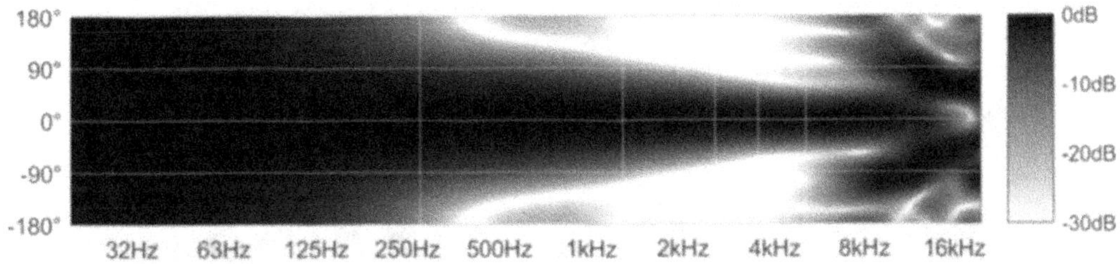

(c) 10 dB noise boost $(280, 1.3k, 2.6k, 3.6k)$ individual max-$\boldsymbol{r}_{\mathrm{E}}$ weights

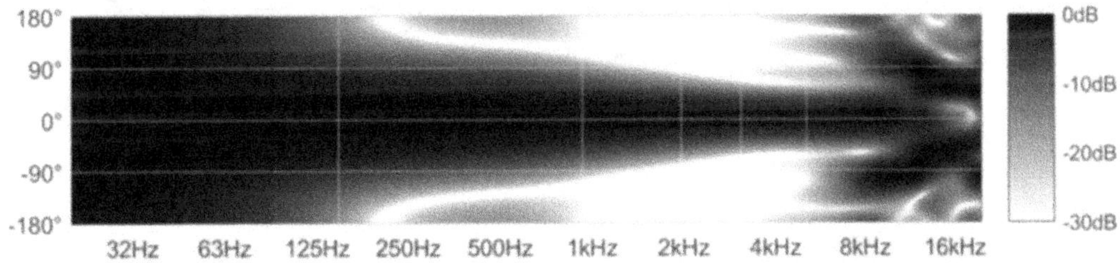

(d) 15 dB noise boost $(150, 950, 2k, 3.15k)$ individual max-$\boldsymbol{r}_{\mathrm{E}}$ weights

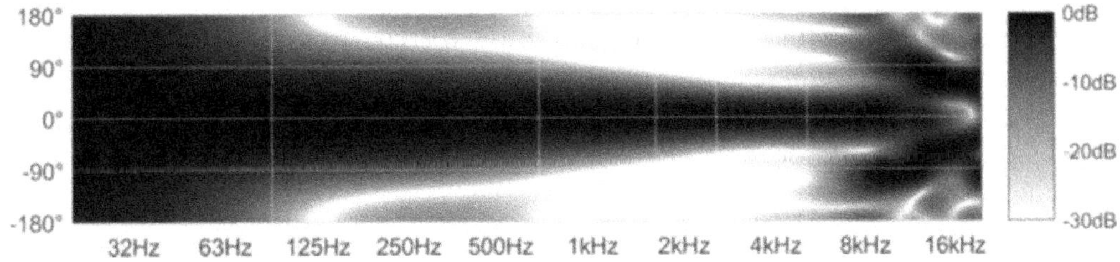

(e) 20 dB noise boost $(90, 680, 1.65k, 2.6k)$ individual max-$\boldsymbol{r}_{\mathrm{E}}$ weights

Fig. 6.16 Direction spread/dB for over frequency/Hz and zenith/degrees of filterbank with different settings to achieve 0, 5, 10, 15, 20 dB noise boosts; the maps show direction spreads normalized to their values at 0° at every frequency as above

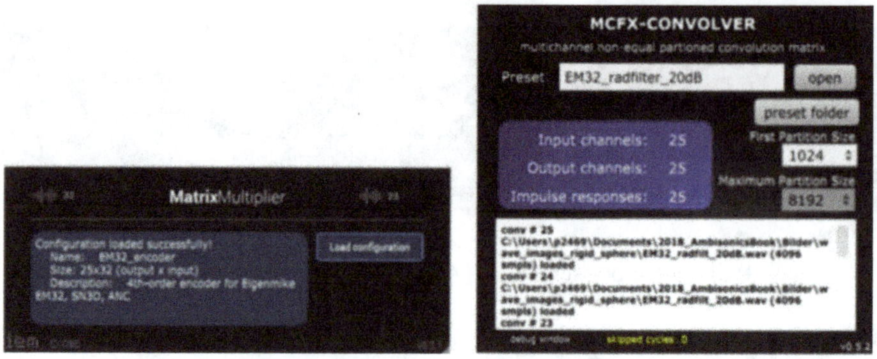

(a) em 32 encoding (b) em 32 radial filters

Fig. 6.17 IEM `MatrixMultiplier` encoding the Eigenmike em32 signals and `mcfx-convolver` applying radial filters to encoded em32 recording

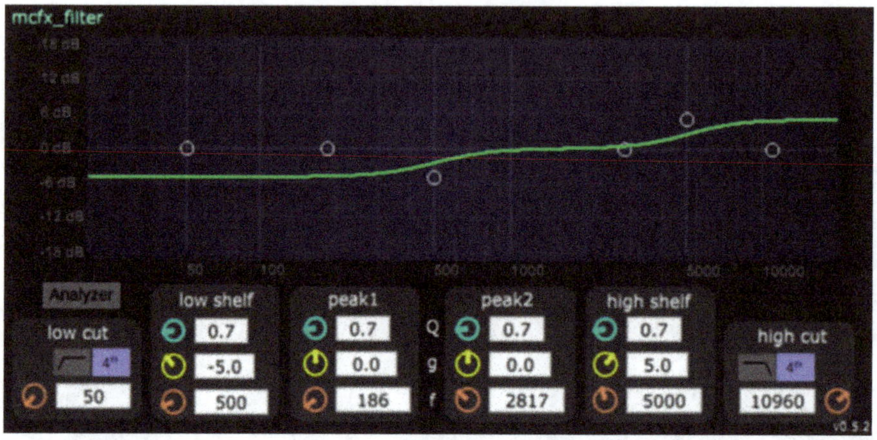

Fig. 6.18 Practical equalization of the em32 transducer characteristics by two parametric shelving filters of the `mcfx_filter`, cf. [28]

files and config files for `mcfx-convolver`) are provided online http://phaidra.kug.ac.at/o:79231 and are available for the different noise boosts 0, 5, 10, 15, 20 dB.

As found in [28], the em32 transducers exhibit a frequency response that favors low frequencies and attenuates high frequencies. This behavior is sufficiently well equalized in practice using two parametric shelving filters, a low shelf at 500 Hz with a gain of -5 dB, and a high shelf at 5 kHz using a gain of $+5$ dB, see Fig. 6.18.

6.11.2 SPARTA Array2SH

The SPARTA suite by Aalto University includes the `Array2SH` plug-in shown in Fig. 6.19 to convert the transducer signals of a microphone array into Ambisonics. It provides both encoding of the signals, as well as calculation and application of radial-focusing filters based on the geometry of the array. It supports rigid and open arrays

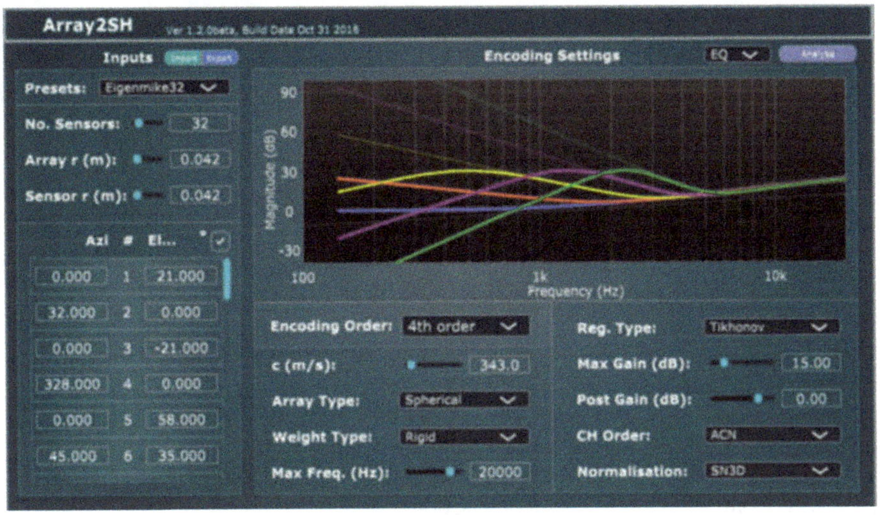

Fig. 6.19 SPARTA `Array2SH` encoding for, e.g., em32

and comes with presets for several arrays, such as the Eigenmike em32. The plug-in allows to adjust the radial filters in terms of regularization type and maximum gain. The `Reg. Type` called `Z-Style` corresponds to the linear-phase design of Sect. 6.9.

References

1. J. Daniel, Evolving views on HOA: from technological to pragmatic concernts, in *Proceedings of the 1st Ambisonics Symposium* (Graz, 2009)
2. G.W. Elko, R.A. Kubli, J. Meyer, Audio system based on at least second-order eigenbeams, in *PCT Patent*, vol. WO 03/061336, no. A1 (2003)
3. G.W. Elko, Superdirectional microphone arrays, in *Acoustic Signal Processing for Telecommunication*, ed. by J. Benesty, S.L. Gay (Kluwer Academic Publishers, Dordrecht, 2000)
4. J. Meyer, G. Elko, A highly scalable spherical microphone array based on an orthonormal decomposition of the soundfield, in *IEEE International Conference on Acoustics, Speech, and Signal Processing, 2002. Proceedings.(ICASSP'02)*, vol. 2 (Orlando, 2002)
5. G.W. Elko, Differential microphone arrays, in *Audio Signal Processing for Next-Generation Multimedia Communication Systems*, ed. by Y. Huang, J. Benesty (Springer, Berlin, 2004)
6. J. Daniel, S. Moreau, Further study of sound ELD coding with higher order ambisonics, in *116th AES Convention* (2004)
7. S.-O. Petersen, Localization of sound sources using 3d microphone array, M. Thesis, University of South Denmark, Odense (2004). www.oscarpetersen.dk/speciale/Thesis.pdf
8. B. Rafaely, Analysis and design of spherical microphone arrays. IEEE Trans. Speech Audio Process. (2005)
9. S. Moreau, Étude et réalisation d'outils avancés d'encodage spatial pour la technique de spatialisation sonore Higher Order Ambisonics: microphone 3d et contrôle de distance, Ph.D. Thesis, Université du Maine (2006)
10. H. Teutsch, *Modal Array Signal Processing: Principles and Applications of Acoustic Wavefield Decomposition* (Springer, Berlin, 2007)
11. Z. Li, R. Duraiswami, Flexible and optimal design of spherical microphone arrays for beamforming. IEEE Trans. ASLP **15**(2) (2007)

12. W. Song, W. Ellermeier, J. Hald, Using beamforming and binaural synthesis for the psychoacoustical evaluation of target sources in noise. J. Acoust. Soc. Am. **123**(2) (2008)
13. B. Rafaely, *Fundamentals of Spherical Array Processing*, 2nd edn. (Springer, Berlin, 2019)
14. ISO 31-11:1978, Mathematical signs and symbols for use in physical sciences and technology (1978)
15. ISO 80000-2, quantities and units? Part 2: Mathematical signs and symbols to be used in the natural sciences and technology (2009)
16. E.G. Williams, *Fourier Acoustics* (Academic, Cambridge, 1999)
17. B. Rafaely, B. Weiss, E. Bachmat, Spatial aliasing in spherical microphone arrays. IEEE Trans. Signal Process. **55**(3) (2007)
18. F. Zotter, Sampling strategies for acoustic holography/holophony on the sphere, in *NAG-DAGA, Rotterdam* (2009)
19. P. Lecomte, P.-A. Gauthier, C. Langrenne, A. Berry, A. Garcia, A fifty-node Lebedev grid and its applications to ambisonics. J. Audio Eng. Soc. **64**(11) (2016)
20. I.H. Sloan, R.S. Womersley, Extremal systems of points and numerical integration on the sphere. Adv. Comput. Math. **21**, 107–125 (2004)
21. B. Bernschütz, C. Pörschmann, S. Spors, Soft-limiting bei modaler amplitudenverstärkung bei sphärischen mikrofonarrays im plane-wave decomposition verfahren, in *Fortschritte der Akustik - DAGA* (2011)
22. T. Rettberg, S. Spors, On the impact of noise introduced by spherical beamforming techniques on data-based binaural synthesis, in *Fortschritte der Akustik - DAGA* (2013)
23. T. Rettberg, S. Spors, Time-domain behaviour of spherical microphone arrays at high orders, in *Fortschritte der Akustik - DAGA* (2014)
24. B. Rafaely, *Fundamentals of Spherical Array Processing*, 1st edn. (Springer, Berlin, 2015)
25. D.L. Alon, B. Rafaely, Spatial decomposition by spherical array processing, in *Parametric Time-Frequency Domain Spatial Audio*, ed. by V. Pulkki, S. Delikaris-Manias, A. Politis (Wiley, New Jersey, 2017)
26. S. Lösler, F. Zotter, Comprehensive radial filter design for practical higher-order ambisonic recording, in *Fortschritte der Akustik – DAGA Nürnberg* (2015)
27. F. Zotter, M. Zaunschirm, M. Frank, M. Kronlachner, A beamformer to play with wall reflections: The icosahedral loudspeaker. Comput. Music J. **41**(3) (2017)
28. F. Zotter, M. Frank, C. Haar, Spherical microphone array equalization for ambisonics, in *Fortschritte der Akustik - DAGA* (Nürnberg, 2015)

Chapter 7
Compact Spherical Loudspeaker Arrays

Le haut-parleur anonymise la source réelle.
Pierre Boulez [1], ICA, 1983.
...adjustable radiation naturalize[s] alien sounds by embedding them in the natural spatiality of the room.
OSIL project [2], InSonic, 2015.

Abstract This chapter introduces auditory objects that can be created by adjustable-directivity sources in rooms. After showing basic positioning properties in distance and direction, we describe physical first- and higher-order spherical loudspeaker arrays and their control, such as the loudspeaker cubes or the icosahedral loudspeaker (IKO). Not only static auditory objects, but such traversing space by their time-varying beam forming are considered here. Signal dependency and different practical setups are discussed and briefly analyzed. This young Ambisonic technology brings new means of expression to sound reinforcement, electroacoustic or computer music.

While surrounding Ambisonic loudspeaker arrays play sound from outside the listening area into the audience, compact spherical loudspeaker arrays play sound into the room from a single position. Directivity adjustable in orientation and shape can be used to steer sound beams in order to excite wall reflections in the given, acoustic environment. The directional shapes and orientations of such beams are all controlled by—guess what—Ambisonic signals. Despite the huge practical difference, both applications do not only share the spherical harmonics that lend their shapes to Ambisonic signals: The control of radiating sound beams employs nearly the same model- or measurement-based radial steering filters as those of compact higher-order Ambisonic microphones.

The works of Warusfel [3], Kassakian [4], Avizienis [5], Zotter [6, 7], Pomberger [8], Pollow [9], Mattioli Pasqual [10] established the electroacoustic background technology required to describe compact spherical loudspeaker arrays built with electrodynamic transducers. The early works on auditory objects were written by Schmeder [11], Sharma, Frank, and Zotter [2, 12, 13]. And some contemporary results were found in the project "Orchestrating the Space by Icosahedral Loudspeaker" (OSIL) between 2015 and 2018 [14–19].

F. Zotter and M. Frank, *Ambisonics*, Springer Topics in Signal Processing 19,
https://doi.org/10.1007/978-3-030-17207-7_7

7.1 Auditory Events of Ambisonically Controlled Directivity

7.1.1 Perceived Distance

Laitinen showed in [20] that increasing the directivity of a listener-facing loudspeaker array from omnidirectional to second order was able to create auditory events that were perceptually closer than the physical distance to the loudspeaker array. The experimental results can be explained by the increase of the direct-to-reverberant energy ratio, as the sound beam of the directional source does not as much excite room reflections.

Wendt extended Laitinen's work by experiments employing a simulation of a third-order directional source in a virtual room (third-order image source model) played back by a loudspeaker ring in an anechoic room [16]. He could show that the perceived distance between the listener and the higher-order directional source could not only be controlled by the order of the directivity pattern but also by the orientation of the source (towards the listener, away from the listener). Beams projecting sounds away from the listener were perceived behind the source, cf. Fig. 7.1. Again, the perceptual results could be modeled by simple measures known from room acoustics.

7.1.2 Perceived Direction

Using a similar room simulation, the study in [21] asked participants to indicate the perceived direction of an auditory event created by a third-order directional source. The results showed that for different source orientations, listeners perceived auditory objects at directions that often did not coincide with the sound source, but with the delayed reflection paths, cf. Fig. 7.2. Perceived directions focused on the direct sound and the three first reflections after 6, 8, and 9 ms. For some orientations, still even

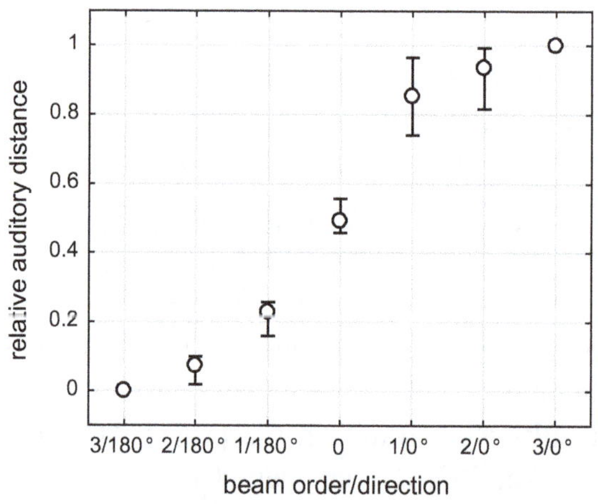

Fig. 7.1 Perceived distance (medians and 95% confidence intervals) depends on order and direction of a static source emitting sound with max-r_E directivity (180° towards listener, 0° away from listener) in a simulated environment

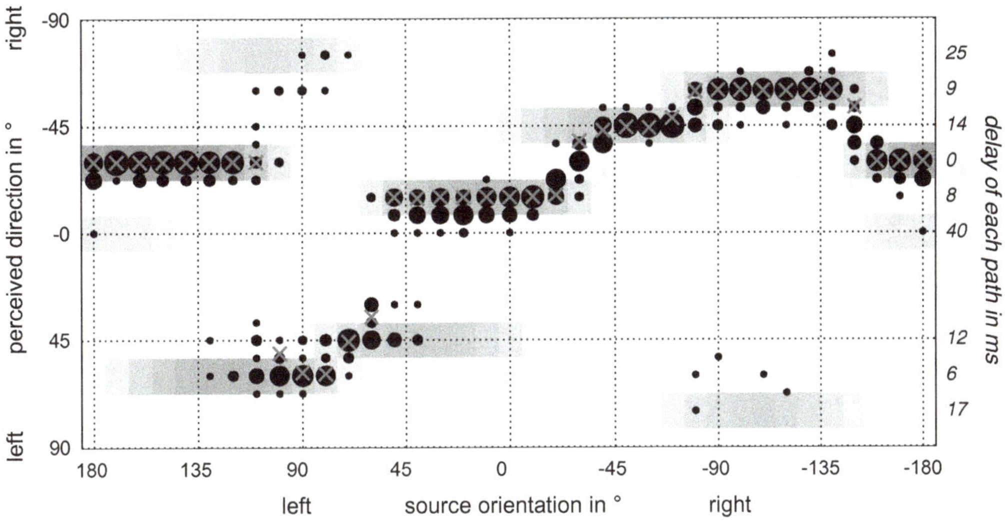

Fig. 7.2 IKO perceived directions (black circles, radii indicate relative amount of answers) and modeling (gray crosses), 3rd-order max-$\boldsymbol{r}_{\mathrm{E}}$ beam, 2nd-order image source model. Gray shading in the background indicates level of each path

the second-order reflections at 12 and 14 ms were dominating localization. However, the influence of later reflections is reduced by the precedence effect. The perceived directions can be modeled by the extended energy vector originally developed for off-center listening positions in surrounding loudspeakers arrangements, as also shown in [17]. Experiments in [22] showed that panning between a reflection and the direct sound creates auditory objects in between. When applying the appropriate delay and gain to the direct sound to compensate for the longer path of the reflection, the localization curves are similar to those of standard stereo using a pair of loudspeakers.

7.2 First-Order Compact Loudspeaker Arrays and Cubes

The simplest way of creating a loudspeaker array with adjustable directivity in a practical sense is a cube with loudspeakers on its plane surfaces, as suggested by Misdariis [23]. Restricting the directivity control to two dimensions reduces the number of loudspeaker drivers to four and facilitates to equip the array with a carrying handle on top and a flange adapter at the bottom, cf. [24] and Fig. 7.3.

Directivity control. First-order Ambisonics utilizes monopole and dipole modes, which directly translate to the corresponding far-field radiation patterns. These modes can easily be created due to the cubic shape by either playing of all four drivers in phase or the opposing drivers out of phase, cf. Fig. 7.4. Nevertheless, the frequency responses of such monopole and dipole modes need to be equalized to enable their phase- and magnitude-aligned superposition in the far field. Filters and measurement data of cube loudspeakers built at IEM [24] are freely available on http://phaidra.kug.ac.at/o:67631.

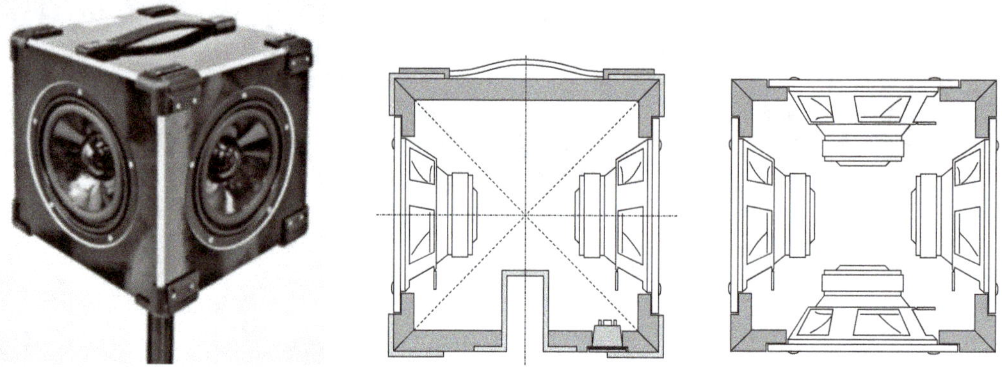

(a) loud speaker cube (b) vertical cross section (c) horizontal cross section

Fig. 7.3 Design of a loudspeaker cube: prototype, and vertical and horizontal cross section plots

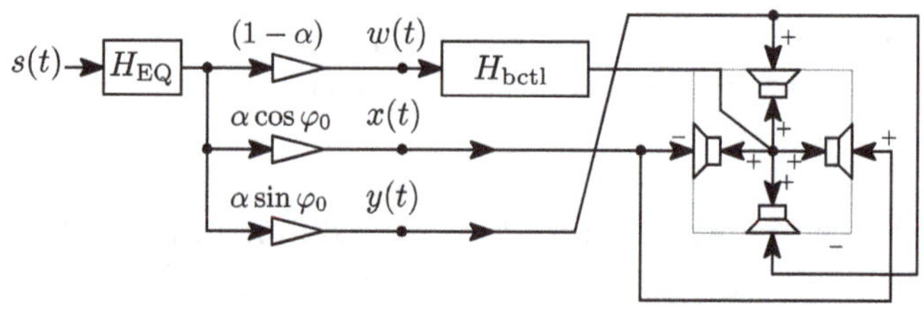

Fig. 7.4 System controlling the monopoles and dipole modes of the loudspeaker cubes, to accomplish first-order beamforming with the shape parameter α and beam direction φ_0

To overcome the compressive effort of interior volume changes at low frequencies, the filter H_{bctl} in Fig. 7.4 equalizes the smaller velocity of the loudspeaker cones when driven omnidirectionally to the velocity when driven in dipoles as a first step, and as a second step, it attenuates the monopole pattern slightly to account for its more efficient radiation at low frequencies. The filter H_{EQ} is a general equalizer required to obtain a flat frequency response, $0 \leq \alpha \leq 1$ is a first-order omni to dipole beam-shape parameter, and φ_0 is the beam direction. The filter H_{bctl} can be specified as a 5th-order IIR filter purely based on geometric and electroacoustic parameters [19].

Direct and indirect sound with two cubes. The study in [19] examined the width of the listening area for the creation of a central auditory object between a pair of loudspeaker cubes cf. Fig. 7.5. Steering the two beams directly at the listener yielded a narrow listening area that increased with the distance to the loudspeakers, similar as known from typical stereo applications, cf. Fig. 2.9. A much wider listening area is achieved by steering the beams to the front wall to excite reflections. To this end, max-$\boldsymbol{r}_{\mathrm{E}}$ (super-cardioid) beams were chosen and oriented in a way to ideally suppress direct sound from the loudspeaker cubes at the listening position. The proposed setup of two loudspeaker cubes can be used to play back stable L, C, R channels of a surround production without the need of an actual center loudspeaker.

Surround with depth: Together with the distance control described by Laitinen [20], the stable in-between auditory image has been used in [19] to establish a *surround-*

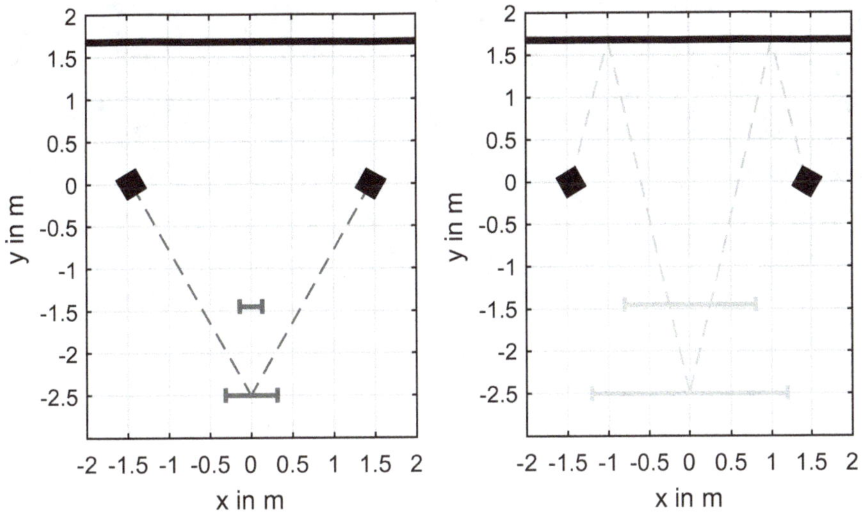

(a) beam pair aiming directly at central listening position at 2.5 m distance

(b) beam pair excting front wall reflections

Fig. 7.5 Width of the listening area for a central auditory object at two distances from a pair of loudspeaker cubes with different orientation of max-$\boldsymbol{r}_{\mathrm{E}}$/super-cardioid beams

with-depth system consisting of a quadraphonic setup of four loudspeaker cubes. As first layer, it uses the direct sounds from the 4 loudspeakers from ±45° and ±135° together with the 4 in-between images at 0°, ±90°, and 180° to obtain 8 directions for third-order Ambisonic surround panning. As a second layer for depth, *surround with depth* uses 4 cardioid beams pointing into the 4 room corners to provide the impression of distant sounds. Blending between those two layer is used to control the distance impression of surround sounds.

7.3 Higher-Order Compact Spherical Loudspeaker Arrays and IKO

With transducers mounted on spheres or polyhedra, higher-order radiators can be built. Typically, those are Platonic solids such as dodecahedra or icosahedra, as they can easily be manufactured from equal-sided polygons cf. Fig. 7.6. Often, the loudspeakers are also mounted onto a common interior volume. Hereby, the higher-order modes can be controlled at reduced impedance of the inner stiffness, however, this also causes acoustic coupling of the transducer motions. Typically, multiple-input-multiple-output (MIMO) crosstalk cancellers are employed to suppress the coupling and to control the velocity of the transducer cones. If this is accomplished, the acoustic radiation can be modeled and equalized by the spherical cap model, cf. [6, 15, 25, 26].

Fig. 7.6 Powerful icosahedral loudspeaker array (IKO by IEM and Sonible) and reflecting baffles in Ligeti concert hall, in preparation of an electroacoustic music concert

Cap model. Higher-order loudspeaker arrays on a compact spherical housing are modeled by the spherical cap model. It assumes for the exterior air that the radial surface velocity is a boundary condition consisting of separated spherical cap shapes of the size α centered around the directions $\{\boldsymbol{\theta}_l\}$, each unity in value. These idealized transducer shapes driven by the transducer velocities v_l compose the surface velocity

$$v(\boldsymbol{\theta}) = \sum_{l=0}^{\mathrm{L}} u(\boldsymbol{\theta}_l^{\mathrm{T}}\boldsymbol{\theta} - \cos\tfrac{\alpha}{2})\, v_l. \tag{7.1}$$

Here, $u(\zeta)$ denotes the unit step function that is unity for $\zeta \geq 0$ and zero otherwise. The surface velocity distribution can be decomposed into spherical harmonics as

$$v(\boldsymbol{\theta}) = \sum_{n=0}^{\infty}\sum_{m=-n}^{n} Y_n^m(\boldsymbol{\theta}) \sum_{l=0}^{\mathrm{L}} w_{nm}^{(l)}\, v_l\,. \tag{7.2}$$

The coefficients $w_{nm}^{(l)}$ of the lth cap are defined by spherical convolution Eq. (A.56) of a Dirac delta $\delta(\boldsymbol{\theta}_l^{\mathrm{T}}\boldsymbol{\theta} - 1)$ pointing to the cap center with a zenithal cap $u(\cos\vartheta - \cos\frac{\alpha}{2})$:

$$w_{nm}^{(l)} = w_n\, Y_n^m(\boldsymbol{\theta}_l), \tag{7.3}$$

where $Y_n^m(\boldsymbol{\theta}_l)$ are the coefficients expressing the Dirac delta, extended to a cap by weighting with w_n. The term $w_n = 2\pi \int_{\cos\frac{\alpha}{2}}^{1} P_n(\zeta)\,\mathrm{d}\zeta$ is derived in Eq. (A.60)

$$w_n = 2\pi \begin{cases} -\frac{P_{n+1}(\cos\frac{\alpha}{2}) - \cos\frac{\alpha}{2}\,P_n(\cos\frac{\alpha}{2})}{n}, & \text{for } n > 0,\\ 1 - \cos\frac{\alpha}{2}, & \text{for } n = 0. \end{cases} \tag{7.4}$$

Decoder. Without radiation control yet, any low-order target spherical harmonic $n \leq \mathrm{N}$ can be synthesized as velocity pattern ϕ_{nm} by superimposing the spherical cap coefficients $w_{nm}^{(l)}$ with suitable transducer velocities v_l, i.e. $\phi_{nm} = \sum_l w_n\, Y_n^m(\boldsymbol{\theta}_l)\, v_l$. We write a matrix/vector notation with the matrix $\boldsymbol{Y} = [\boldsymbol{y}(\boldsymbol{\theta}_1), \ldots, \boldsymbol{y}(\boldsymbol{\theta}_\mathrm{L})]$ containing the spherical harmonics $\boldsymbol{y}(\boldsymbol{\theta}) = [Y_n^m(\boldsymbol{\theta})]_{nm}$ sampled at the transducer positions $\{\boldsymbol{\theta}_l\}$ to represent Dirac deltas pointing there, and $\boldsymbol{w} = [w_n]_{nm}$ to represent the cap shape,

$$\boldsymbol{\phi} = \mathrm{diag}\{\boldsymbol{w}\}\boldsymbol{Y}\,\boldsymbol{v}. \tag{7.5}$$

As long as the order N up to which coefficients are controlled is low enough $\mathrm{L} \geq (\mathrm{N}+1)^2$ and transducers are well-distributed, perfect control is feasible. The corresponding velocities are found by solving a least-squares problem, see Appendix A.4, Eq. (A.63), yielding the right inverse of the Nth-order cap-coefficient matrix,

$$\boldsymbol{v} = \boldsymbol{Y}_\mathrm{N}^\mathrm{T}(\boldsymbol{Y}_\mathrm{N}\boldsymbol{Y}_\mathrm{N}^\mathrm{T})^{-1}\,\mathrm{diag}\{\boldsymbol{w}_\mathrm{N}\}^{-1}\boldsymbol{\phi}_\mathrm{N} = \boldsymbol{D}\,\mathrm{diag}\{\boldsymbol{w}_\mathrm{N}\}^{-1}\boldsymbol{\phi}_\mathrm{N}. \tag{7.6}$$

The right inverse $\boldsymbol{D} = \boldsymbol{Y}_\mathrm{N}^\mathrm{T}(\boldsymbol{Y}_\mathrm{N}\boldsymbol{Y}_\mathrm{N}^\mathrm{T})^{-1}$ is a mode-matching decoder, cf. Eq. (4.40).

Exterior problem. The radiated sound pressure is described by the exterior problem denoted by the coefficients c_{nm} in Eq. (6.13) and the spherical Hankel functions $h_n^{(2)}(kr)$. To relate it to a time-derived surface velocity at the array radius $r = \mathrm{a}$, we derive the exterior solution with regard to radius $\frac{\partial p}{\partial r} = k\frac{\partial p}{\partial kr} = -\mathrm{i}kc\,\rho v$, cf. Eq. (6.2),

$$v(\boldsymbol{\theta}) = \frac{\mathrm{i}}{\rho c}\sum_{n=0}^{\infty}\sum_{m=-n}^{n} h_n^{\prime(2)}(k\mathrm{a})\; Y_n^m(\boldsymbol{\theta})\; c_{nm}. \tag{7.7}$$

Comparing Eq. (7.2) to Eq. (7.7) yields $c_{nm} = \rho c[\mathrm{i}h_n^{\prime(2)}(k\mathrm{a})]^{-1}\sum_{l=0}^{\mathrm{L}} w_n\, Y_n^m(\boldsymbol{\theta}_l)\, v_l$, the coefficients to calculate the radiated pressure. Far away, we replace the spherical Hankel function that approaches $h_n^{(2)}(kr) \rightarrow \mathrm{i}^{n+1}k^{-1}e^{-\mathrm{i}kr}$ by the term $\mathrm{i}^{n+1}k^{-1}$ in Eq. (6.13) so that the radiated far-field sound pressure $p \propto \sum \mathrm{i}^{n+1}k^{-1}Y_n^m c_{nm}$ becomes

$$p(\boldsymbol{\theta}) \propto \sum_{n=0}^{\infty}\sum_{m=-n}^{n} Y_n^m(\boldsymbol{\theta})\;\frac{\mathrm{i}^n\, w_n}{k\, h_n^{\prime(2)}(k\mathrm{a})}\sum_{l=0}^{\mathrm{L}} Y_n^m(\boldsymbol{\theta}_l)\, v_l. \tag{7.8}$$

7.3.1 Directivity Control

The spherical harmonics coefficients of the far-field sound pressure pattern in Eq. (7.8) are controlled by the cap velocities v_l

$$\psi_{nm} = \frac{\mathrm{i}^n\, w_n}{k\, h_n^{\prime(2)}(k\mathrm{a})}\sum_{l=0}^{\mathrm{L}} Y_n^m(\boldsymbol{\theta}_l)\, v_l, \tag{7.9}$$

and we desire to form the directional sound beam they represent according to a max-$\boldsymbol{r}_\mathrm{E}$ pattern $a_n\,Y_n^m(\boldsymbol{\theta}_0)$ yielding radiation focused towards $\boldsymbol{\theta}_0$

$$\psi_{nm} = a_n\,Y_n^m(\boldsymbol{\theta}_0). \tag{7.10}$$

To find suitable cap velocities v_l, we equate the model Eqs. (7.9) and (7.10). In matrix/vector notation never used the equation is

$$\mathrm{diag}\{[\mathrm{i}^n w_n\,k^{-1}/\,h_n'^{(2)}(k\mathrm{a})]_{nm}\}\;\boldsymbol{Y}\boldsymbol{v} = \mathrm{diag}\{[a_n]_{nm}\}\boldsymbol{y}(\boldsymbol{\theta}_0). \tag{7.11}$$

The diagonal matrix on the left is easy to invert, and for patterns up to the order $n \leq \mathrm{N}$, the mode-matching decoder $\boldsymbol{D}$ of Eq. (7.6) already gives us a way to define velocities inverting the matrix $\boldsymbol{Y}_\mathrm{N}$ from the right. The preliminary solution becomes

$$\Rightarrow \boldsymbol{v} = \boldsymbol{D}\,\mathrm{diag}\{[\mathrm{i}^{-n}w_n^{-1}\,k\,h_n'^{(2)}(k\mathrm{a})\,a_n]_{nm}\}\;\boldsymbol{y}_\mathrm{N}(\boldsymbol{\theta}_0). \tag{7.12}$$

On-axis equalized, sidelobe-suppressing directivity control limiting the excursion. The inverse cap shape coefficient w_n^{-1} and the max-$\boldsymbol{r}_\mathrm{E}$ weight a_n can be regarded as a part of the radiation control filters $\mathrm{i}^{-n}\,k\,h_n'^{(2)}(k\mathrm{a})$. The expression $\mathrm{i}^{-n-1}(k\mathrm{a})^2\,h_n'^{(2)}(k\mathrm{a})$ of compact spherical microphone arrays (Sect. 6.6) qualitatively differs by a factor k. Practical implementation of radiation control filters and their regularization is therefore quite similar to radial filters of spherical microphone arrays. There are three main differences, as explained in [15]:

- With loudspeaker arrays, it is rather the excursion that is limited, which primarily entails a different strategy of adjusting the filter bank cut-on frequencies, which due to size are at lower frequencies where group-delay distortions are less disturbing, and linear-phase implementations would cause avoidably long delays.
- Moreover, instead of cut-on filter slopes of $(n+1)$th order required for noise removal in signals obtained from spherical microphone arrays, limited excursion requires cut-on slopes of at least $(n+3)$th order, i.e. 4th order to cut on the 1st-order Ambisonic signals. Thereof, one additional order is caused by the qualitative difference of k^{-1} in radial filters, and another order by the conversion of velocity to excursion by a factor $(\mathrm{i}\omega)^{-1}$.
- Finally, instead of diffuse-field equalization that is useful for surround sound playback of spherical microphone array signals, it is more useful to equalize spherical sound beams on-axis (free field).

On-axis equalization yields a different scaling of the sub-band max-$\boldsymbol{r}_\mathrm{E}$ weights

$$a_{n,b} = \begin{cases} P_n\!\left(\cos\frac{137.9^\circ}{b+1.51}\right)\frac{\sum_{n=0}^{\mathrm{N}}(2n+1)P_n\left(\cos\frac{137.9^\circ}{\mathrm{N}+1.51}\right)}{\sum_{n=0}^{b}(2n+1)P_n\left(\cos\frac{137.9^\circ}{b+1.51}\right)}, & \text{for}\,n \leq b,\\ 0, & \text{otherwise.}\end{cases} \tag{7.13}$$

Typically, cut-on frequencies for compact spherical loudspeaker arrays are low, and linear-phase filterbanks would require long pre-delays. It is useful to employ Linkwitz-Riley filters for the crossovers, to get a low-latency implementation. To emphasize the similarity to Eq. (6.22), we write Linkwitz-Riley filters [27] as combination of an all-pass A^m with twice the phase response of an mth-order Butterworth low-pass combined either with the magnitude-squared low-pass response $[1+(\omega/\omega_c)^{2m}]^{-1}$ or high-pass response $(\omega/\omega_c)^{2m}[1+(\omega/\omega_c)^{2m}]^{-1}$ per crossover. Such a minimum-phase crossover is of even order, so that the minimum-order cut-on slope must be rounded up to the next even order $2\lceil\frac{b+3}{2}\rceil$. Plain high/low crossovers would be in-phase unless combined with further crossovers to form narrower bands. However, an in-phase filterbank is obtained after inserting the product of all all-passes in every band, cf. [28]. Although non-minimum-phase, this is still low-latency. For the band b containing Ambisonic orders $0 \le n \le b$, the modified filterbank is

$$H_b(\omega) = \frac{\left(\frac{\omega}{\omega_b}\right)^{2\lceil\frac{b+3}{2}\rceil}}{1+\left(\frac{\omega}{\omega_b}\right)^{2\lceil\frac{b+3}{2}\rceil}} \frac{1}{1+\left(\frac{\omega}{\omega_{b+1}}\right)^{2\lceil\frac{b+4}{2}\rceil}} \prod_{b'=0}^{\mathrm{N}} A_{\omega_{b'}}^{\lceil\frac{b'+3}{2}\rceil}(\omega). \tag{7.14}$$

The sum $\sum_b H_b(\omega)$ is considered to be sufficiently flat, so that the radial filters for compact spherical loudspeaker arrays using Eqs. (7.8), (7.13), (7.14) become

$$\rho_n(\omega) = \left[\sum_{b=n}^{\mathrm{N}} a_{n,b} H_b(\omega)\right] \mathrm{i}^{-n}\, w_n^{-1}\, h_n^{\prime(2)}(k\mathrm{a})\, e^{\mathrm{i}k\mathrm{a}}. \tag{7.15}$$

Figure 7.7 shows the block diagram to control compact spherical loudspeaker arrays by Ambisonic input signals, including the radiation control filters, the decoder, and a voltage-equalizing crosstalk canceller feeding the loudspeakers.

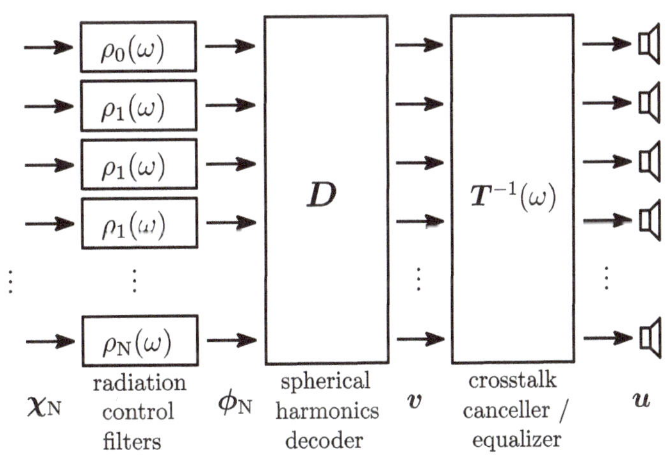

Fig. 7.7 Signal processing for higher-order compact spherical loudspeaker array control: the Ambisonic directivity signals $\boldsymbol{\chi}_{\mathrm{N}}(t)$ run through radiation control filters $\rho_n(\omega)$, a decoder yielding desired velocities $\boldsymbol{v}(t)$, and a crosstalk canceller/equalizer provides suitable output voltages $\boldsymbol{u}(t)$

7.3.2 Control System and Verification Based on Measurements

Velocity equalization/crosstalk cancellation. In the frequency domain, laser vibrometer measurements, cf. Fig. 7.8a, characterize the physical multiple-input-multiple-output (MIMO) system of transducer input voltages $u_l(\omega)$ to transducer velocities $v_l(\omega)$

$$\boldsymbol{v}(\omega) = \boldsymbol{T}(\omega)\,\boldsymbol{u}(\omega), \tag{7.16}$$

including the effect of acoustic coupling through the common enclosure. Corresponding open measurement data sets[1] can be found online, as described in [18]. Theoretically, the frequency-domain inverse of the matrix $\boldsymbol{T}(\omega)$ can be used to equalize and control the transducer velocities with acoustic crosstalk cancelled, as indicated in Fig. 7.7,

$$\boldsymbol{u}(\omega) = \boldsymbol{T}^{-1}(\omega)\,\boldsymbol{v}(\omega). \tag{7.17}$$

In practice, this is only useful up to the frequency at which the loudspeaker cone vibration breaks up into modes, so typically below 1 kHz.

Control system: The entire control system with Ambisonic signals $\boldsymbol{\chi}_\mathrm{N}(\omega)$ as inputs uses Eqs. (7.6), (7.15), (7.17)

$$\boldsymbol{u}(\omega) = \boldsymbol{T}^{-1}(\omega)\,\boldsymbol{D}\,\mathrm{diag}\{\boldsymbol{\rho}(\omega)\}\,\boldsymbol{\chi}_\mathrm{N}(\omega). \tag{7.18}$$

Directivity measurement. It is useful to characterize the directivity obtained by measurements to verify the results; high-resolution 648×20 measurements $\boldsymbol{G}(\omega)$ of the IKO are found online[1]. The sound pressure can be decomposed with the known directional sampling by left-inversion of a spherical harmonics matrix $\boldsymbol{Y}_{17}^\mathrm{T}$, see Appendix A.4, Eq. (A.65, which can be up to 17th order on a $10° \times 10°$ grid in azimuth and zenith:

$$\boldsymbol{p}(\omega) = \boldsymbol{G}(\omega)\,\boldsymbol{u}(\omega), \qquad \Rightarrow \boldsymbol{\psi}_{17}(\omega) = (\boldsymbol{Y}_{17}^\mathrm{T})^\dagger\,\boldsymbol{p}(\omega). \tag{7.19}$$

With the highly resolved spherical harmonics coefficients, polar diagrams or balloon diagrams can be evaluated at any direction

$$p(\boldsymbol{\theta}, \omega) = \boldsymbol{y}_{17}(\boldsymbol{\theta})^\mathrm{T}\boldsymbol{\psi}_{17}(\omega), \tag{7.20}$$

given any control system delivering suitable voltages $\boldsymbol{u}$ for beamforming, as e.g. obtained by Eq. (7.18).

[1] http://phaidra.kug.ac.at/o:67609.

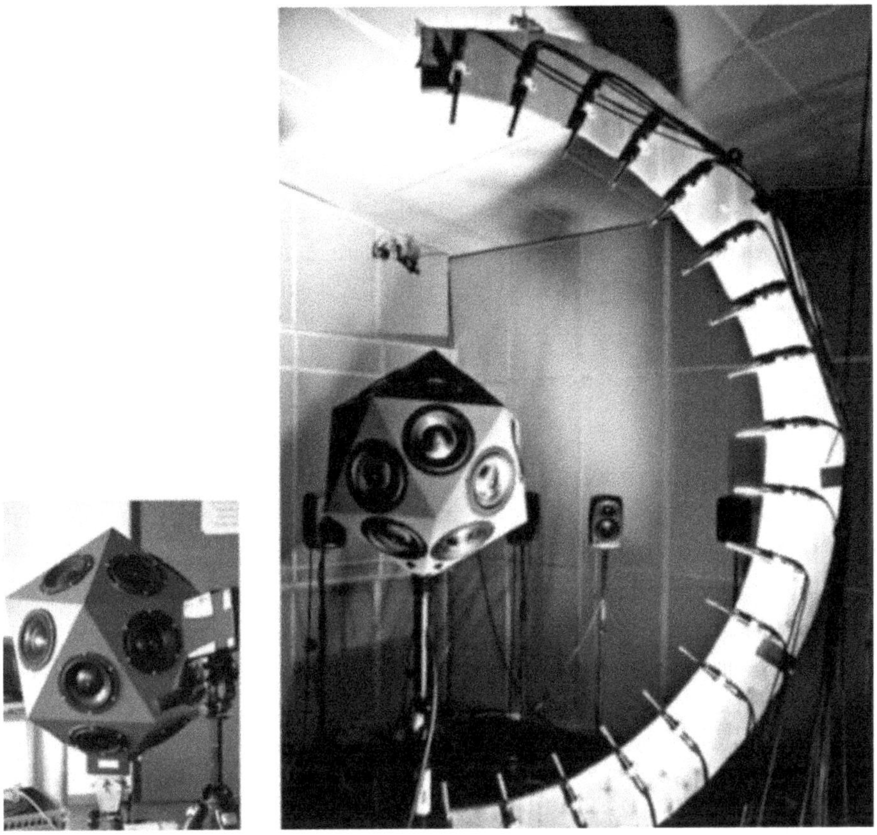

(a) velocities (b) directivities

Fig. 7.8 Measurements on the IKO as a MIMO system in terms of transducer output velocities (left) and radiation patterns (right) depending on the transducer input voltages

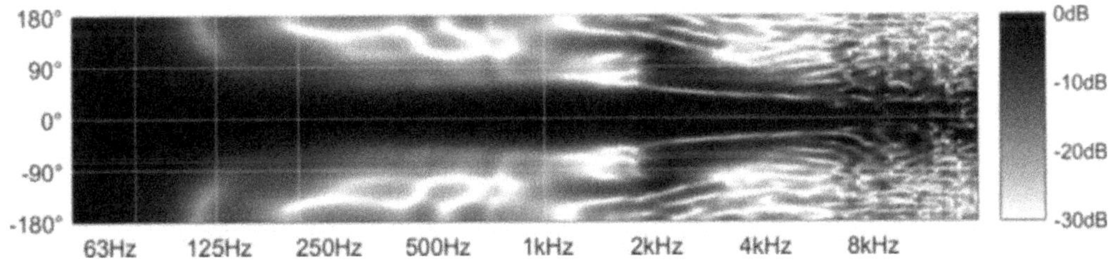

Fig. 7.9 Horiontal cross section of the IKO's directivity/dB over frequency/Hz and azimuth/degrees when beamforming to 0° azimuth on the horizon, with radiation control filters above, with filterbank frequencies (38, 75, 125, 210) Hz

To inspect the frequency-dependent directivity, a horizontal cross section is shown in Fig 7.9. The beamforming gets effective above 100 Hz and a beam width of $\pm 30°$ is held until 2 kHz. The filterbank starts the 0th order above 38 Hz, and with 75, 125, 210 Hz, 1st, 2nd, and 3rd order are successively added including on-axis equalized max-$\boldsymbol{r}_{\mathrm{E}}$ weightings. Above 2 kHz both spatial aliasing and modal breakup of the transducer cones affect directivity. However, these beamforming-direction-dependent distortions are often negligible in typical rooms.

7.4 Auditory Objects of the IKO

7.4.1 Static Auditory Objects

The study in [16] showed that distance control by changing the directivity and its orientation can also be achieved with the IKO in a real room, cf. Fig. 7.10. The experiments used stationary pink noise and could create auditory objects nearly 2 m behind the IKO, which corresponds to the distance between the IKO and the front wall of the playback room.

The maximum distance of auditory objects created by the IKO is strongly signal-dependent. Experiments in [14] showed that the auditory distance of pink noise bursts decreased for shorter fade-in times, while the fade-out time had no influence, cf. Fig. 7.11. A transient click sound was perceived even closer to the IKO. This can be explained by the precedence effect, that favors the earlier direct sound over the reflected sound from the walls. While this effect is strong for transient sounds, it is inhibited for stationary sounds with long fade-in times.

However, the precedence effect can even be reduced for transient click sounds by simultaneous playback of a masker sound that reduces the influence of the direct sound [29]. In comparison to no masker, playing a pink noise masker doubles the auditory distance, cf. Fig. 7.12. Using the room noise as a masker by playing the

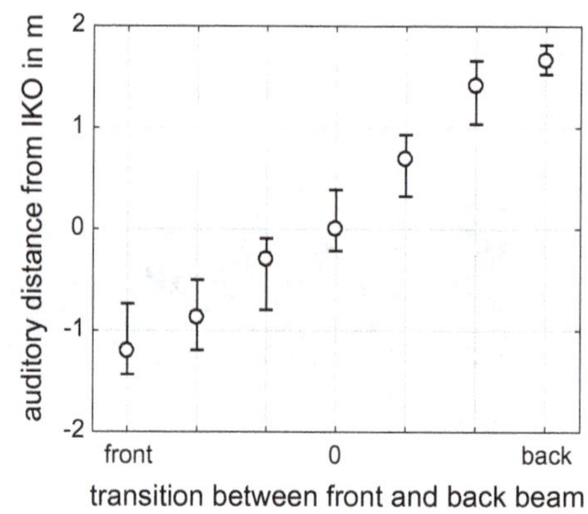

Fig. 7.10 Distance control using the IKO in the IEM CUBE

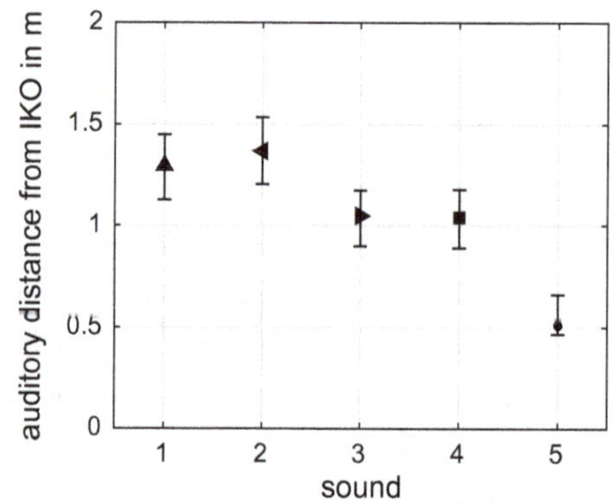

Fig. 7.11 Signal-dependent distance of auditory objects created by the IKO: markers indicate short or long fade-in and fade-out of pink noise bursts, · indicates transient click sound

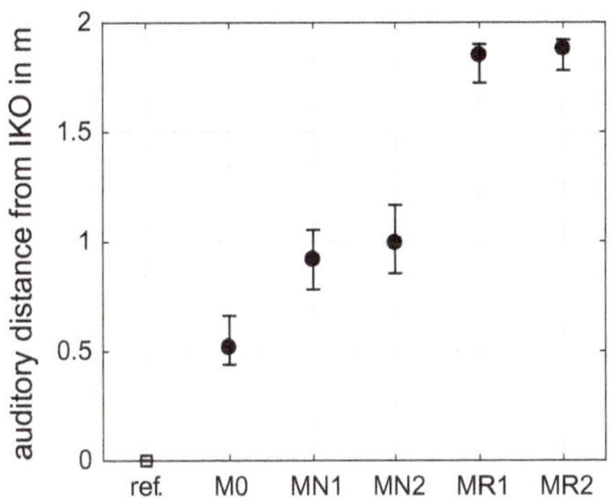

Fig. 7.12 Distance of auditory objects created by the IKO playing a transient click sound in dependence of maskers: no masker (M0), noise at different levels (M1, M2) and playback at very low level (room masker MR1 and MR2)

target sound very softly further increases the distance and yield a perception that is detached from the IKO.

7.4.2 *Moving Auditory Objects*

The studies in [14, 15] extended the previous listening experiments towards simple time-varying beam directions, such as from the left to the right, front/back or circles. To report the perceived locations of the moving auditory objects, listeners used a touch screen that showed a floor plan of the room, including the listening position and the position of the IKO. They had to indicate the location of the auditory object's trajectory every 500 ms. The perceived trajectories depend on the listening position, but they can always be recognized, cf. Fig. 7.13. The empirical knowledge was

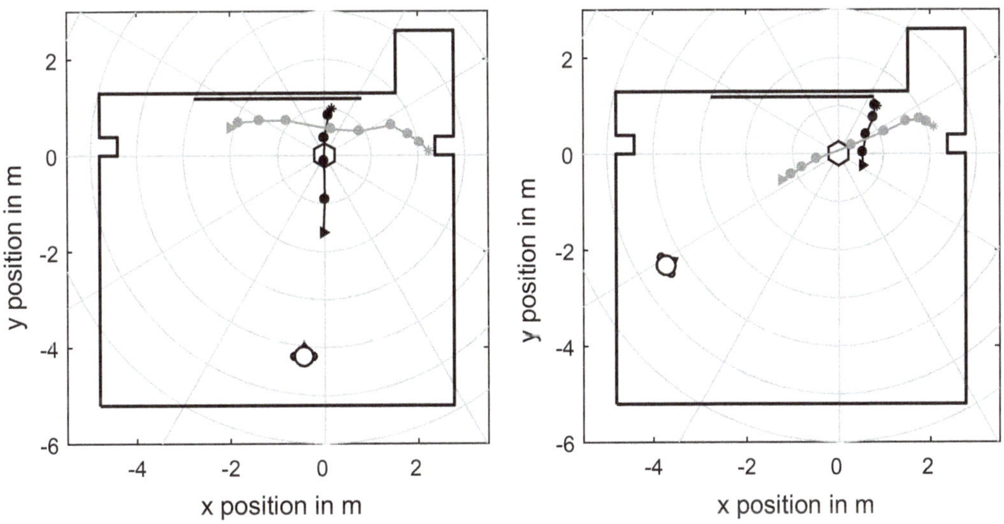

Fig. 7.13 Average perceived locations for each 500 ms step during front/back-movement (dark gray) and left/right-movement (light gray) at two listening positions, triangle indicates start and asterisk end of the trajectory

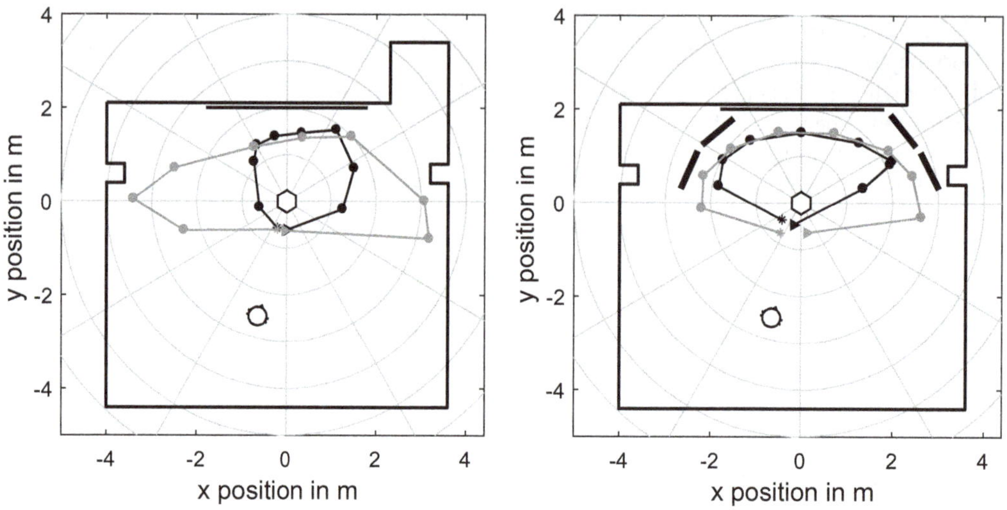

Fig. 7.14 Average perceived locations for each 500 ms step during circular movement of transient sound (dark gray) and stationary noise (light gray) without and with additional reflectors, triangle indicates start and asterisk end of the trajectory

applied in the artistic study in [14] about body-space relations, composing sounds that are spatialized with different static directions and simple movements.

For concerts, the artistic practice evolved to set the IKO up together with reflector baffles, cf. Fig. 7.14. A recent study in [30] investigated their effect on the perception of moving transient and stationary sounds. The baffles obviously reduce the signal-dependency by contributing more additional reflection paths, contrasting the direct sound.

7.5 Practical Free-Software Examples

7.5.1 IEM Room Encoder and Directivity Shaper

The IEM `Room Encoder` VST plug-in, cf. Fig. 4.36, can not only be used to simulate the room reflections of an omnidirectional sound source based on the image-source method, but it also supports directional sound sources. As format, it employs Ambisonics with ACN ordering and adjustable normalization up to seventh order. Thus, it enables to utilize data from directivity measurements or even directional recordings done with a surrounding spherical microphone, e.g. to put real instrument recordings into the virtual room.

As an alternative, the IEM `Directivity Shaper`, cf. Fig. 7.15 provides simple means to generate a frequency-dependent directivity pattern from scratch and to apply it on a mono input signal. This is useful to generate the typical rotary speaker effect of a Leslie cabinet.

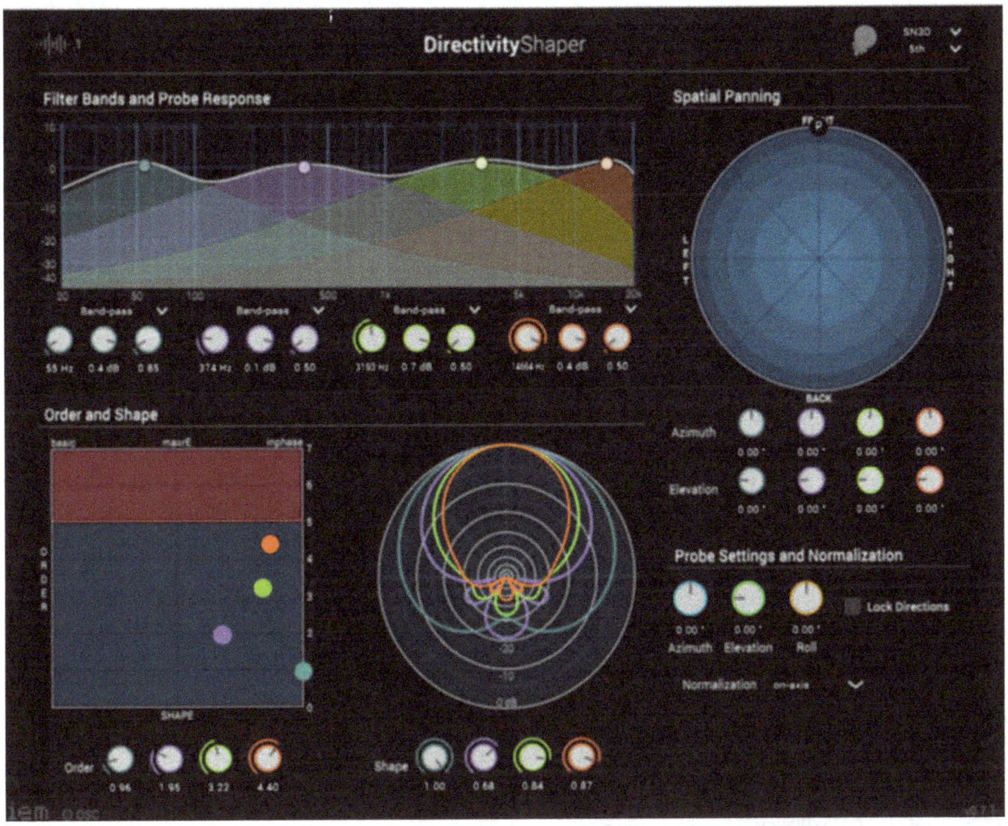

Fig. 7.15 IEM `Directivity Shaper` plug-in

7.5.2 IEM Cubes 5.1 Player and Surround with Depth

As shown in Fig. 7.5, a pair of loudspeaker cubes can create a stable auditory event in between them to replace an actual center loudspeaker. In order to play back an entire 5.1 production, the IEM `cubes 5.1 Player` plug-in extends this approach by two additional beams to the side walls for the surround channels, cf. Fig. 7.16. The plug-in provides a control of the shape, direction, and level of all beams, as well as a delay compensation for the reflection paths.

Surround sound with depth can be realized with a quadraphonic setup of four loudspeaker cubes and a combination of the `cubes Surround Decoder` and multiple `Distance Encoder` plug-ins, cf. Fig. 7.16. For each source, the `Distance Encoder` controls position and distance, i.e. the blending between the two layers. The output of the plug-in is a 10-channel audio stream including 7 channels for third-order (inner layer) and 3 for first-order 2D Ambisonics (outer depth layer). The `cubes Surround Decoder` plug-in decodes the 10-channel audio stream and distributes the signals to the 16 drivers of four loudspeaker cubes. For each loudspeaker cube, the directions to excite direct and reflected sound of the inner layer and the diffuse sound of the depth layer can be adjusted in order to adapt to the playback environment. Additionally, the directivity patterns for direct, reflected, and diffuse sound beams

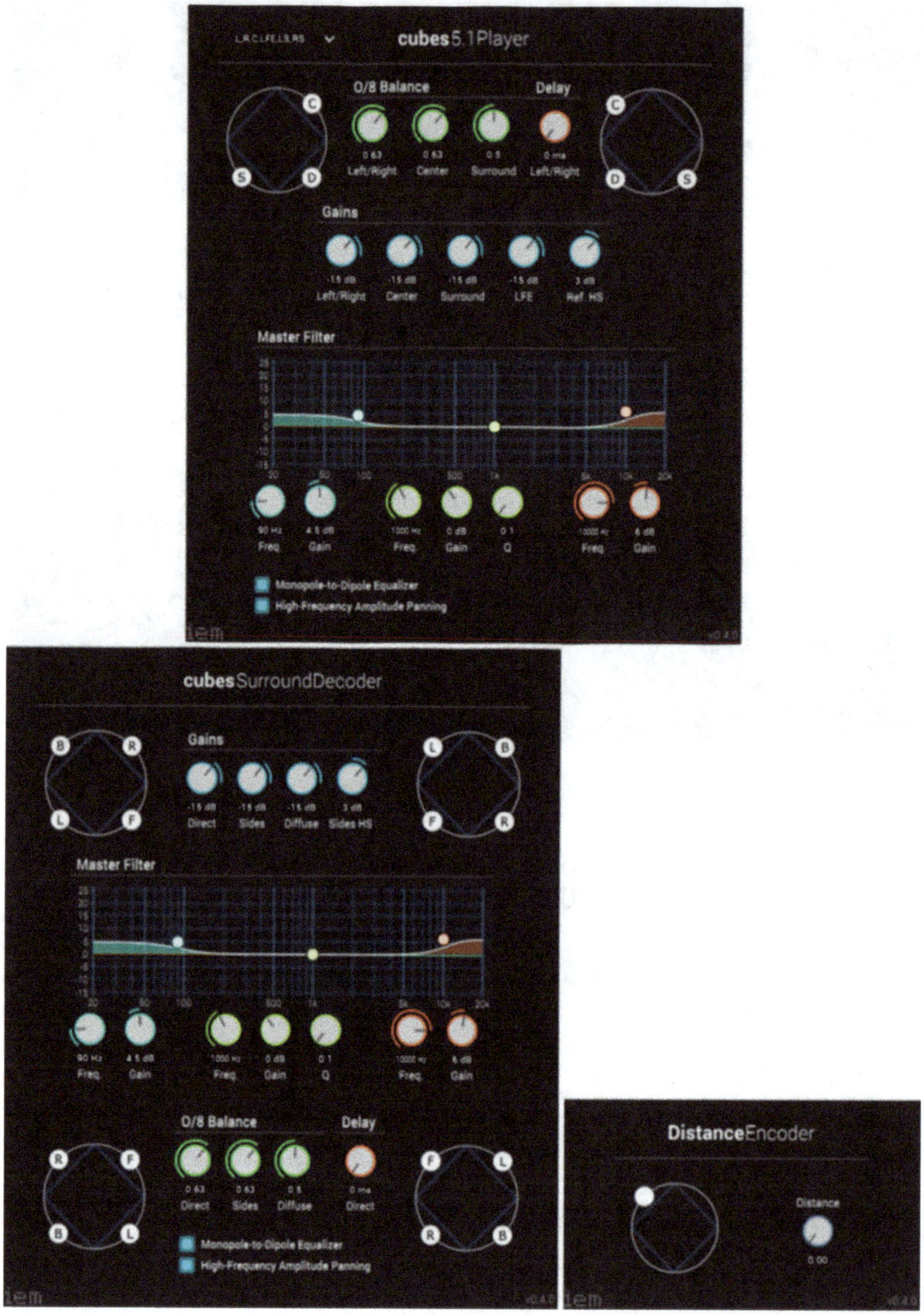

Fig. 7.16 IEM `cubes 5.1 Player`, `cubes Surround Decoder` and `Distance Encoder` plug-ins

can be controlled, as well as a delay to compensate for the longer propagation paths of the reflected sound. The plug-ins are available under https://git.iem.at/audioplugins/CubeSpeakerPlugins.

7.5.3 IKO

Spatialization using the IKO can use a similar infrastructure of plug-ins as surrounding loudspeaker arrays. Ambisonic encoder plug-ins, such as the `ambix_encoder` or the IEM `StereoEncoder` or `MultiEncoder`, create the third-order Ambisonic signals that are subsequently fed to a decoder. Decoding to the IKO requires the processing steps as shown in Fig. 7.7: radiation control filters in the spherical harmonic domain, decoding from spherical harmonics to transducer signals, as well as crosstalk cancellation and equalization of the transducers. This processing can be summarized in a 16 (spherical harmonics up to third order) $\times$ 20 (transducers) filter matrix. Convolution can be done efficiently using the `mcfx_convolver` plug-in. Filter presets for the IKO can be found under http://phaidra.kug.ac.at/o:79235.

References

1. P. Boulez, L'acoustique et la musique contemporaine, in *Proceedings of the 11th International Congress on Acoustics*, vol. 8 (Paris, 1983), https://www.icacommission.org/Proceedings/ICA1983Paris/ICA11%20Proceedings%20Vol8.pdf
2. M. Frank, G.K. Sharma, F. Zotter, What we already know about spatialization with compact spherical arrays as variable-directivity loudspeakers, in *Proceedings of the inSONIC2015* (Karlsruhe, 2015)
3. O. Warusfel, P. Derogis, R. Causse, Radiation synthesis with digitally controlled loudspeakers, in *103rd AES Convention* (Paris, 1997)
4. P. Kassakian, D. Wessel, Characterization of spherical loudspeaker arrays, in *Convention paper #1430* (San Francisco, 2004)
5. R. Avizienis, A. Freed, P. Kassakian, D. Wessel, A compact 120 independent element spherical loudspeaker array with programmable radiation patterns, in *120th AES Convention* (France, Paris, 2006)
6. F. Zotter, R. Höldrich, Modeling radiation synthesis with spherical loudspeaker arrays, in *Proceedings of the ICA* (Madrid, 2007), http://iaem.at/Members/zotter
7. F. Zotter, Analysis and synthesis of sound-radiation with spherical arrays, PhD. Thesis, University of Music and Performing Arts, Graz (2009)
8. H. Pomberger, Angular and radial directivity control for spherical loudspeaker arrays, M. Thesis, Institut für Elektronische Musik und Akustik, Kunstuni Graz, Technical University Graz, Graz, A (2008)
9. M. Pollow, G.K. Behler, Variable directivity for platonic sound sources based on spherical harmonics optimization. Acta Acust. United Acust. **95**(6), 1082–1092 (2009)
10. A.M. Pasqual, P. Herzog, J.R. Arruda, Theoretical and experimental analysis of the behavior of a compact spherical loudspeaker array for directivity control. J. Acoust. Soc. Am. **128**(6), 3478–3488 (2010)
11. A. Schmeder, An exploration of design parameters for human-interactive systems with compact spherical loudspeaker arrays, in *1st Ambisonics Symposium* (Graz, 2009)
12. G.K. Sharma, F. Zotter, M. Frank, Orchestrating wall reflections in space by icosahedral loudspeaker: findings from first artistic research exploration, in *Proceedings of the ICMC/SMC* (Athens, 2014), pp. 830–835
13. F. Zotter, M. Frank, A. Fuchs, D. Rudrich, Preliminary study on the perception of orientation-changing directional sound sources in rooms, in *Proceedings of the Forum Acusticum* (Kraków, 2014)

14. F. Wendt, G.K. Sharma, M. Frank, F. Zotter, R. Höldrich, Perception of spatial sound phenomena created by the icosahedral loudspeaker. Comput. Music J. **41**(1) (2017)
15. F. Zotter, M. Zaunschirm, M. Frank, M. Kronlachner, A beamformer to play with wall reflections: the icosahedral loudspeaker. Comput. Music J. **41**(3) (2017)
16. F. Wendt, F. Zotter, M. Frank, R. Höldrich, Auditory distance control using a variable-directivity loudspeaker. MDPI Appl. Sci. **7**(7) (2017)
17. F. Wendt, M. Frank, On the localization of auditory objects created by directional sound sources in a virtual room, in *VDT Tonmeistertagung* (Cologne, 2018)
18. F. Schultz, M. Zaunschirm, F. Zotter, Directivity and electro-acoustic measurements of the IKO, in *Audio Engineering Society Convention 144* (2018), http://www.aes.org/e-lib/browse.cfm?elib=19557
19. T. Deppisch, N. Meyer-Kahlen, F. Zotter, M. Frank, Surround with depth on first-order beam-controlling loudspeakers, in *Audio Engineering Society Convention 144* (2018), http://www.aes.org/e-lib/browse.cfm?elib=19494
20. M.-V. Laitinen, A. Politis, I. Huhtakallio, V. Pulkki, Controlling the perceived distance of an auditory object by manipulation of loudspeaker directivity. J. Acoust. Soc. Am. **137**(6) EL462–EL468 (2015)
21. F. Zotter, M. Frank, Investigation of auditory objects caused by directional sound sources in rooms. Acta Phys. Pol. A **128**(1-A) (2015), http://przyrbwn.icm.edu.pl/APP/PDF/128/a128z1ap01.pdf
22. F. Zagala, J. Linke, F. Zotter, M. Frank, Amplitude panning between beamforming-controlled direct and reflected sound, in *Audio Engineering Society Convention 142* (Berlin, 2017), http://www.aes.org/e-lib/browse.cfm?elib=18679
23. N. Misdariis, F. Nicolas, O. Warusfel, R. Caussee, Radiation control on multi-loudspeaker device : La timée," in *International Computer Music Conference* (La Habana, 2001)
24. N. Meyer-Kahlen, F. Zotter, K. Pollack, Design and measurement of a first-order, in *Audio Engineering Society Convention 144* (2018), http://www.aes.org/e-lib/browse.cfm?elib=19559
25. F. Zotter, H. Pomberger, A. Schmeder, Efficient directivity pattern control for spherical loudspeaker arrays, in *ACOUSTICS08* (2008)
26. F. Zotter, A. Schmeder, M. Noisternig, Crosstalk cancellation for spherical loudspeaker arrays, in *Fortschritte der Akustik - DAGA* (Dresden, 2008)
27. S.P. Lipshitz, J. Vanderkoy, In-phase crossover network design. J. Audio Eng. Soc. **34**(11), 889–894 (1986)
28. S. Lösler, F. Zotter, Comprehensive radial filter design for practical higher-order ambisonic recording, in *Fortschritte der Akustik – DAGA Nürnberg* (2015)
29. J. Linke, F. Wendt, F. Zotter, M. Frank, How masking affects auditory objects of beamformed sounds, in *Fortschritte der Akustik, DAGA* (Munich, 2018)
30. J. Linke, F. Wendt, F. Zotter, M. Frank, How the perception of moving sound beams is influenced by masking and reflector setup, in *VDT Tonmeistertagung* (Cologne, 2018)

Appendix

Many have written of the experience of mathematical beauty as being comparable to that derived from the greatest art.

S. Zeki, J.P. Romaya, D.M.T. Benincasa, M.F. Atiyah [1]
Frontiers in Human Neuroscience, Feb. 2014.

A.1 Harmonic Functions

The Laplacian is defined in the D-dimensional Cartesian space as

$$\triangle = \sum_{j=1}^{\mathrm{D}} \frac{\partial^2}{\partial x_j^2}.$$

The Laplacian eigenproblem

$$\triangle f = -\lambda f$$

is solved by harmonics, on a finite interval.

A.2 Laplacian in Orthogonal Coordinates

In general, coordinates can be expressed by n-tuples of values. For instance, the Cartesian coordinates are $(x_1, x_2, \dots)$ and coordinates of another coordinate systems are $(u_1, u_2, \dots)$, and both describe the location of a point in a space depending on a finite number of dimensions. Each location of the space should be accessible by both

F. Zotter and M. Frank, *Ambisonics*, Springer Topics in Signal Processing 19,
https://doi.org/10.1007/978-3-030-17207-7

coordinate systems and there should be a bijective mapping between both systems, e.g. $u_j = u_j(x_1, x_2, \dots)$. A single differentiation with regard to the component x_i, for instance, is described by the chain rule and consists of the sum of weighted partial differentials with regard to u_j:

$$\frac{\partial}{\partial x_i} = \sum_j \frac{\partial u_j}{\partial x_i} \frac{\partial}{\partial u_j}. \tag{A.1}$$

Written in terms of vectors, the (Cartesian) gradient $\nabla = \frac{\partial}{\partial \boldsymbol{x}}$ with $\frac{\partial}{\partial \boldsymbol{u}}$ yields:

$$\nabla = \frac{\partial \boldsymbol{u}^{\mathrm{T}}}{\partial \boldsymbol{x}} \frac{\partial}{\partial \boldsymbol{u}} := \boldsymbol{J}_{\partial u/\partial x} \frac{\partial}{\partial \boldsymbol{u}}, \tag{A.2}$$

for which the Jacobian matrix $\boldsymbol{J}_{\partial u/\partial x} = \left[\frac{\partial u_j}{\partial x_i}\right]_{ij}$ that is either written in dependency of $\boldsymbol{x}$ or $\boldsymbol{u}$ represents all the partial derivatives of the mapping between the coordinate systems. For bijective mappings also Jacobian of the inverse mapping exists $\boldsymbol{J}_{\partial x/\partial u} = \left[\frac{\partial x_i}{\partial u_j}\right]_{ji}$. The coordinate systems are equivalent if the determinant of the Jacobian is non-zero $|\boldsymbol{J}| \neq 0$.

Orthogonal coordinate systems have the interesting property that the rows of the Jacobian (or its columns) are orthogonal, so that $\boldsymbol{J}^{\mathrm{T}}\boldsymbol{J}$ yields a diagonal matrix. With $\boldsymbol{J}_{\partial x/\partial u} = \frac{\partial \boldsymbol{x}^{\mathrm{T}}}{\partial \boldsymbol{u}}$ the meaning of this property becomes easier to understand: the differential changes in the location $\partial \boldsymbol{x}/\partial u_j$ of each Cartesian coordinate into the direction of each individual non-Cartesian coordinate u_j describes an orthogonal set of motion directions in space, whose orientation depends on the location and whose individual lengths may vary.

To obtain a description of the Laplăcian in the Helmholtz equation $\triangle = \sum_i \frac{\partial^2}{\partial x_i^2}$ is our goal here, and it can be obtained with the chain rule, now calculating from x_i to u_j,

$$\begin{aligned}\triangle &= \sum_i \frac{\partial}{\partial x_i}\left(\frac{\partial}{\partial x_i}\right) = \sum_i \frac{\partial}{\partial x_i}\left(\sum_j \frac{\partial u_j}{\partial x_i} \frac{\partial}{\partial u_j}\right) \\ &= \sum_{i,j} \frac{\partial^2 u_j}{\partial x_i^2} \frac{\partial}{\partial u_j} + \sum_{i,j,k} \frac{\partial u_j}{\partial x_i} \frac{\partial u_k}{\partial x_i} \frac{\partial^2}{\partial u_j \partial u_k}, \\ \text{with} \sum_{i,j,k} \frac{\partial u_j}{\partial x_i} \frac{\partial u_k}{\partial x_i} \frac{\partial^2}{\partial u_j \partial u_k} &= \boldsymbol{1}^{\mathrm{T}} \left[\underbrace{(\boldsymbol{J}^{\mathrm{T}}\boldsymbol{J})}_{\text{ortho:diag}} \circ \left(\frac{\partial}{\partial \boldsymbol{u}^{\mathrm{T}}} \frac{\partial}{\partial \boldsymbol{u}}\right)\right] = \sum_{i,j} \left(\frac{\partial u_j}{\partial x_i}\right)^2 \frac{\partial^2}{\partial u_j^2},\end{aligned} \tag{A.3}$$

with $\circ$ denoting the element-wise, i.e. Hadamard product. Orthogonal coordinates largely simplify the Laplacian (see last line) and make it consist of first- and second-order differentials with regard to the new coordinates, individually, with all mixed derivatives canceling. Both first- and second-order differentials are weighted by the partial derivatives of the coordinate mapping. For each u_j, the Laplacian is composed of those two expressions

$$\triangle = \sum_j \triangle_{u_j}, \quad \text{where } \triangle_{u_j} = \left[\sum_i \frac{\partial^2 u_j}{\partial x_i^2}\right] \frac{\partial}{\partial u_j} + \left[\sum_i \left(\frac{\partial u_j}{\partial x_i}\right)^2\right] \frac{\partial^2}{\partial u_j^2}. \tag{A.4}$$

A.3 Laplacian in Spherical Coordinates

The right-handed spherical coordinate systems in ISO31-11, ISO80000-2, [2, 3], uses a radius r, an azimuth angle φ, and a zenigh angle ϑ, mapping to Cartesian coordinates $x = r\,\cos\varphi\,\sin\vartheta$, $y = r\,\sin\varphi\,\sin\vartheta$, $z = r\,\cos\vartheta$, or inversely $r = \sqrt{x^2 + y^2 + z^2}$, $\varphi = \arctan\frac{y}{x}$, $\vartheta = \arctan\frac{\sqrt{x^2+y^2}}{z}$, see Fig. 4.11.

Re-expressing the zenith angle coordinate by $\zeta = \cos\vartheta = \frac{z}{r}$ reduces the effort in calculation and yields $x = r\,\cos\varphi\,\sqrt{1-\zeta^2}$, $y = r\,\sin\varphi\,\sqrt{1-\zeta^2}$, $z = r\,\zeta$.

In order to obtain solutions along the angular dimensions azimuth and zenith, we first need to re-write the Laplacian from Cartesian to spherical coordinates. For first-order derivative along the x axis, we get the generalized differential

$$\frac{\partial}{\partial x} = \left[\frac{\partial r}{\partial x}\right] \frac{\partial}{\partial r} + \left[\frac{\partial \varphi}{\partial x}\right] \frac{\partial}{\partial \varphi} + \left[\frac{\partial \zeta}{\partial x}\right] \frac{\partial}{\partial \zeta}.$$

As the Cartesian and spherical coordinates are orthogonal, therefore any mixed second-order derivatives in Cartesian or spherical coordinates vanish. We may derive a second time wrt. x:

$$\begin{aligned} \frac{\partial}{\partial x}\left[\frac{\partial}{\partial x}\right] = &\left[\frac{\partial^2 r}{\partial x^2} + \left(\frac{\partial r}{\partial x}\right)^2 \frac{\partial}{\partial r}\right] \frac{\partial}{\partial r} \\ &+ \left[\frac{\partial^2 \varphi}{\partial x^2} + \left(\frac{\partial \varphi}{\partial x}\right)^2 \frac{\partial}{\partial \varphi}\right] \frac{\partial}{\partial \varphi} + \left[\frac{\partial^2 \zeta}{\partial x^2} + \left(\frac{\partial \zeta}{\partial x}\right)^2 \frac{\partial}{\partial \zeta}\right] \frac{\partial}{\partial \zeta}. \end{aligned}$$

Obviously, we require all first-order derivatives squared, and all second-order derivatives of the spherical coordinates.

A.3.1 *The Radial Part*

With $r = \sqrt{x^2 + y^2 + z^2}$ we obtain for the radial part $\triangle_\mathrm{r} = \left[\frac{\partial^2 r}{\partial x^2} + \left(\frac{\partial r}{\partial x}\right)^2 \frac{\partial}{\partial r}\right] \frac{\partial}{\partial r}$ of the Laplacian

$$\left[\frac{\partial r}{\partial x}\right]^2=\left[\frac{\partial\sqrt{x^2+y^2+z^2}}{\partial x}\right]^2=\left[\frac{1}{2\sqrt{x^2+y^2+z^2}}2x\right]^2=\left[\frac{x}{r}\right]^2=\frac{x^2}{r^2},$$

$$\frac{\partial^2 r}{\partial x^2}=\frac{\partial}{\partial x}\left[\frac{x}{r}\right]=\frac{1}{r}\frac{\partial x}{\partial x}+x\frac{\partial r}{\partial x}\frac{\partial}{\partial r}\frac{1}{r}=\frac{1}{r}-x\frac{x}{r}\frac{1}{r^2}=\frac{r^2-x^2}{r^3}.$$

For x, y, z altogether, this is for:

$$\Delta_{\mathrm{r}}=\left[\frac{3r^2-x^2-y^2-z^2}{r^3}\right]\frac{\partial}{\partial r}+\left[\frac{x^2}{r^2}+\frac{y^2}{r^2}+\frac{z^2}{r^2}\right]\frac{\partial^2}{\partial r^2}=\frac{2}{r}\frac{\partial}{\partial r}+\frac{\partial^2}{\partial r^2}. \quad \text{(A.5)}$$

2D. In two dimensions, there is no z coordinate, therefore there is just one term fewer:

$$\Delta_{\mathrm{r,2D}}=\left[\frac{2r^2-x^2-y^2}{r^3}\right]\frac{\partial}{\partial r}+\left[\frac{x^2}{r^2}+\frac{y^2}{r^2}\right]\frac{\partial^2}{\partial r^2}=\frac{1}{r}\frac{\partial}{\partial r}+\frac{\partial^2}{\partial r^2}. \quad \text{(A.6)}$$

A.3.2 The Azimuthal Part

With $\varphi=\arctan\frac{y}{x}$ and $\frac{\mathrm{d}}{\mathrm{d}x}\arctan x=\frac{1}{1+x^2}$, the azimuthal part $\Delta_\varphi=\left[\frac{\partial^2\varphi}{\partial x^2}+\left(\frac{\partial\varphi}{\partial x}\right)^2\frac{\partial}{\partial\varphi}\right]\frac{\partial}{\partial\varphi}$ of the Laplacian becomes

$$\left[\frac{\partial\varphi}{\partial x}\right]^2=\left[\frac{\partial\arctan\frac{y}{x}}{\partial x}\right]^2=\left[\frac{1}{1+\frac{y^2}{x^2}}\frac{\partial}{\partial x}\frac{y}{x}\right]^2=\left[-\frac{x^2}{x^2+y^2}\frac{y}{x^2}\right]^2=\left[-\frac{y}{r_{\mathrm{xy}}^2}\right]^2=\frac{y^2}{r_{\mathrm{xy}}^4},$$

$$\left[\frac{\partial\varphi}{\partial y}\right]^2=\left[\frac{\partial\arctan\frac{y}{x}}{\partial y}\right]^2=\left[\frac{x^2}{x^2+y^2}\frac{\partial}{\partial y}\frac{y}{x}\right]^2=\left[\frac{x^2}{x^2+y^2}\frac{1}{x}\right]^2=\left[\frac{x}{r_{\mathrm{xy}}^2}\right]^2=\frac{x^2}{r_{\mathrm{xy}}^4},$$

$$\frac{\partial^2\varphi}{\partial x^2}=\frac{\partial}{\partial x}\left[-\frac{y}{r_{\mathrm{xy}}^2}\right]=\frac{2xy}{r_{\mathrm{xy}}^3},$$

$$\frac{\partial^2\varphi}{\partial y^2}=\frac{\partial}{\partial y}\left[\frac{x}{r_{\mathrm{xy}}^2}\right]=-\frac{2xy}{r_{\mathrm{xy}}^3},\qquad\frac{\partial\varphi}{\partial z}=0.$$

It only depends on x and y, and altogether, we obtain

$$\Delta_\varphi=\left[\frac{2xy-2xy}{r_{\mathrm{xy}}^3}\right]\frac{\partial}{\partial\varphi}+\left[\frac{x^2+y^2}{r_{\mathrm{xy}}^4}\right]\frac{\partial^2}{\partial\varphi^2}=\frac{1}{r_{\mathrm{xy}}^2}\frac{\partial^2}{\partial\varphi^2}=\frac{1}{r^2(1-\zeta^2)}\frac{\partial^2}{\partial\varphi^2}. \quad \text{(A.7)}$$

2D. In two dimensions, $\zeta=0$, therefore

$$\Delta_{\varphi,2D} = \frac{1}{r^2}\frac{\partial^2}{\partial \varphi^2}. \tag{A.8}$$

A.3.3 The Zenithal Part

The zenith angle is actually ϑ, and we define $\zeta = \cos\vartheta$ as a variable to express it in order to simplify the derivation. With $\zeta = \frac{z}{\sqrt{x^2+y^2+z^2}} = \frac{z}{r}$, the zenithal part $\Delta_\zeta = \left[\frac{\partial^2\zeta}{\partial x^2} + \left(\frac{\partial\zeta}{\partial x}\right)^2 \frac{\partial}{\partial\zeta}\right]\frac{\partial}{\partial\zeta}$ becomes

$$\begin{aligned}
\left[\frac{\partial\zeta}{\partial x}\right]^2 &= \left[\frac{\partial}{\partial x}\frac{z}{r}\right]^2 = \left[-\frac{z}{r^2}\frac{x}{r}\right]^2 = \left[-\frac{xz}{r^3}\right]^2 = \frac{x^2z^2}{r^6},\\
\left[\frac{\partial\zeta}{\partial z}\right]^2 &= \left[\frac{\partial}{\partial z}\frac{z}{r}\right]^2 = \left[\frac{1}{r} - \frac{z}{r^2}\frac{z}{r}\right]^2 = \left[\frac{r^2 - z^2}{r^3}\right]^2 = \left[\frac{r_{\mathrm{xy}}^2}{r^3}\right]^2 = \frac{r_{\mathrm{xy}}^4}{r^6},\\
\frac{\partial^2\zeta}{\partial x^2} &= \frac{\partial}{\partial x}\left[-\frac{xz}{r^3}\right] = -\frac{z}{r^3} + 3xz\frac{1}{r^4}\frac{x}{r} = z\frac{3x^2 - r^2}{r^5},\\
\frac{\partial^2\zeta}{\partial z^2} &= \frac{\partial}{\partial z}\left[\frac{r_{\mathrm{xy}}^2}{r^3}\right] = -3\frac{r_{\mathrm{xy}}^2}{r^4}\frac{z}{r} = -z\frac{3r_{\mathrm{xy}}^2}{r^5}.
\end{aligned}$$

For x, y, and z altogether, we get

$$\begin{aligned}
\Delta_\zeta &= z\frac{3x^2+3y^2-2r^2-3r_{\mathrm{xy}}^2}{r^5}\frac{\partial}{\partial\zeta} + \frac{(x^2+y^2)z^2 + r_{\mathrm{xy}}^4}{r^6}\frac{\partial^2}{\partial\zeta^2} = -\frac{2r^2}{r^5}\frac{\partial}{\partial\zeta} + \frac{r^2 r_{\mathrm{xy}}^2}{r^6}\frac{\partial^2}{\partial\zeta^2}\\
&= -z\frac{2}{r^3}\frac{\partial}{\partial\zeta} + \frac{r^2(1-\zeta^2)}{r^4}\frac{\partial^2}{\partial\zeta^2} = -\frac{2}{r^2}\zeta\frac{\partial}{\partial\zeta} + \frac{1-\zeta^2}{r^2}\frac{\partial^2}{\partial\zeta^2}.
\end{aligned} \tag{A.9}$$

2D. This part does not exist in 2D.

A.3.4 Azimuthal Solution in 2D and 3D

The azimuth harmonics are found by solving $\Delta_\varphi \Phi = -\lambda r_{\mathrm{xy}}^2\,\Phi$

$$\frac{\mathrm{d}^2}{\mathrm{d}\varphi^2}\Phi = -\lambda\, r_{\mathrm{xy}}^2\,\Phi. \tag{A.10}$$

We know that $\cos'' x = -\cos x$ and $\sin'' x = -\sin x$, therefore we can insert the solutions

$$\hat{\Phi} = \begin{cases} \cos(a\varphi), & \text{for}\, a \geq 0, \\ \sin(|a|\varphi), & \text{for}\, a < 0, \end{cases}$$

and obtain with $\frac{\mathrm{d}^2}{\mathrm{d}\varphi^2}\hat{\Phi} = -a^2\hat{\Phi}$ the characteristic equation that fixes a

$$-a^2 = -\lambda\, r_{\mathrm{xy}}^2.$$

Geometrically, we desire that $\hat{\Phi}(\varphi) = \hat{\Phi}(\varphi + 2\pi\, l)$ with $l \in \mathbb{Z}$. This is only possible with $\lambda\, r_{\mathrm{xy}}^2 = m^2$, and $m \in \mathbb{Z}$.

We can therefore define for $-\infty \leq m \leq \infty$ the terms of a normalized Fourier series

$$\Phi_m(\varphi) = \frac{1}{\sqrt{2\pi}} \begin{cases} \sqrt{2}\sin(|m|\varphi), & \text{for}\, m < 0, \\ 1, & \text{for}\, m = 0, \\ \sqrt{2}\cos(m\varphi), & \text{for}\, m > 0. \end{cases} \tag{A.11}$$

The azimuth harmonics are orthogonal: none of the products $\cos(i\varphi)\sin(j\varphi)$, $\cos(i\varphi)\cos(j\varphi)$, or $\sin(i\varphi)\sin(j\varphi)$ produces a constant component unless $i = j$, excluding the mixed cosine and sine product. Normalization ensures that the non-zero result is unity

$$\int_0^{2\pi} \Phi_i\, \Phi_j\, \mathrm{d}\varphi = \delta_{ij} = \begin{cases} 1, & \text{for}\, i = j, \\ 0, & \text{else}, \end{cases} \tag{A.12}$$

by $\int_0^{2\pi} \frac{1}{\sqrt{2\pi}^2}\mathrm{d}\varphi = \int_0^{2\pi} \frac{\cos^2 m\varphi}{\sqrt{\pi}^2}\,\mathrm{d}\varphi = \int_0^{2\pi} \frac{\sin^2 m\varphi}{\sqrt{\pi}^2}\,\mathrm{d}\varphi = 1$.

2D functions. With unbounded $|m| \to \infty$, the circular harmonics are complete in the Hilbert space of square-integrable circular polynomials. By their orthonormality, we can derive a transformation integral of a function $g(\varphi)$ that should be represented as series

$$g = \sum_{j=-\infty}^{\infty} \gamma_j\, \Phi_j \tag{A.13}$$

by integration of g over $\Phi_m \int \mathrm{d}\varphi$

$$\int_{-\pi}^{\pi} g(\varphi)\, \Phi_m\, \mathrm{d}\varphi = \sum_{m=-\infty}^{\infty} \gamma_m \underbrace{\int_{-\pi}^{\pi} \Phi_m \Phi_j\, \mathrm{d}\varphi}_{=\delta_{mj}} = \gamma_m. \tag{A.14}$$

2D panning functions. To decompose an infinitely narrow unit-surface Dirac delta function that represents an infinite-order panning function towards the direction φ_{s},

$$\delta(\varphi - \varphi_s) = \begin{cases} \lim_{\varepsilon\to 0}\frac{1}{2\varepsilon}, & \text{for } |\varphi - \varphi_s| \le \varepsilon, \\ 0 & \text{otherwise,} \end{cases} \tag{A.15}$$

we the obtain the coefficients from the transformation integral

$$\begin{aligned} \gamma_m &= \int_{-\pi}^{\pi} \delta(\varphi - \varphi_s)\, \Phi_m \,\mathrm{d}\varphi = \Phi_m(\varphi_s) \lim_{\varepsilon\to 0} \int_{-\varepsilon}^{\varepsilon} \frac{1}{2\varepsilon} \mathrm{d}\varphi \\ &= \Phi_m(\varphi_s). \end{aligned} \tag{A.16}$$

Typically, a finite-order series will be employed as a panning function

$$g_\mathrm{N} = \sum_{m=-\mathrm{N}}^{\mathrm{N}} a_m\, \Phi_m(\varphi_s)\Phi_m(\varphi), \tag{A.17}$$

involving a weight $a_m = a_{|m|}$ controlling the side lobes. To evaluate the E measure of loudness, we can write

$$\begin{aligned} E &= \int_{-\pi}^{\pi} g_\mathrm{N}^2 \,\mathrm{d}\varphi = \int_{-\pi}^{\pi} \left[\sum_{i=0}^{\mathrm{N}} a_i\, \Phi_i\right] \left[\sum_{j=0}^{\mathrm{N}} a_j\, \Phi_j\right] \mathrm{d}\varphi = \sum_{i,j} a_i\, a_j \int_{-\pi}^{\pi} \Phi_i \Phi_j \mathrm{d}\varphi \\ &= \sum_{i=0}^{\mathrm{N}} \frac{2-\delta_i}{2\pi}\, a_i^2. \end{aligned} \tag{A.18}$$

For $\varphi_s = 0$, we obtain an axisymmetric function in terms of a pure cosine series, as $\sin 0 = 0$,

$$g_\mathrm{N}(\varphi) = \sum_{m=0}^{\mathrm{N}} a_m\, \frac{2-\delta_m}{2\pi} \cos(m\varphi), \tag{A.19}$$

with $\delta_m = 1$ for 1 for $m = 0$ and 0 elsewhere. The axisymmetric panning function is easier to design.

2D max-$\boldsymbol{r}_\mathrm{E}$. For the narrowest-possible spread, we maximize the length of $\boldsymbol{r}_\mathrm{E}$,

$$\begin{aligned} r_\mathrm{E} &= \frac{\int_{-\pi}^{\pi} g_\mathrm{N}^2 \cos\varphi \,\mathrm{d}\varphi}{\int_{-\pi}^{\pi} g_\mathrm{N}^2 \,\mathrm{d}\varphi} = \frac{\int \sum_{i,j=0}^{\mathrm{N}} a_i a_j (2-\delta_i)(2-\delta_j) \cos(i\varphi)\cos(j\varphi)\, \cos(\varphi)\,\mathrm{d}\varphi}{(2\pi)^2 E} \\ &= \frac{\sum_{i=1}^{\mathrm{N}} a_i a_{i-1}}{\pi E} =: \frac{\hat{r}_\mathrm{E}}{E} \end{aligned} \tag{A.20}$$

where we used $\cos(i\varphi)\cos(\varphi) = \frac{\cos[(i+1)\varphi] + \cos[(i-1)\varphi]}{2-\delta_i}$, inserted the orthogonality of the cosine $\int_{-\pi}^{\pi} \frac{(2-\delta_i)\cos(i\varphi)\cos(j\varphi)}{2\pi} \mathrm{d}\varphi = \delta_{ij}$, and combined $\sum(a_i a_{i-1} + a_i a_{i+1}) = 2\sum a_i a_{i-1}$. To maximize, we zero the derivative to a_m

$$r_{\mathrm{E}}' = \frac{\hat{r}_{\mathrm{E}}'}{E} - \frac{\hat{r}_{\mathrm{E}}}{E^2}E' = \frac{1}{E}\left[\hat{r}_{\mathrm{E}}' - E'\, r_{\mathrm{E}}\right] = 0$$
$$= a_{m-1} + a_{m+1} - (2 - \delta_m) a_m\, r_{\mathrm{E}} = 0.$$

If we assume that $a_m = \cos(m\alpha)$, and we insert this for $\frac{a_{m+1}+a_{m-1}}{2-\delta_m} = a_m\, r_{\mathrm{E}}$, we recognize by inserting the above theorem $\cos(m\alpha)\cos(\alpha) = \frac{\cos[(m+1)\alpha]+\cos[(m-1)\alpha]}{2-\delta_m}$

$$\frac{\cos[(m+1)\alpha] + \cos[(m-1)\alpha]}{2 - \delta_m} = \cos(m\alpha)\, r_{\mathrm{E}} = \cos(m\alpha)\cos(\alpha)$$

that $r_{\mathrm{E}} = \cos\alpha$. And to maximize r_{E} by constraining that $a_{\mathrm{N}+1} = 0$, we get the smallest-possible spread $\alpha = \pm\frac{\pi}{2}\frac{1}{\mathrm{N}+1} = \pm\frac{90^\circ}{\mathrm{N}+1}$

$$a_m = \begin{cases} \cos\left(\frac{\pi}{2}\frac{m}{\mathrm{N}+1}\right), & \text{for } 0 \le m \le \mathrm{N}, \\ 0, & \text{elsewhere.} \end{cases} \tag{A.21}$$

The max-$\boldsymbol{r}_{\mathrm{E}}$ panning function in 2D consequently is

$$g_{\mathrm{N}}(\varphi) = \sum_{m=-\mathrm{N}}^{\mathrm{N}} a_m\, \Phi_m(\varphi_{\mathrm{s}})\, \Phi_m(\varphi) = \sum_{m=-\mathrm{N}}^{\mathrm{N}} \cos\left(\frac{\pi}{2}\frac{m}{\mathrm{N}+1}\right)\, \Phi_m(\varphi_{\mathrm{s}})\, \Phi_m(\varphi). \tag{A.22}$$

A.3.5 Towards Spherical Harmonics (3D)

The spherical harmonics are harmonics depending only on angular terms. We may superimpose both parts $\Delta_{\varphi\zeta} = \Delta_\varphi + \Delta_\zeta$ of the Laplacian and solve the eigenproblem $r^2\, \Delta_{\varphi\zeta}\, Y = -\lambda Y$

$$\frac{1}{1-\zeta^2}\frac{\partial^2}{\partial\varphi^2}Y - 2\zeta\frac{\partial}{\partial\zeta}Y + (1-\zeta^2)\frac{\partial^2}{\partial\zeta^2}Y = -\lambda Y.$$

We assume Y to be a product of the azimuth harmonics $\Phi_m(\varphi)$ from above and undefined zenith harmonics $\Theta(\zeta)$

$$Y = \Phi_m\, \Theta, \tag{A.23}$$

which yields a differential equation ($\partial \to \mathrm{d}$) only in ζ after inserting $\frac{\mathrm{d}^2}{\mathrm{d}^2\varphi}\Phi_m = -m^2\Phi_m$

$$\Theta\frac{-m^2}{1-\zeta^2}\Phi_m - 2\zeta\,\Phi_m\frac{\mathrm{d}}{\mathrm{d}\zeta}\Theta + (1-\zeta^2)\Phi_m\frac{\mathrm{d}^2}{\mathrm{d}\zeta^2}\Theta = -\lambda\Phi_m\Theta.$$

And after dividing by Φ_m, we obtain the *associated Legendre differential equation*

$$\frac{-m^2}{1-\zeta^2}\Theta - 2\zeta\frac{\mathrm{d}}{\mathrm{d}\zeta}\Theta + (1-\zeta^2)\frac{\mathrm{d}^2}{\mathrm{d}\zeta^2}\Theta = -\lambda\Theta,$$

$$\left[(1-\zeta^2)\frac{\mathrm{d}^2}{\mathrm{d}\zeta^2} - 2\zeta\frac{\mathrm{d}}{\mathrm{d}\zeta} + \lambda - \frac{m^2}{1-\zeta^2}\right]\Theta = 0. \tag{A.24}$$

A.3.6 Zenithal Solution: Associated Legendre Differential Equation

The associated Legendre differential equation (written in x and y for mathematical simplicity) is

$$(1-x^2)y'' - 2xy' + \left[\lambda - \frac{m^2}{1-x^2}\right]y = 0,$$

or after gathering the derivatives

$$\left[(1-x^2)y'\right]' + \left[\lambda - \frac{m^2}{1-x^2}\right]y = 0.$$

Simplifying the differential equation by $\frac{1}{1-x^2}$. In the associated Legendre differential equation, we would like to get rid of the denominator $\frac{1}{1-x^2}$. In this case, it is typical to substitute $y = (1-x^2)^\alpha v$ and try out which α succeeds. For insertion into the differential equation, the derivative of y is

$$y' = -\alpha(1-x^2)^{\alpha-1}2x\,v + (1-x^2)^\alpha v' = -2\alpha(1-x^2)^{\alpha-1}\,x\,v + (1-x^2)^\alpha\,v',$$

and the second-order derivative term is

$$\begin{aligned}
[(1-x^2)y']' &= \left[-2\alpha(1-x^2)^\alpha\,x\,v + (1-x^2)^{\alpha+1}\,v'\right]' \\
&= 4\alpha^2(1-x^2)^{\alpha-1}\,x^2\,v - 2\alpha(1-x^2)^\alpha\,v - 2\alpha(1-x^2)^\alpha\,x\,v' \\
&\qquad - 2(\alpha+1)(1-x^2)^\alpha\,x\,v' + (1-x^2)^{\alpha+1}\,v'' \\
&= (1-x^2)^\alpha\left[\frac{4\alpha^2}{1-x^2}x^2\,v - 2\alpha\,v - 2(2\alpha+1)\,x\,v' + (1-x^2)\,v''\right].
\end{aligned}$$

Together with the term $\left[\lambda - \frac{m^2}{1-x^2}\right]y$, the associated Legendre differential equation becomes

$$(1-x^2)^\alpha\left[\frac{4\alpha^2}{1-x^2}x^2\,v-2\alpha\,v+\left(\lambda-\frac{m^2}{1-x^2}\right)v-2(2\alpha+1)\,x\,v'+(1-x^2)\,v''\right]=0$$

$$-m^2\frac{1-\frac{4\alpha^2}{m^2}x^2}{1-x^2}v+(\lambda-2\alpha)\,v-2(2\alpha+1)\,x\,v'+(1-x^2)\,v''=0.$$

We see that the term $\frac{1}{1-x^2}$ entirely cancels by $\alpha=\frac{m}{2}$, which fixes the substitution

$$y=\sqrt{1-x^2}^m\,v. \tag{A.25}$$

Note that for rotational symmetric solutions around the Cartesian z coordinate, the choice of $m=0$ would ensure a constant azimuthal part $\Phi_m=\text{const}$. Reinserting $x=\zeta=\cos\vartheta$, the preceding term $\sqrt{1-\cos^2\vartheta}^m=\sin^m\vartheta$ is understandably required to represent shapes that aren't rotationally symmetric around z, but any other, freely rotated axis, for which we also required the sinusoids in 2D. The differential equation for $v=v(\cos\vartheta)$ is

$$(1-x^2)\,v''-2(m+1)\,x\,v'+[\lambda-m(m+1)]\,v=0. \tag{A.26}$$

Still, the above equation is singular at $x\pm 1$, which means that the second-derivative term multiplied by $(1-x^2)$ vanishes there, rendering the differential equation into a first-order differential equation, locally. Instead of the more comprehensive Frobenius method we keep it simple: Desired spherical polynomials $Y_n^m=\Phi_m\Theta_n^m=\mathcal{P}_n(\theta_\mathrm{x},\theta_\mathrm{y},\theta_\mathrm{z})$ with $\Phi_m(\varphi)\propto\frac{1}{\sqrt{1-\theta_\mathrm{z}^2}^m}\begin{pmatrix}\theta_\mathrm{x} & -\theta_\mathrm{y}\\ \theta_\mathrm{y} & \theta_\mathrm{x}\end{pmatrix}^{m-1}\begin{pmatrix}\theta_\mathrm{x}\\ \theta_\mathrm{y}\end{pmatrix}=\frac{\mathcal{P}_m(\theta_\mathrm{x},\theta_\mathrm{y})}{\sqrt{1-\theta_\mathrm{z}^2}^m}$ imply that Θ_n^m must contain $\sqrt{1-\theta_\mathrm{z}^2}^m\mathcal{P}_{n-m}(\theta_\mathrm{z})$ to be polynomial and nth-order: in condensed notation this is $y=\sqrt{1-x^2}^m\sum_{k=0}^{n-m}a_k\,x^k$, see also [4].

Power-series for v. With $v=\sum_{k=0}^{\infty}a_k\,x^k$, we get after inserting and deriving

$$(1-x^2)\sum_{k=2}^{\infty}k(k-1)\,a_k\,x^{k-2}-2(m+1)x\sum_{k=1}^{\infty}k\,a_k\,x^{k-1}$$
$$+\left[\lambda-m(m+1)\right]\sum_{k=0}^{\infty}a_k\,x^k=0,$$

$$\sum_{k=2}^{\infty}k(k-1)\,a_k\,x^{k-2}-\sum_{k=2}^{\infty}k(k-1)\,a_k\,x^k-2(m{+}1)\sum_{k=1}^{\infty}k\,a_k\,x^k$$
$$+\left[\lambda-m(m+1)\right]\sum_{k=0}^{\infty}a_k\,x^k=0.$$

For $k\geq 2$, all sum terms are present and the comparison of coefficients for the kth power yields:

$$(k+1)(k+2)\,a_{k+2} = \left[k(k-1) + 2(m+1)k - \left[\lambda - m(m+1)\right]\right] a_k$$
$$a_{k+2} = \frac{k(k+2m+1) + m(m+1) - \lambda}{(k+1)(k+2)}\, a_k.$$

Typically for such a two-step recurrence, two starting conditions $a_0 = 1, a_1 = 0$ and $a_0 = 0, a_1 = 1$ yield a pair of linearly independent solutions (even and odd).

If the series in x should converge, it will most certainly do so when v is *polynomial* and stops at some order. To design y to be of some arbitrary finite order $n \in \mathbb{Z}$, we take into account that $\sqrt{1-x^2}^m$ is of mth order already, so the polynomial v must be $(n-m)$th order, and $|m| \leq n$. The series is forced to stop the coefficient a_k for $k = n-m$ if the numerator is forced to become zero by a suitably chosen λ, thus $\lambda = (n-m)(n+m+1) + m(m+1) = n(n+1)$. Corresponding to the termination either at an even or odd $k = n-m$, even $a_0 = 1, a_1 = 0$ or odd $a_0 = 0, a_1 = 1$ starting conditions must be chosen. The otherwise wrong-parity solution is an infinite series [5, Eq. 3.2.45] whose convergence radius R indicates singularities at $x = \pm 1$,

$$R = \lim_{k\to\infty} \frac{a_k}{a_{k+2}} = \lim_{k\to\infty} \frac{(k+1)(k+2)}{k(k+2m+1) - m(m+1) - n(n+1)} = \lim_{k\to\infty} \frac{k^2 + \cdots}{k^2 + \cdots} = 1. \tag{A.27}$$

Using $\lambda = n(n+1)$ and writing the differentials in condensed form, the defining differential equations for associated Legendre functions P_n^m (m is no exponent but a second index) and their polynomial part v_n^m become

$$\frac{\mathrm{d}}{\mathrm{d}x}\left[(1-x^2)\frac{\mathrm{d}}{\mathrm{d}x}P_n^m\right] + \left[n(n+1) - \frac{m(m+1)}{1-x^2}\right]P_n^m = 0, \tag{A.28}$$
$$(1-x^2)^{-m}\frac{\mathrm{d}}{\mathrm{d}x}\left[(1-x^2)^{m+1}\frac{\mathrm{d}}{\mathrm{d}x}v_n^m\right] + \left[n(n+1) - m(m+1)\right]v_n^m = 0. \tag{A.29}$$

Orthogonality of associated Legendre functions. The resulting associated Legendre differential equation

$$\left[(1-x^2)\left[P_n^m\right]'\right]' + \left[n(n+1) - \frac{m^2}{1-x^2}\right]P_n^m = 0,$$

yields a sequence of finite-order functions P_n^m with the order $n \in \mathbb{N}_0$ and $|m| \leq n$. Before even defining these functions, we can prove their orthogonality $\int_{-1}^{1} P_n^m P_l^m \,\mathrm{d}x = 0$ for $n \neq l$. This means no product of a pair of associated Legendre functions of different indices $n \neq l$ produces any constant part on $x \in [-1; 1]$, and P_n^m and P_l^m do not contain shapes of the respective other function. This is important to uniquely decompose shapes and to define transformation integrals. We multiply the differential equation with P_l^m and integrate it over x

$$\int_{-1}^{1}\left[(1-x^2)[P_n^m]'\right]' P_l^m \,\mathrm{d}x + \int_{-1}^{1}\left[n(n+1) - \frac{m^2}{1-x^2}\right] P_n^m P_l^m \,\mathrm{d}x = 0.$$

Integration by parts of the first integral yields

$$\int_{-1}^{1}\left[(1-x^2)[P_n^m]'\right]' P_l^m \,\mathrm{d}x = \underbrace{(1-x^2)[P_n^m]' P_l^m\Big|_{-1}^{1}}_{=0} - \int_{-1}^{1}(1-x^2)[P_n^m]'[P_l^m]' \,\mathrm{d}x, \tag{A.30}$$

where the vanishing part is because of $(1-x^2)=0$ at the endpoints $x=\pm1$ where $[P_n^m]'$ and P_l^m are finite. We get

$$\int_{-1}^{1}(1-x^2)[P_n^m]'[P_l^m]' \,\mathrm{d}x = \int_{-1}^{1}\left[n(n+1) - \frac{m^2}{1-x^2}\right] P_n^m P_l^m \,\mathrm{d}x.$$

We could have arrived at an alternative expression, with the only difference in $l(l+1)$ instead of $n(n+1)$,

$$\int_{-1}^{1}(1-x^2)[P_l^m]'[P_n^m]' \,\mathrm{d}x = \int_{-1}^{1}\left[l(l+1) - \frac{m^2}{1-x^2}\right] P_l^m P_n^m \,\mathrm{d}x,$$

if we had started integrating the differential equation of P_l^m over P_n^m, instead. The difference of both equations is

$$\left[n(n+1) - l(l+1)\right]\int_{-1}^{1} P_n^m P_l^m \,\mathrm{d}x = 0,$$

and the scalar in brackets only vanishes for $n=l$. For the equation to hold at other $n \neq l$, we conclude that the associated Legendre functions $\int_{-1}^{1} P_n^m P_l^m \mathrm{d}x = 0$ must be orthogonal. (Orthogonality needs not hold for different m, as Φ_m achieves this orthogonality in azimuth.)

Solving for polynomial part of associated Legendre functions. To solve the differential equation for the polynomial part v_n^m in a way to arrive at the elegant Rodrigues formula, we first play with a test function

$$u_n = (1-x^2)^n, \quad \text{differentiated} \quad u_n' = -2n\,x\,(1-x^2)^{n-1} = -2n\,(1-x^2)^{-1}\,x\,u_n.$$

We may write its derivative as differential equation

$$(1-x^2)u_n' + 2n\,x\,u_n = 0$$

and derive it l times by the Leibniz rule $(f\,g)^{(n)} = \sum_{k=0}^{n} \binom{n}{k} f^{(k)}\, g^{(n-k)}$ for repeated differentiation of products, with the binomial coefficient $\binom{n}{k} = \frac{n!}{k!(n-k)!}$ and $f^{(k)} = \frac{\mathrm{d}^k f}{\mathrm{d}x^k}$ for simplicity. The few non-zero derivatives of x and $(1-x^2)$ simplify differentiation, $x' = 1$, $[(1-x^2)' = -2x]' = -2$,

$$(1-x^2)u_n^{(l+1)} + l\,(-2x)\,u_n^{(l)} + \frac{(l-1)l}{2}(-2)\,u_n^{(l-1)} \quad + 2n\,x\,u_n^{(l)} + 2n\,l\,u_n^{(l-1)} = 0,$$
$$(1-x^2)\,u_n^{(l+1)} - 2(l-n)\,x\,u_n^{(l)} + l(2n-l+1)\,u_n^{(l-1)} = 0.$$

This equation matches $(1-x^2)\,v_n''^m - 2(m+1)\,x\,v_n'^m + [n(n+1) - m(m+1)]\, v_n^m = 0$ by matching the coefficients $l - n = m + 1$, hence $l = m + n + 1$, which nicely implies $l(2n - l + 1) = n(n+1) - m(m+1)$,

$$(1-x^2)\,u_n^{(m+n+2)} - 2(m+1)\,x\,u_n^{(m+n+1)} + \big[n(n+1) - m(m+1)\big]\,u_n^{(m+n)} = 0.$$

We therefore find the solutions $v_n^m = u_n^{(n+m)} = \frac{\mathrm{d}^{n+m}}{\mathrm{d}x^{n+m}}(1-x^2)^n$ yielding $y_n^m = \sqrt{1-x^2}^m \frac{\mathrm{d}^{n+m}}{\mathrm{d}x^{n+m}}(1-x^2)^n$.

Rodrigues formula. By of the above, the Rodrigues formula for the associated Legendre functions P_n^m becomes

$$P_n^m = \frac{(-1)^{n+m}}{2^n n!}\sqrt{1-x^2}^m \frac{\mathrm{d}^{n+m}}{\mathrm{d}x^{n+m}}(1-x^2)^n \tag{A.31}$$
$$\text{or} \quad P_n^m = (-1)^m \sqrt{1-x^2}^m \frac{\mathrm{d}^m}{\mathrm{d}x^m} P_n, \qquad \text{with} \quad P_n = \frac{(-1)^n}{2^n n!}\frac{\mathrm{d}^n}{\mathrm{d}x^n}(1-x^2)^n.$$

and $P_n = P_n^0$ are the Legendre polynomials. The Legendre polynomials are normalized to $P_n(1) = 1$ by the factor $\frac{(-1)^n}{2^n n!}$. Because $(1-x^2)$ is zero at $x = 1$ with any positive integer exponent, only the part of its n-fold derivative that exclusively affects the power of $(1-x^2)^n$ for n times is responsible for its value there: $n!(-2x)^n(1-x^2)^0|_{x=1} = n!2^n(-1)^n$. The scaling of the associated Legendre functions with $m > 0$ is somewhat more arbitrary in sign and value.

Indies n and m. The boundaries for the index m of the Legendre functions $m \in \mathbb{Z}$ are typically $-n \le m \le m$, however due to the shift of the eigenvalue by $\frac{m^2}{1-x^2}$, functions for positive and negative m are linearly dependent. We observe this by inspecting the highest-order terms in [6]

$$2^n n! \sqrt{1-x^2}^m P_n^m = (-1)^{n+m}(1-x^2)^m \frac{\mathrm{d}^{n+m}(1-x^2)^n}{\mathrm{d}x^{n+m}}$$
$$= x^{2m}\frac{\mathrm{d}^{n+m}}{\mathrm{d}x^{n+m}}\left[x^{2n} - \cdots\right] = x^{2m}\left[\frac{(2n)!}{(n-m)!}x^{n-m} - \cdots\right]$$
$$2^n n! \sqrt{1-x^2}^m P_n^{-m} = (-1)^{n+m}\frac{\mathrm{d}^{n-m}(1-x^2)^n}{\mathrm{d}x^{n-m}}$$

$$= (-1)^m \frac{\mathrm{d}^{n-m}}{\mathrm{d}x^{n-m}}\left[x^{2n} - \cdots\right] = (-1)^m \left[\frac{(2n)!}{(n+m)!} x^{n+m} - \cdots\right]$$

$$\Longrightarrow P_n^{-m} = (-1)^m \frac{(n-m)!}{(n+m)!} P_n^m \tag{A.32}$$

and to avoid confusion, it convenient to only use $m \geq 0$, or $|m|$ to evaluate the associated Legendre functions.

Alternative definition: three-term recurrence. Any polynomial $\mathcal{P}_n$ of the order n can be decomposed into Legendre polynomials $\mathcal{P}_n = \sum_{i=0}^n c_i P_i$, and the Legendre polynomial P_j is orthogonal to all those Legendre polynomials $\int_{-1}^1 \mathcal{P}_n P_j \mathrm{d}x = 0$ if $j > n$. With this knowledge it is interesting to describe $\int_{-1}^1 (x P_i) P_j \mathrm{d}x$. As $(x P_i)$ is of $(i+1)$th order, the integral must vanish for $j > i + 1$. Because of commutativity, $\int_{-1}^1 P_i (x P_j) \mathrm{d}x$, and $(x P_j)$ being $(j+1)$th order, it also vanishes for $i > j + 1$. Hereby, re-expansion of $x P_n$ can maximally use three terms, $x P_n = \alpha\, P_{n-1} + \gamma\, P_n + \beta\, P_{n+1}$. In fact only two terms remain as P_{2k} are even functions on $x \in [-1; 1]$ and P_{2k+1} are odd, thus orthogonal. The product $x P_n$ changes the parity of P_n, leaving $x P_n = \alpha\, P_{n-1} + \beta\, P_{n+1}$. At $x = 1$ all polynomials were normalized to $P_i(1) = 1$, therefore evaluation at $x = 1$ leaves $1 = \alpha + \beta$, so $\alpha = 1 - \beta$, hence

$$x\, P_n = \beta_n\, P_{n+1} + (1 - \beta_n)\, P_{n-1}.$$

As also the associated Legendre functions P_n^m for a specific m are orthogonal, the recurrence is more general

$$x\, P_n^m = \beta_n^m\, P_{n+1}^m + (1 - \beta_n^m)\, P_{n-1}^m.$$

To determine the coefficient β_n^m, we only need to find out how the highest-power coefficients x^{n-m+1} of the polynomial parts in $x P_n^m$ and P_{n+1}^m are related. We see this after inserting $P_n = \sqrt{1-x^2}^m \frac{(-1)^{n+m}}{2^n n!} \frac{\mathrm{d}^{n+m}}{\mathrm{d}x^{n+m}} (1-x^2)^n$ and division by $\sqrt{1-x^2}^m \frac{(-1)^{n+m}}{2^n n!}$, which leaves a recurrence for the polynomial part

$$\underbrace{x\, v_n^m}_{O=n-m+1} = -\underbrace{\frac{\beta_n^m}{2(n+1)} v_{n+1}^m}_{O=n-m+1} - \underbrace{2(n+1)(1-\beta_n^m) v_{n-1}^m}_{O=n-m-1}.$$

Of the highest powers x^{n-m+1} in both $x\, v_n^m$ and v_{n+1}^m the coefficients $c_{n,n-m}^m$ and $c_{n+1,n-m+1}^m$ define

$$\beta_n^m = -2(n+1) \frac{c_{n,n-m}^m}{c_{n+1,n-m+1}^m}.$$

To find it, we binomially expand $(1-x^2)^n$ to $(1-x^2)^n = (-1)^n \sum_{k=0}^n (-1)^k \binom{n}{k}$ $x^{2(n-k)}$

$$\frac{v_n^m}{(-1)^n} = \frac{\mathrm{d}^{n+m}}{\mathrm{d}x^{n+m}} \sum_{k=0}^{n} \binom{n}{k} (-1)^k x^{2(n-k)} = \sum_{k=0}^{\lfloor \frac{n-m}{2} \rfloor} \binom{n}{k} \frac{(2n-2k)!(-1)^k}{(n-m-2k)!} x^{n-m-2k},$$

so that with $k = 0$ we can find $c_{n,n-m}^m = \frac{(-1)^n n!}{n!} \frac{(2n)!}{(n-m)!} = \frac{(-1)^n (2n)!}{(n-m)!}$ for the highest-power coefficient of v_n^m. Accordingly, coefficient of the recurrence is

$$\beta_n^m = 2(n+1) \frac{(n-m+1)}{(2n+1)(2n+2)} = \frac{n-m+1}{2n+1},$$

hence with $1 - \beta_n^m = \frac{n+m}{2n+1}$, and $x P_n^m = \frac{n-m+1}{2n+1} P_{n+1}^m + \frac{n+m}{2n+1} P_{n-1}^m$, we can construct P_n^m recursively by

$$P_{n+1}^m = \frac{2n+1}{n-m+1} x P_n^m - \frac{n+m}{n-m+1} P_{n-1}^m. \tag{A.33}$$

The start value is $P_n^n = \frac{(-1)^{2n}}{2^n n!} \sqrt{1-x^2}^n \frac{\mathrm{d}^{2n}}{\mathrm{d}x^{2n}} (1-x^2)^n = \frac{(-1)^n (2n)!}{2^n n!} \sqrt{1-x^2}^n$, and for $n = m$, the term P_{n-1}^m is excluded.

Normalization. A unity square integral (orthonormalization) simplifies the definition of transform integrals. We would like to obtain the corresponding factor N_n^m with

$$\int_{-1}^{1} (P_n^m N_n^m)^2 \, \mathrm{d}x = 1.$$

Normalization for $m = 0$ is easy to find by repeated integration by parts

$$\begin{aligned}
\frac{2^{2n} n!^2}{(N_n)^2} &= 2^{2n} n!^2 \int_{-1}^{1} (P_n)^2 \mathrm{d}x = \underbrace{\big[(1-x^2)^n\big]^{(n-1)} \big[(1-x^2)^n\big]^{(n)} \big|_{-1}^{1}}_{=0} \\
&\quad - \int_{-1}^{1} \big[(1-x^2)^n\big]^{(n-1)} \big[(1-x^2)^n\big]^{(n)} \mathrm{d}x \\
&= \cdots = (-1)^n \int_{-1}^{1} (1-x^2)^n \big[(1-x^2)^n\big]^{(2n)} \mathrm{d}x \\
&= \int_{-1}^{1} (1-x^2)^n \, (2n)! \, \mathrm{d}x = (2n)! \int_{0}^{\pi} \sin^{2n} \vartheta \, \sin \vartheta \, \mathrm{d}\vartheta.
\end{aligned}$$

With the integral $\int_0^\pi \sin^{2n+1} \vartheta \, \mathrm{d}\vartheta = 2 \frac{(2n)!!}{(2n+1)!!} = 2 \frac{2^{2n} n!^2}{(2n+1)!}$, this is $(N_n)^{-2} = \frac{2\,(2n)!}{(2n+1)!} = \frac{2}{2n+1}$. For N_n^m, a trick to insert the relation between P_n^m and P_n^{-m} is used [6], and integration by parts until the differentials are of the same order

$$\frac{1}{(N_n^m)^2} = \int_{-1}^{1} P_n^m P_n^m \, \mathrm{d}x = \int_{-1}^{1} P_n^m \frac{(-1)^m (n-m)!}{(n+m)!} P_n^{-m} \, \mathrm{d}x$$

$$= \frac{1}{2^{2n}n!^2}\frac{(-1)^m(n-m)!}{(n+m)!}\int_{-1}^{1}\left[(1-x^2)\right]^{(n+m)}\left[(1-x^2)\right]^{(n-m)}\mathrm{d}x = \cdots$$

$$= \frac{(n-m)!}{(n+m)!}\underbrace{\int_{-1}^{1}\frac{1}{2^{2n}n!^2}\left[(1-x^2)\right]^{(n-m)}\left[(1-x^2)\right]^{(n-m)}\mathrm{d}x}_{=1/(N_n)^2} = \frac{2}{2n+1}\frac{(n+m)!}{(n-m)!}$$

$$\Rightarrow N_n^m = (-1)^m\sqrt{\frac{(2n+1)}{2}\frac{(n-m)!}{(n+m)!}} \tag{A.34}$$

The $(-1)^m$ can be excluded if not used in the Rodrigues formula. (*It is always a wise idea to check and compare signs as conventions may differ ... in practice* $(-1)^m$ *is a rotation around* z *by* 180°.)

A.3.7 Spherical Harmonics

With all the above definitions, we obtain the fully normalized spherical harmonics

$$Y_n^m(\varphi, \vartheta) = N_n^{|m|}\, P_n^{|m|}(\cos\vartheta)\, \Phi_m(\varphi) \tag{A.35}$$

Orthonormality. They are *orthonormal* when integrated over the sphere

$$\int_{\mathbb{S}^2} Y_n^m Y_{n'}^{m'}\, \mathrm{d}\cos\theta\, \mathrm{d}\varphi = \delta_{nn'}\delta_{mm'}. \tag{A.36}$$

Transform integral. Because of their completeness in the Hilbert space, any square-integrable function $g(\varphi, \vartheta)$ can be decomposed by

$$g(\varphi, \vartheta) = \sum_{n=0}^{\infty}\sum_{m'=-n'}^{n'} \gamma_{n'm'}\, Y_{n'}^{m'}(\varphi, \vartheta). \tag{A.37}$$

From a known function $g(\varphi, \vartheta)$, the coefficients are obtained by integrating g with another spherical harmonic Y_n^m over the unit sphere $\mathbb{S}^2$, $\int_{-1}^{1}\mathrm{d}\cos\vartheta \int_0^{2\pi}\mathrm{d}\varphi$. For a simple notation, we gather the two variables in a direction vector $\boldsymbol{\theta} = [\cos\varphi\ \sin\vartheta,\ \sin\varphi\ \sin\vartheta,\ \cos\vartheta]^{\mathrm{T}}$ and write

$$\int_{\mathbb{S}^2} g(\boldsymbol{\theta})\, Y_n^m\, \mathrm{d}\boldsymbol{\theta} = \sum_{n'=0}^{\infty}\sum_{m'=-n'}^{n'} \gamma_{n'm'} \underbrace{\int_{\mathbb{S}^2} Y_{n'}^{m'}\, Y_n^m\, \mathrm{d}\boldsymbol{\theta}}_{\delta_{nn'}\delta_{mm'}} = \gamma_{nm}. \tag{A.38}$$

Parseval's theorem. Due to orthonormality, the integral norm of any pattern $g(\boldsymbol{\theta})$ composed as $\sum_{n=0}^{\infty}\sum_{m=-n}^{n}\gamma_{nm}\,Y_n^m(\boldsymbol{\theta})$ is equivalent to

$$\int_{\mathbb{S}^2}|g(\boldsymbol{\theta})|^2\,\mathrm{d}\boldsymbol{\theta}=\sum_{n=0}^{\infty}\sum_{m=-n}^{n}|\gamma_{nm}|^2 \tag{A.39}$$

because $\int_{\mathbb{S}^2}\sum_{n,n',m,m'}\gamma_{nm}\gamma_{n'm'}^{*}\,Y_n^m(\boldsymbol{\theta})\,Y_{n'}^{m'}(\boldsymbol{\theta})\,\mathrm{d}\boldsymbol{\theta}=\sum_{n,n',m,m'}\gamma_{nm}\gamma_{n'm'}^{*}\delta_{nn'}\delta_{mm'}$.

3D panning functions: Dirac delta on the sphere. An infinitely narrow range around the desired direction $\boldsymbol{\theta}_\mathrm{s}$ can be described by limiting the dot product $\boldsymbol{\theta}_\mathrm{s}^\mathrm{T}\boldsymbol{\theta}>\cos\varepsilon\rightarrow 1$. A unit-surface Dirac delta distribution $\delta(1-\boldsymbol{\theta}_\mathrm{s}^\mathrm{T}\boldsymbol{\theta})$ can be described as

$$\delta(1-\boldsymbol{\theta}_\mathrm{s}^\mathrm{T}\boldsymbol{\theta})=\frac{1}{2\pi}\begin{cases}\lim_{\varepsilon\rightarrow 0}\frac{1}{1-\cos\varepsilon}, & \text{for } \arccos\boldsymbol{\theta}_\mathrm{s}^\mathrm{T}\boldsymbol{\theta}<\varepsilon\\ 0, & \text{otherwise.}\end{cases} \tag{A.40}$$

And its coefficients are found by the transformation integral

$$\gamma_{nm}=\int_{\mathbb{S}^2}Y_n^m(\boldsymbol{\theta})\,\mathrm{d}\boldsymbol{\theta}=Y_n^m(\boldsymbol{\theta}_\mathrm{s})\int_0^{2\pi}\mathrm{d}\varphi\,\lim_{\varepsilon\rightarrow 0}\int_{\cos\varepsilon}^{1}\mathrm{d}\zeta\;=Y_n^m(\boldsymbol{\theta}_\mathrm{s}). \tag{A.41}$$

Typically, a finite-order panning function with $n\leq\mathrm{N}$ employs a weight a_n to reduce side lobes

$$g_\mathrm{N}(\boldsymbol{\theta})=\sum_{n=0}^{\mathrm{N}}\sum_{m=-n}^{n}a_n\,Y_n^m(\boldsymbol{\theta}_\mathrm{s})\,Y_n^m(\boldsymbol{\theta}). \tag{A.42}$$

Assuming the panning direction is $\boldsymbol{\theta}_\mathrm{s}=[0,0,1]^\mathrm{T}$, we get the axisymmetric panning function, with $Y_n^0=\sqrt{\frac{2n+1}{4\pi}}P_n$,

$$g_\mathrm{N}(\vartheta)=\frac{1}{2\pi}\sum_{n=0}^{\mathrm{N}}\frac{2n+1}{2}\,a_n\,P_n(\cos\vartheta). \tag{A.43}$$

We can evaluate its E measure by integrating g_N^2 over the sphere

$$\begin{aligned}E&=\underbrace{\int_0^{2\pi}\mathrm{d}\varphi}_{2\pi}\int_{-1}^{1}g_\mathrm{N}(\zeta)^2\,\mathrm{d}\zeta=\sum_{i,j}\frac{2i+1}{2}\frac{2j+1}{2}\,a_i\,a_j\underbrace{\int_{-1}^{1}P_i\,P_j\,\mathrm{d}\zeta}_{\delta_{ij}\frac{2}{2i+1}}\\&=\sum_{n=0}^{\mathrm{N}}\frac{2n+1}{2}a_n^2.\end{aligned} \tag{A.44}$$

The $\boldsymbol{r}_\mathrm{E}$ measure is, because of the axisymmetry, perfectly aligned with z, therefore, its length is calculated by

$$r_\mathrm{E} = \frac{\int_{-1}^{1} g_\mathrm{N}(\zeta)\,\zeta\,\mathrm{d}\zeta}{E} = \frac{\sum_{i,j} \frac{2i+1}{2}\frac{2j+1}{2}\,a_i\,a_j \int_{-1}^{1} P_i \overbrace{\zeta\,P_j}^{\frac{j+1}{2j+1}P_{j+1}+\frac{j}{2j+1}P_{j-1}} \mathrm{d}\zeta}{E}$$
$$= \frac{\sum_{i,j} \frac{2i+1}{4}\,a_i\,a_j \int_{-1}^{1} P_i\,[(j+1)P_{j+1} + j\,P_{j-1}]\mathrm{d}\zeta}{E}$$
$$= \frac{\sum_{n=0}^{\mathrm{N}} [n\,a_n\,a_{n-1} + (n+1)\,a_n\,a_{n+1}]}{2E} = \frac{\sum_{n=1}^{\mathrm{N}} n\,a_n\,a_{n-1}}{E}. \tag{A.45}$$

3D max-$\boldsymbol{r}_\mathrm{E}$. For the narrowest-possible spread, we maximize r_E, which we decompose into $r_\mathrm{E} = \frac{\hat{r}_\mathrm{E}}{E}$ and we zero its derivative, as for 2D,

$$r'_\mathrm{E} = \frac{\hat{r}'_\mathrm{E}}{E} - \frac{\hat{r}_\mathrm{E}}{E^2}E' = \frac{1}{E}[\hat{r}'_\mathrm{E} - E'\,r_\mathrm{E}] = 0$$
$$n\,a_{n-1} + (n+1)\,a_{n+1} - (2n+1)\,a_n\,r_\mathrm{E} = 0. \tag{A.46}$$

If we assume that $a_n = P_n(\zeta)$, we see by $(n+1)P_{n+1} + n\,P_{n-1} = (2n+1)\,P_n\,\zeta$

$$(n+1)\;P_{n+1} + n\;P_{n-1} = (2n+1)\;P_n\,\zeta\; = P_n\,r_\mathrm{E} \tag{A.47}$$

that $r_\mathrm{E} = \zeta$ and $a_n = P_n(\zeta) = P_n(r_\mathrm{E})$. We maximize r_E under the constraint that $P_{\mathrm{N}+1}(r_\mathrm{E}) = 0$. Therefore, r_E must be as close to 1 as possible, and be a zero of the Legendre polynomial $P_{\mathrm{N}+1}$. It can be discovered by a root-finding algorithm in MATLAB, e.g. Newton–Raphson, when the function P_n is implemented. In [7], the useful approximation $r_\mathrm{E} = \cos\frac{2.4062}{\mathrm{N}+1.51} = \cos\frac{137.9^\circ}{\mathrm{N}+1.51}$ was given.

Squared norm mirror/rotation invariance. The norm of any pattern $a(\boldsymbol{\theta})$ is invariant under orthogonal coordinate transform (rotation/mirror) $\hat{\boldsymbol{\theta}} = \boldsymbol{R}\,\boldsymbol{\theta}$ with $\boldsymbol{R}^\mathrm{T}\boldsymbol{R} = \boldsymbol{I}$,

$$\int_{\mathbb{S}^2} a^2(\boldsymbol{\theta})\,\mathrm{d}\boldsymbol{\theta} = \int_{\mathbb{S}^2} b^2(\boldsymbol{\theta})\,\mathrm{d}\boldsymbol{\theta}, \qquad \text{with } b(\boldsymbol{\theta}) = a(\boldsymbol{R}\,\boldsymbol{\theta}). \tag{A.48}$$

The norm equivalence of the corresponding spherical harmonics coefficients α_{nm} and β_{nm} follows from Parseval's theorem

$$\sum_{n=0}^{\infty}\sum_{m=-n}^{n} |\alpha_{nm}|^2 = \sum_{n=0}^{\infty}\sum_{m=-n}^{n} |\beta_{nm}|^2. \tag{A.49}$$

In vector notation of the coefficients, i.e. $\boldsymbol{\alpha} = [\alpha_{00}, \ldots, \alpha_{\mathrm{NN}}]^\mathrm{T}$ and $\boldsymbol{\beta} = [\beta_{00}, \ldots, \beta_{\mathrm{NN}}]^\mathrm{T}$, this is $\|\boldsymbol{\alpha}\|^2 = \|\boldsymbol{\beta}\|^2$. To fulfill this equivalence, both vectors are related by an orthogonal matrix $\boldsymbol{\beta} = \boldsymbol{Q}\,\boldsymbol{\alpha}$ with $\boldsymbol{Q}^\mathrm{T}\boldsymbol{Q} = \boldsymbol{I}$, and hence $\|\boldsymbol{b}\|^2 = \boldsymbol{\beta}^\mathrm{T}\boldsymbol{\beta} = \boldsymbol{\alpha}^\mathrm{T}\boldsymbol{Q}^\mathrm{T}\boldsymbol{Q}\boldsymbol{\alpha} =$

$\boldsymbol{\alpha}^{\mathrm{T}}\boldsymbol{\alpha} = \|\boldsymbol{\alpha}\|^2$. Moreover, rotation/mirroring neither creates components of higher nor lower orders, so that $\boldsymbol{Q}$ must be block structured

$$\boldsymbol{Q} = \begin{bmatrix} \boldsymbol{Q}_0 & \mathbf{0} & \dots & \mathbf{0} \\ \mathbf{0} & \boldsymbol{Q}_1 & \mathbf{0} & \vdots \\ \vdots & \ddots & \ddots & \ddots \end{bmatrix}. \tag{A.50}$$

The order subspaces in n therefore stay de-coupled, so that the coefficient vectors for every order-subspace n are related and norm equivalent under mirror/rotation operations

$$\|\boldsymbol{\alpha}_n\|^2 = \|\boldsymbol{\beta}_n\|^2, \qquad \boldsymbol{\beta}_n = \boldsymbol{Q}_n\,\boldsymbol{\alpha}_n, \qquad \boldsymbol{Q}_n^{\mathrm{T}}\boldsymbol{Q}_n = \boldsymbol{I}_{2n+1}. \tag{A.51}$$

Pseudo all-pass character of the Dirac delta. Dirac delta distributions $\delta(\boldsymbol{\theta}^{\mathrm{T}}\boldsymbol{\theta}_{\mathrm{s}} - 1)$ yield the coefficients $Y_n^m(\boldsymbol{\theta}_{\mathrm{s}})$, and due to rotation invariance they yield constant energy in every spherical harmonic order n, regardless of the aiming $\boldsymbol{\theta}_{\mathrm{s}}$. One can determine the norm for zenithal aiming $\vartheta = 0$, i.e. $\boldsymbol{\theta}_{\mathrm{z}} = 0, 0, 1^{\mathrm{T}}$, yielding a non-zero coefficient for $m = 0$

$$Y_n^m(\boldsymbol{\theta}_{\mathrm{z}}) = \sqrt{\tfrac{2n+1}{4\pi}}\,\overbrace{P_n(1)}^{=1}\,\delta_m = \sqrt{\tfrac{2n+1}{4\pi}}\,\delta_m.$$

Because of the rotation invariance we recognize a pseudo-allpass character (Unsöld theorem) of the spherical harmonics of any order n

$$\sum_{m=-n}^{n} |Y_n^m(\boldsymbol{\theta}_{\mathrm{s}})|^2 = \sum_{m=-n}^{n} |Y_n^m(\boldsymbol{\theta}_{\mathrm{z}})|^2 = \tfrac{2n+1}{4\pi} = (2n+1)\,|Y_0^0(\boldsymbol{\theta}_{\mathrm{s}})|^2.$$

For encoded single-direction Ambisonic signals $\alpha_{nm}(t)$, this implies

$$\sum_{m=-n}^{n} |\alpha_{nm}(t)|^2 = (2n+1)\,|\alpha_{00}(t)|^2. \tag{A.52}$$

Expected norm in the diffuse field. An ideal diffuse sound field is composed of directional signals $a(\boldsymbol{\theta}_{\mathrm{s}}, t)$ from all directions $\boldsymbol{\theta}$, with no correlation for signals from different directions $\mathcal{E}\{a(\boldsymbol{\theta}_1, t)\,a(\boldsymbol{\theta}_2, t)\} = \sigma_{\mathrm{a}}^2\,\delta(\boldsymbol{\theta}_1^{\mathrm{T}}\boldsymbol{\theta}_2 - 1)$. Its coefficients are obtained by the integral over the directions

$$\alpha_{nm}(t) = \int_{\mathbb{S}^2} a(\boldsymbol{\theta}_{\mathrm{s}}, t)\,Y_n^m(\boldsymbol{\theta}_{\mathrm{s}})\,\mathrm{d}\boldsymbol{\theta}_{\mathrm{s}}, \tag{A.53}$$

and we can show that not only the expected directional signals, but also the spherical harmonic coefficients are orthogonal by $\mathcal{E}\{a(\boldsymbol{\theta}_1, t)\,a(\boldsymbol{\theta}_2, t)\} = \sigma_{\mathrm{a}}^2\,\delta(\boldsymbol{\theta}_1^{\mathrm{T}}\boldsymbol{\theta}_2 - 1)$ and orthonormality $\int_{\mathbb{S}^2} Y_n^m Y_{n'}^{m'}\,\mathrm{d}\boldsymbol{\theta} = \delta_{nn'}\delta_{mm'}$ of the spherical harmonics

$$\begin{aligned}\mathcal{E}\{\alpha_{nm}(t)\,\alpha_{n'm'}(t)\} &= \int_{\mathbb{S}^2}\int_{\mathbb{S}^2} \mathcal{E}\{a(\boldsymbol{\theta}_1,t)\,a(\boldsymbol{\theta}_2,t)\}\,Y_n^m(\boldsymbol{\theta}_1)\,Y_{n'}^{m'}(\boldsymbol{\theta}_2)\,\mathrm{d}\boldsymbol{\theta}_1\mathrm{d}\boldsymbol{\theta}_2 \\ &= \sigma_{\mathrm{a}}^2 \int_{\mathbb{S}^2} Y_n^m(\boldsymbol{\theta})\,Y_{n'}^{m'}(\boldsymbol{\theta})\,\mathrm{d}\boldsymbol{\theta} = \sigma_{\mathrm{a}}^2\delta_{nn'}\delta_{mm'}. \end{aligned} \tag{A.54}$$

In a perfectly diffuse field, we therefore expect the same norm in every spherical harmonic component per frequency band. We could reformulate this to

$$\mathcal{E}\{|\alpha_{nm}(t)|^2\} = \mathcal{E}\{|\alpha_{00}(t)|^2\},$$

however the temporal disjointness assumption of SDM only invents a drastically thinned out content in the individual higher-order spherical harmonics. To cover the available temporal information from all $(2n+1)$ spherical harmonic signals within each order n and for a similar formulation as for a single-direction component, we may re-formulate

$$\sum_{m=-n}^{n} \mathcal{E}\{|\alpha_{nm}(t)|^2\} = (2n+1)\,\mathcal{E}\{|\alpha_{00}(t)|^2\}. \tag{A.55}$$

Spherical convolution. By the argumentation used above to prove rotation invariance, we can argue that isotropic filtering of spherical patterns is invariant under rotation, and must therefore depend only on the order n. Spherical convolution of is defined in [8] by the coefficients β_{nm} of a function $b(\boldsymbol{\theta})$ convolved with the coefficients α_n of a rotationally symmetric shape $a(\boldsymbol{\theta}) = a(\theta_{\mathrm{z}})$

$$\gamma_{nm} = \alpha_n\,\beta_{nm}. \tag{A.56}$$

Spherical cap function. A rotationally symmetric spherical cap function at $\pm\frac{\alpha}{2}$ centered around $\vartheta = 0$, briefly $\boldsymbol{\theta}_{\mathrm{z}}$, can be written in terms of a unit-step. We find the shape coefficients w_n for its spherical harmonic decomposition by

$$u(\cos\vartheta - \cos\tfrac{\alpha}{2}) = \sum_{n=0}^{\infty}\sum_{m=-n}^{n} w_n\,Y_n^m(\boldsymbol{\theta})\,Y_n^m(\boldsymbol{\theta}_{\mathrm{z}}) = \sum_{n=0}^{\infty} w_n\,P_n(\cos\vartheta)\,\tfrac{2n+1}{4\pi}, \tag{A.57}$$

where $Y_n^m(\boldsymbol{\theta}_{\mathrm{z}}) = \sqrt{\frac{2n+1}{4\pi}}P_n^m(1) = \sqrt{\frac{2n+1}{4\pi}}\delta_m$ and $P_n^0 = P_n$ was used. The coefficients w_n are obtained by integration over Legendre polynomials $\mathrm{d}\cos\vartheta \int_{-1}^{1} P_{n'}(\cos\vartheta)$ and for the right hand-side using their orthogonality $\int_{-1}^{1} P_{n'}(\zeta)P_n(\zeta)\,\mathrm{d}\zeta = \frac{2}{2n+1}\delta_{nn'}$, leaving $w_n\,\frac{2}{2n+1}\,\frac{2n+1}{4\pi}$, so that

$$w_n = 2\pi \int_{\cos\frac{\alpha}{2}}^{1} P_n(x)\,\mathrm{d}x.$$

The integral is solved by $2^n n! P_n = \frac{\mathrm{d}^n v^n}{\mathrm{d}x^n}$ with $v = (x^2 - 1)$ after replacing the inner-most derivative by $\frac{\mathrm{d}}{\mathrm{d}x} = \frac{\mathrm{d}v}{\mathrm{d}x}\frac{\mathrm{d}}{\mathrm{d}v} = 2x\frac{\mathrm{d}}{\mathrm{d}v}$, and Leibniz' rule for repeated derivatives

$$\begin{aligned}\frac{\mathrm{d}^n v^n}{\mathrm{d}x^n} &= \frac{\mathrm{d}^{n-1} 2x \frac{\mathrm{d}v^n}{\mathrm{d}v}}{\mathrm{d}x^{n-1}} = \frac{\mathrm{d}^{n-1}(2xnv^{n-1})}{\mathrm{d}x^{n-1}} \\ &= (2n)\left[\binom{0}{n-1}\frac{\mathrm{d}^0 x}{\mathrm{d}x^0}\frac{\mathrm{d}^{n-1}v^{n-1}}{\mathrm{d}x^{n-1}} + \binom{1}{n-1}\frac{\mathrm{d}^1 x}{\mathrm{d}x^1}\frac{\mathrm{d}^{n-2}v^{n-1}}{\mathrm{d}x^{n-2}}\right] \\ &= (2n)\left[x\,\frac{\mathrm{d}^{n-1}v^{n-1}}{\mathrm{d}x^{n-1}} + (n-1)\frac{\mathrm{d}^{n-2}v^{n-1}}{\mathrm{d}x^{n-2}}\right].\end{aligned}$$

We may increase n by one, observe that the last expression has one fewer differential, thus is an integrated version, and obtain after re-inserting $2^n n! P_n = \frac{\mathrm{d}^n v^n}{\mathrm{d}x^n}$

$$\begin{aligned}2^{n+1}(n+1)!\, P_{n+1} &= 2(n+1)\left[2^n n!\, x P_n + n 2^n n! \int P_n \mathrm{d}x - nC\right] \\ \int P_n \mathrm{d}x &= \frac{P_{n+1} - x P_n}{n} + C \qquad \text{(A.58)}\end{aligned}$$

With the definite integration limits $x_0 = \cos\frac{\alpha}{2}$ and 1 and $\int_{x_0}^1 P_n \mathrm{d}x$ only depends on lower boundary as $P_{n+1}(1) - 1 \cdot P_n(1) = 0$,

$$\int_{x_0}^1 P_n(x)\mathrm{d}x = -\frac{P_{n+1}(x_0) - x_0 P_n(x_0)}{n} \qquad \text{(A.59)}$$

$$w_n = -2\pi\,\frac{P_{n+1}\left(\cos\frac{\alpha}{2}\right) - \cos\frac{\alpha}{2} P_n\left(\cos\frac{\alpha}{2}\right)}{n}, \qquad \text{for } n > 0, \qquad \text{(A.60)}$$

and $w_0 = 2\pi \int_{\cos\frac{\alpha}{2}}^1 \mathrm{d}x = 2\pi(1 - \cos\frac{\alpha}{2})$ for $n = 0$.

The recurrence $(2n+1)xP_n - (n+1)P_{n+1} = nP_{n-1}$ yields alternatively for $n > 0$

$$w_n = 2\pi\,\frac{P_{n+1}\left(\cos\frac{\alpha}{2}\right) + P_{n-1}\left(\cos\frac{\alpha}{2}\right)}{2n+1} = 2\pi\,\frac{P_{n-1}\left(\cos\frac{\alpha}{2}\right) - \cos\frac{\alpha}{2} P_n\left(\cos\frac{\alpha}{2}\right)}{n+1}. \qquad \text{(A.61)}$$

A.4 Encoding to SH and Decoding to SH

Mode-matching decoder: L loudspeakers driven by the weights g_l and given by their directions $\{\boldsymbol{\theta}_l\}$ produce a pattern $f(\boldsymbol{\theta})$ linearly composed of Dirac deltas

$$f(\boldsymbol{\theta}) = \sum_{l=1}^{\mathrm{L}} \delta(\boldsymbol{\theta}^{\mathrm{T}}\boldsymbol{\theta}_l - 1)\, g_l \tag{A.62}$$

$$= \sum_{n=0}^{\infty} \sum_{m=-n}^{n} Y_n^m(\boldsymbol{\theta}) \sum_{l=1}^{\mathrm{L}} Y_n^m(\boldsymbol{\theta}_l)\, g_l = \boldsymbol{y}(\boldsymbol{\theta})^{\mathrm{T}} \boldsymbol{Y} \boldsymbol{g},$$

in an order-unlimited representation. The vector $\boldsymbol{y} = [Y_n^m(\boldsymbol{\theta})]_{nm}$ contains all the spherical harmonics $0 \le n \le \infty$, $-n \le m \le n$, in a suitable order, e.g. Ambisonic Channel Number (ACN) $n^2 + m + n$, and the matrix $\boldsymbol{Y} = [\boldsymbol{y}(\boldsymbol{\theta}_l)]_l$ contains the spherical harmonic coefficient vectors of every loudspeaker. Obviously, the spherical harmonic coefficients synthesized by the loudspeakers are $\boldsymbol{\phi} = \boldsymbol{Y}\,\boldsymbol{g}$, so that

$$f(\boldsymbol{\theta}) = \boldsymbol{y}(\boldsymbol{\theta})^{\mathrm{T}} \boldsymbol{Y}\, \boldsymbol{g} = \boldsymbol{y}(\boldsymbol{\theta})^{\mathrm{T}} \boldsymbol{\phi}.$$

With L loudspeakers, at most $(\mathrm{N}+1)^2 \le \mathrm{L}$ spherical harmonics can be controlled. Therefore control typically restricts to the under-determined Nth-order subspace

$$\boldsymbol{\phi}_{\mathrm{N}} = \boldsymbol{Y}_{\mathrm{N}}\, \boldsymbol{g},$$

in which we can synthesize any coefficient vector $\boldsymbol{\phi}_{\mathrm{N}}$. To get a finite and well-determined solution with the exceeding and arbitrary degrees of freedom in $\boldsymbol{g}$, the least-squares solution for $\boldsymbol{g}$ is searched under the constraint

$$\begin{aligned} &\min \|\boldsymbol{g}\|^2 \\ &\text{subject to: } \boldsymbol{\phi}_{\mathrm{N}} = \boldsymbol{Y}_{\mathrm{N}}\, \boldsymbol{g}, \end{aligned} \tag{A.63}$$

yielding the cost function with the Lagrange multipliers $\boldsymbol{\lambda}$

$$J(\boldsymbol{g}, \boldsymbol{\lambda}) = \boldsymbol{g}^{\mathrm{T}} \boldsymbol{g} + (\boldsymbol{\phi}_{\mathrm{N}} - \boldsymbol{Y}_{\mathrm{N}}\, \boldsymbol{g})^{\mathrm{T}} \boldsymbol{\lambda}.$$

For the optimum in $\boldsymbol{g}$, its derivative to $\boldsymbol{g}$ is zero, and in $\boldsymbol{\lambda}$ the corresponding derivative:

$$\frac{\partial J}{\partial \boldsymbol{g}} = 2\boldsymbol{g}_{\mathrm{opt}} - \boldsymbol{Y}_{\mathrm{N}}^{\mathrm{T}} \boldsymbol{\lambda} = 0, \qquad \frac{\partial J}{\partial \boldsymbol{\lambda}} = \boldsymbol{\phi}_{\mathrm{N}} - \boldsymbol{Y}_{\mathrm{N}}\, \boldsymbol{g} = 0.$$

For $\boldsymbol{g}$ the equation yields $\boldsymbol{g}_{\mathrm{opt}} = \frac{1}{2}\, \boldsymbol{Y}_{\mathrm{N}}^{\mathrm{T}}\, \boldsymbol{\lambda}$, and for $\boldsymbol{\lambda}$ the original constraint $\boldsymbol{\phi}_{\mathrm{N}} = \boldsymbol{Y}_{\mathrm{N}}\, \boldsymbol{g}$ that only allows to insert the optimal $\boldsymbol{g}$ yielding $\boldsymbol{\phi}_{\mathrm{N}} = \boldsymbol{Y}_{\mathrm{N}}\, (\frac{1}{2}\, \boldsymbol{Y}_{\mathrm{N}}^{\mathrm{T}}\, \boldsymbol{\lambda}_{\mathrm{opt}})$. Inversion of $(\frac{1}{2}\boldsymbol{Y}_{\mathrm{N}}\boldsymbol{Y}_{\mathrm{N}}^{\mathrm{T}})^{-1}$ from the left yields the multipliers $2(\boldsymbol{Y}_{\mathrm{N}}\boldsymbol{Y}_{\mathrm{N}}^{\mathrm{T}})^{-1}\boldsymbol{\phi}_{\mathrm{N}} = \boldsymbol{\lambda}_{\mathrm{opt}}$, so that

$$\boldsymbol{g} = \boldsymbol{Y}_{\mathrm{N}}^{\mathrm{T}}\, (\boldsymbol{Y}_{\mathrm{N}}\boldsymbol{Y}_{\mathrm{N}}^{\mathrm{T}})^{-1} \boldsymbol{\phi}_{\mathrm{N}}. \tag{A.64}$$

The solution is right-inverse to $\boldsymbol{Y}_{\mathrm{N}}$, i.e. $\boldsymbol{Y}_{\mathrm{N}}\, [\boldsymbol{Y}_{\mathrm{N}}^{\mathrm{T}}\, (\boldsymbol{Y}_{\mathrm{N}}\boldsymbol{Y}_{\mathrm{N}}^{\mathrm{T}})^{-1}] = \boldsymbol{I}$.

Best-fit encoder by MMSE: When given M samples of a spherical function $g(\boldsymbol{\theta})$ at the locations $\{\boldsymbol{\theta}_l\}$, we can minimize the means-square error (MMSE)

$$\min \sum_{l=1}^{\mathrm{L}} \left[g(\boldsymbol{\theta}_l) - \sum_{n=0}^{\mathrm{N}} \sum_{m=-n}^{n} Y_n^m(\boldsymbol{\theta}_l)\, \gamma_{nm} \right]^2$$

to find suitable spherical harmonic coefficients γ_{nm}. Using the matrix notation from above, this is

$$\min \|\boldsymbol{e}\|^2 = \min \|\boldsymbol{g} - \boldsymbol{Y}_{\mathrm{N}}^{\mathrm{T}} \boldsymbol{\gamma}_{\mathrm{N}}\|^2 \tag{A.65}$$

and we find by zeroing the derivative

$$\frac{\partial}{\partial \boldsymbol{\gamma}_{\mathrm{N}}} \boldsymbol{e}^{\mathrm{T}} \boldsymbol{e} = 2 \left(\frac{\partial \boldsymbol{e}}{\partial \boldsymbol{\gamma}_{\mathrm{N}}} \right)^{\mathrm{T}} \boldsymbol{e} = 2 \boldsymbol{Y}_{\mathrm{N}} \boldsymbol{e} = 2 \boldsymbol{Y}_{\mathrm{N}} \boldsymbol{Y}_{\mathrm{N}}^{\mathrm{T}} \boldsymbol{\gamma}_{\mathrm{N}} - 2 \boldsymbol{Y}_{\mathrm{N}} \boldsymbol{g} = 0,$$
$$\Longrightarrow \boldsymbol{\gamma}_{\mathrm{N}} = (\boldsymbol{Y}_{\mathrm{N}}^{\mathrm{T}} \boldsymbol{Y}_{\mathrm{N}})^{-1} \boldsymbol{Y}_{\mathrm{N}}\, \boldsymbol{g}. \tag{A.66}$$

The resulting matrix is left-inverse to the thin matrix $\boldsymbol{Y}_{\mathrm{N}}^{\mathrm{T}}$, and can be written in terms of the more general pseudo inverse $(\boldsymbol{Y}_{\mathrm{N}}^{\mathrm{T}})^{\dagger}$.

A.5 Covariance Constraint for Binaural Ambisonic Decoding

The interaural covariance matrix is related to the expectation value of the auto/cross-covariances of the left and right HRTFs:

$$\boldsymbol{R}^* = \int_{\mathbb{S}^2} \begin{bmatrix} h_{\mathrm{left}}(\boldsymbol{\theta}, \omega) \\ h_{\mathrm{right}}(\boldsymbol{\theta}, \omega) \end{bmatrix} \left[h_{\mathrm{left}}(\boldsymbol{\theta}, \omega)^* \; h_{\mathrm{right}}(\boldsymbol{\theta}, \omega)^* \right] \mathrm{d}\boldsymbol{\theta} = \int_{\mathbb{S}^2} \boldsymbol{h}(\boldsymbol{\theta}, \omega) \boldsymbol{h}(\boldsymbol{\theta}, \omega)^{\mathrm{H}}\, \mathrm{d}\boldsymbol{\theta}. \tag{A.67}$$

When specified in terms of spherical harmonic coefficients $h = \boldsymbol{y}^{\mathrm{T}} \boldsymbol{h}_{\mathrm{SH}}$, the integral $\int_{\mathbb{S}^2} h_1^* h_2 \mathrm{d}\boldsymbol{\theta}$ of any of $\boldsymbol{R}$'s entries vanishes by the orthogonality of the spherical harmonics $\boldsymbol{h}_{\mathrm{SH1}}^{\mathrm{H}} \int_{\mathbb{S}^2} \boldsymbol{y} \boldsymbol{y}^{\mathrm{T}}\, \mathrm{d}\boldsymbol{\theta}\, \boldsymbol{h}_{\mathrm{SH2}} = \boldsymbol{h}_{\mathrm{SH1}}^{\mathrm{H}} \boldsymbol{h}_{\mathrm{SH2}}$, and we obviously only need the inner product between the spherical-harmonic coefficients of the HRTFs.

A very-high-order spherical harmonics HRTF dataset $\boldsymbol{H}_{\mathrm{SH}}^{\mathrm{H}}$ of dimensions $2 \times (\mathrm{M}+1)^2$ with the order ($\mathrm{M} \gg \mathrm{N}$) yields a covariance matrix at every frequency

$$\boldsymbol{R} = \boldsymbol{H}_{\mathrm{SH}}^{\mathrm{H}} \boldsymbol{H}_{\mathrm{SH}} = \boldsymbol{X}^{\mathrm{H}} \boldsymbol{X}$$

that can be factored into a quadratic form of a 2×2 matrix $\boldsymbol{X}$ by Cholesky factorization, which reduces the degrees of freedom involved to the minimum required size.

The Nth-order Ambisonically reproduced, high-frequency modified HRTF dataset $\tilde{\boldsymbol{H}}_{\mathrm{SH}}$ of dimensions $2 \times (\mathrm{M}+1)^2$ also has a 2×2 covariance matrix $\hat{\boldsymbol{R}}$ that will differ from $\boldsymbol{R}$, and which we also decompose in Cholesky factors $\hat{\boldsymbol{X}}$,

$$\hat{\boldsymbol{R}} = \hat{\boldsymbol{H}}_{\mathrm{SH}}^{\mathrm{H}} \hat{\boldsymbol{H}}_{\mathrm{SH}} = \hat{\boldsymbol{X}}^{\mathrm{H}} \hat{\boldsymbol{X}}. \tag{A.68}$$

To equalize $\boldsymbol{R} = \hat{\boldsymbol{R}}$, the reproduced HRTF set is corrected by a 2×2 filter matrix $\boldsymbol{M}$,

$$\hat{\boldsymbol{H}}_{\mathrm{SH,corr}} = \hat{\boldsymbol{H}}_{\mathrm{SH}} \boldsymbol{M}. \tag{A.69}$$

This is done properly as soon as

$$\boldsymbol{X}^{\mathrm{H}} \boldsymbol{X} = \boldsymbol{M}^{\mathrm{H}} \hat{\boldsymbol{X}}^{\mathrm{H}} \hat{\boldsymbol{X}} \boldsymbol{M} = \boldsymbol{M}^{\mathrm{H}} \hat{\boldsymbol{X}}^{\mathrm{H}} \overbrace{\boldsymbol{Q}^{\mathrm{H}} \boldsymbol{Q}}^{\boldsymbol{I}} \hat{\boldsymbol{X}} \boldsymbol{M}, \tag{A.70}$$

and the orthogonal matrix $\boldsymbol{Q}$ is used to compensate for degrees of freedom that the Cholesky factors $\boldsymbol{X}$ and $\hat{\boldsymbol{X}}$ have in sign, phase, mixing, with regard to each other. We recognize the root and hereby the preliminary solution for $\boldsymbol{M}$

$$\boldsymbol{X} = \boldsymbol{Q} \hat{\boldsymbol{X}} \boldsymbol{M}, \qquad \Rightarrow \boldsymbol{M} = \hat{\boldsymbol{X}}^{-1} \boldsymbol{Q}^{\mathrm{H}} \boldsymbol{X}. \tag{A.71}$$

This leaves $\hat{\boldsymbol{H}}_{\mathrm{SH,corr}} = \hat{\boldsymbol{H}}_{\mathrm{SH}} \hat{\boldsymbol{X}}^{-1} \boldsymbol{Q}^{\mathrm{H}} \boldsymbol{X}$ depending on an unspecific orthogonal 2×2 matrix $\boldsymbol{Q}$. To obtain a corrected-covariance HRTFs $\hat{\boldsymbol{H}}_{\mathrm{corr.SH}}$ of highest-possible phase-alignment and correlation to its uncorrected counterpart $\hat{\boldsymbol{H}}_{\mathrm{SH}}$, we maximize the trace, i.e. the sum of diagonal elements

$$\max \Re\mathrm{e}\,\mathrm{Tr}\{\hat{\boldsymbol{H}}_{\mathrm{SH}}^{\mathrm{H}} \hat{\boldsymbol{H}}_{\mathrm{corr,SH}}\} = \max \Re\mathrm{e}\,\mathrm{Tr}\{\hat{\boldsymbol{H}}_{\mathrm{SH}}^{\mathrm{H}} \hat{\boldsymbol{H}}_{\mathrm{SH}} \hat{\boldsymbol{X}}^{-1} \boldsymbol{Q}^{\mathrm{H}} \boldsymbol{X}\} =$$
$$\max \Re\mathrm{e}\,\mathrm{Tr}\{\hat{\boldsymbol{X}}^{\mathrm{H}} \hat{\boldsymbol{X}} \hat{\boldsymbol{X}}^{-1} \boldsymbol{Q}^{\mathrm{H}} \boldsymbol{X}\} = \max \Re\mathrm{e}\,\mathrm{Tr}\{\hat{\boldsymbol{X}}^{\mathrm{H}} \boldsymbol{Q}^{\mathrm{H}} \boldsymbol{X}\} = \max \Re\mathrm{e}\,\mathrm{Tr}\{\boldsymbol{Q}^{\mathrm{H}} \hat{\boldsymbol{X}}^{\mathrm{H}} \boldsymbol{X}\}.$$

For the last expression, the property $\mathrm{Tr}\{\boldsymbol{AB}\} = \mathrm{Tr}\{\boldsymbol{BA}\}$ was used. An orthogonal matrix $\boldsymbol{Q}^{\mathrm{H}} = \boldsymbol{V}\boldsymbol{U}^{\mathrm{H}}$ composed of two orthogonal matrices $\boldsymbol{U}$ and $\boldsymbol{V}$ would yield $\mathrm{Tr}\{\hat{\boldsymbol{X}}^{\mathrm{H}} \boldsymbol{X} \boldsymbol{V} \boldsymbol{U}^{\mathrm{H}}\} = \mathrm{Tr}\{\boldsymbol{U}^{\mathrm{H}} \hat{\boldsymbol{X}}^{\mathrm{H}} \boldsymbol{X} \boldsymbol{V}\}$, and it would maximize the trace if $\boldsymbol{U}$ and $\boldsymbol{V}^{\mathrm{H}}$ diagonalized $\hat{\boldsymbol{X}}^{\mathrm{H}} \boldsymbol{X}$. This is accomplished by singular-value decomposition (SVD) $\hat{\boldsymbol{X}}^{\mathrm{H}} \boldsymbol{X} = \boldsymbol{U} \boldsymbol{S} \boldsymbol{V}^{\mathrm{H}}$, when singular values $\boldsymbol{S} = \mathrm{diag}\{[s_1,\ s_2]\}$ are real and positive, as in most SVD implementations. Using $\boldsymbol{U}$ and $\boldsymbol{V}$, the desired solution is:

$$\hat{\boldsymbol{H}}_{\mathrm{corr,SH}} = \hat{\boldsymbol{H}}_{\mathrm{SH}} \hat{\boldsymbol{X}}^{-1} \boldsymbol{V} \boldsymbol{U}^{\mathrm{H}} \boldsymbol{X}. \tag{A.72}$$

If the SVD delivers negative or complex-valued singular values, the complex/negative factor just need to be pulled out and factored into either the corresponding left or right singular vector.

A.6 Physics of the Helmholtz Equation

A.6.1 Adiabatic Compression

We search for a physical compression equation relating pressure p and volume V.

Ideal gas. The gas pressure p inside the volume V obeys the ideal gas law [9]

$$p\,V = n\,R\,T, \tag{A.73}$$

with n measuring the amount of substance in moles, R is the gas constant, and T is the temperature. This would yield a valid compression equation if the medium of sound propagation was isothermal. However, this is not the case $T \neq \text{const}$, and local temperature fluctuations happen too fast to be equalized by thermal dissipation. Isothermal compression would be too soft, the resulting speed of sound off by -15%. A compression law involving fluctuations of all three quantities (p, V, T) needs an additional equation.

First law of thermodynamics. In thermodynamics [10–12], the enthalpy H describes the energy required to heat up a freely expanding gas under constant pressure p. The enthalpy goes to the internal energy U required to heat up the gas in a constant volume, which is easier, plus the ideal-gas volume work $p\,V$ taken by the gas to expand under the constant external pressure

$$\begin{aligned} H &= U + p\,V, \\ \text{specifically } n\,c_\mathrm{p}\,T &= n\,c_\mathrm{V}\,T + n\,R\,T, \qquad \Rightarrow R = c_\mathrm{p} - c_\mathrm{V}. \end{aligned} \tag{A.74}$$

The quantities c_p and c_V are the specific heat capacities for heating up a gas that is expanding ($p = \text{const.}$) or confined in a fixed volume ($V = \text{const.}$) to a temperature T, which can be accurately measured or modeled. Obviously, the gas constant R is the difference between the two. To make sound propagation isenthalpic, the energy must fluctuate between internal energy U and volume work pV.

Adiabatic process. The above steady-state equations are not useful yet to describe short-term fluctuations of p, V, and T in time and space. A differential formulation related to the change in enthalpy, internal energy, and volume work $\mathrm{d}H = \mathrm{d}U + p\,\mathrm{d}V$ is more useful. Moreover, we regard packages of a constant amount of substance whose internal heat up is just due to compression and not due to external enthalpy sources, it is therefore isenthalpic $\mathrm{d}H = 0$, see [10, Sect. 3.12.2], [13]

$$0 = n\, c_\mathrm{V}\, \mathrm{d}T + \frac{n\, R\, T}{V}\, \mathrm{d}V.$$

We may divide by $n\, c_\mathrm{V}\, T$, replace $R = c_\mathrm{p} - c_\mathrm{V}$, and obtain $\frac{\mathrm{d}T}{T} + (\frac{c_\mathrm{p}}{c_\mathrm{V}} - 1)\frac{\mathrm{d}V}{V} = 0$, whose integration yields $\ln T + (\frac{c_\mathrm{p}}{c_\mathrm{V}} - 1)\ln V = \ln\big(T\, V^{\frac{c_\mathrm{p}}{c_\mathrm{V}}-1}\big) = 0$, hence $T\, V^{\frac{c_\mathrm{p}}{c_\mathrm{V}}-1} = 1$, and with the ideal gas equation inserted as $T = \frac{p\, V}{n\, R}$, the *adiabatic process law* becomes

$$p\, V^{\frac{c_\mathrm{p}}{c_\mathrm{V}}} = n\, R = \text{const}, \tag{A.75}$$

for which the adiabatic exponent is frequently expressed as $\gamma = \frac{c_\mathrm{p}}{c_\mathrm{V}}$. For air, the exponent is $\gamma = 1.4$, and we may express a state change as $(p_0, V_0) \rightarrow (p_0 + p, V_0 + V)$. The equation $p_0\, V_0^\gamma = (p_0 + p)(V_0 + V)^\gamma$ yields after division by $p_0\, V_0^\gamma$ and by $(1 + \frac{V}{V_0})^\gamma$:

$$1 + \frac{p}{p_0} = \left(1 + \frac{V}{V_0}\right)^{-\gamma} \approx 1 - \gamma\, \frac{V}{V_0}, \qquad \text{hence } p = -\gamma\, p_0\, \frac{V}{V_0}.$$

Assuming the Cartesian coordinates x, y, z measured in the resting gas to define its volume $V_0 = \Delta x \Delta y \Delta z$, as well as its deflected coordinates $\xi(x)$, $\eta(y)$, $\zeta(z)$ after a volume change to V, we can approximate the volume change well-enough by the three independent volume changes $\Delta\xi \Delta y \Delta z$, $\Delta x \Delta\eta \Delta z$, and $\Delta x \Delta y \Delta\zeta$, resulting from the superimposed individual elongation into the three coordinates' directions,

$$\lim_{V_0 \to 0} \frac{\Delta V}{V_0} = \lim_{V_0 \to 0} \frac{\Delta\xi \Delta y \Delta z + \Delta x \Delta\eta \Delta z + \Delta x \Delta y \Delta\zeta}{\Delta x \Delta y \Delta z} = \frac{\partial\xi}{\partial x} + \frac{\partial\eta}{\partial y} + \frac{\partial\zeta}{\partial z} = \boldsymbol{\nabla}^\mathrm{T}\boldsymbol{\xi}.$$

Replacing the bulk modulus $K = \gamma p_0 = \rho\, c^2$ of air by more common constants,[1] where $c = \sqrt{K/\rho}$ and applying the derivative in time $\frac{\partial}{\partial t}$, we get the equation of compression in its typical form using the velocities $\frac{\partial \boldsymbol{\xi}}{\partial t} = \boldsymbol{v}$,

$$\frac{\partial p}{\partial t} = -\rho\, c^2\, \boldsymbol{\nabla}^\mathrm{T}\boldsymbol{v}. \tag{A.76}$$

A.6.2 *Potential and Kinetic Sound Energies, Intensity, Diffuseness*

The *potential* energy density or *volume work* stored in the elastic medium that gets compressed by a deformation $\mathrm{d}V$ increases with $\mathrm{d}w_\mathrm{p} = p\, \mathrm{d}V$, while deformation also increases the pressure by $\mathrm{d}p = K\, \mathrm{d}V$. We may substitute for $\mathrm{d}V = K^{-1}\mathrm{d}p$

[1] Typical constants are $\gamma = 1.4$, $p_0 = 10^5$ Pa, $\rho = 1.2\,\mathrm{kg/m^3}$, $c = 343\,\mathrm{m/s}$.

yielding $\mathrm{d}w_\mathrm{p} = K^{-1} p\,\mathrm{d}p$. The volume work stored by a pressure increase from 0 to p is

$$w_\mathrm{p} = \int_0^p \frac{p\,\mathrm{d}p}{K} = \frac{p^2}{2K} = \frac{p^2}{2\rho c^2}. \tag{A.77}$$

The *kinetic* energy density stored in the motion of the medium along any axis, e.g. x, increases by acceleration against its mass, $\mathrm{d}w_{\mathrm{v}_\mathrm{x}} = \rho\,\frac{\mathrm{d}v_\mathrm{x}}{\mathrm{d}t}\mathrm{d}x$. The velocity is $v_\mathrm{x} = \frac{\mathrm{d}x}{\mathrm{d}t}$ so that we substitute for $\mathrm{d}x = v_\mathrm{x}\mathrm{d}t$ to get $\mathrm{d}w_{\mathrm{v}_\mathrm{x}} = \rho\, v_\mathrm{x}\,\mathrm{d}v_\mathrm{x}$. The total kinetic energy density stored in velocities increasing from 0 to v_x, v_y, v_z is

$$w_\mathrm{v} = \rho\left[\int_0^{v_\mathrm{x}} v_\mathrm{x}\,\mathrm{d}v_\mathrm{x} + \int_0^{v_\mathrm{y}} v_\mathrm{y}\,\mathrm{d}v_\mathrm{y} + \int_0^{v_\mathrm{z}} v_\mathrm{z}\,\mathrm{d}v_\mathrm{z}\right] = \rho\frac{v_\mathrm{x}^2 + v_\mathrm{y}^2 + v_\mathrm{z}^2}{2} = \frac{\rho\|\boldsymbol{v}\|^2}{2}. \tag{A.78}$$

Total energy density and intensity. The total energy density therefore becomes

$$w = w_\mathrm{p} + w_\mathrm{v} = \frac{p^2}{2\rho c^2} + \frac{\rho\|\boldsymbol{v}\|^2}{2}, \tag{A.79}$$

and derived with regard to time, it becomes

$$\frac{\partial w}{\partial t} = \frac{p}{\rho c^2}\frac{\partial p}{\partial t} + \rho\boldsymbol{v}^\mathrm{T}\frac{\partial \boldsymbol{v}}{\partial t} = p\nabla^\mathrm{T}\boldsymbol{v} + \boldsymbol{v}^\mathrm{T}\nabla p = \nabla^\mathrm{T}(p\boldsymbol{v}) = \nabla^\mathrm{T}\boldsymbol{I}, \tag{A.80}$$

and defines the (time-domain) intensity vector $\boldsymbol{I} = p\boldsymbol{v}$ that describes the energy flow in space. Hereby, $\frac{\partial w}{\partial t} = \nabla^\mathrm{T}\boldsymbol{I}$ expresses that only a non-zero *divergence* of the intensity causes energy increase (source) or loss (absorption) in the lossless medium.

Direction of arrival and diffuseness: The intensity vector carries a meaning in its own right: it displays into which direction the energy flows (direction of emission). In the frequency domain, it becomes $\boldsymbol{I} = \mathfrak{Re}\{p^*\boldsymbol{v}\}$, and for a plane-wave sound field $p = e^{\mathrm{i}k\boldsymbol{\theta}_\mathrm{s}^\mathrm{T}\boldsymbol{r}}$, where $\boldsymbol{v} = -\frac{\nabla p}{\mathrm{i}k\rho c} = -\frac{\boldsymbol{\theta}_\mathrm{s}\,p}{\rho c}$, it indicates the direction of arrival (DOA)

$$\boldsymbol{r}_\mathrm{DOA} = -\frac{\rho c\,\boldsymbol{I}}{|p|^2} = -\frac{\rho c\,\mathfrak{Re}\{p^*\boldsymbol{v}\}}{|p|^2} = \frac{\rho c\,|p|^2\boldsymbol{\theta}_\mathrm{s}}{\rho c\,|p|^2} = \boldsymbol{\theta}_\mathrm{s}. \tag{A.81}$$

An ideal, uniformly enveloping diffuse field is composed of uncorrelated plane waves $E\{\frac{a(\boldsymbol{\theta}_1)^* a(\boldsymbol{\theta}_2)}{4\pi}\} = \frac{a^2}{4\pi}\delta(1 - \boldsymbol{\theta}_1^\mathrm{T}\boldsymbol{\theta}_2)$ resulting in the sound pressure $p = \int \frac{a(\boldsymbol{\theta}_\mathrm{s})}{\sqrt{4\pi}}\,e^{\mathrm{i}k\boldsymbol{\theta}_\mathrm{s}^\mathrm{T}\boldsymbol{r}}\mathrm{d}\boldsymbol{\theta}_\mathrm{s}$. While the expected sound pressure is non-zero as before $E\{|p|^2\} = \frac{a^2}{4\pi} = |p|^2$, the expected intensity of the uniformly surrounding waves vanishes $-\rho c\,E\{\boldsymbol{I}\} = \frac{a^2}{4\pi}\int_{\mathbb{S}^2}\boldsymbol{\theta}_\mathrm{s}\,\mathrm{d}\boldsymbol{\theta}_\mathrm{s} = \boldsymbol{0}$.

Assuming stochastic interference of all sources, the intensity-based DOA estimator $\boldsymbol{r}_\mathrm{DOA} = -\frac{\rho c\,\mathfrak{Re}\{p^*\boldsymbol{v}\}}{|p|^2}$ is therefore the physical equivalent to the $\boldsymbol{r}_\mathrm{E}$ vector measure.

A typical diffuseness measure $0 \leq \psi \leq 1$ relies on its length between 0 and 1

$$\psi = 1 - \|\boldsymbol{r}_{\mathrm{DOA}}\|^2. \tag{A.82}$$

The signals $W = p$ and $[X, Y, Z]^{\mathrm{T}} = \frac{\sqrt{2}\nabla p}{\mathrm{i}k} = -\rho c\,\sqrt{2}\boldsymbol{v}$ of a first-order Ambisonic microphone allow to describe a time-domain estimator $\boldsymbol{r}_{\mathrm{DOA}}$

$$\boldsymbol{r}_{\mathrm{DOA}} = -\frac{\rho c\, E\{\boldsymbol{I}\}}{E\{p^2\}} = -\frac{\rho c\, E\{p\,\boldsymbol{v}\}}{E\{p^2\}} = \frac{E\{W\,[X, Y, Z]^{\mathrm{T}}\}}{\sqrt{2}\,E\{W^2\}}. \tag{A.83}$$

A.6.3 Green's Function in 3 Cartesian Dimensions

We may compose the Green's function, the solution to the inhomogeneous wave equation

$$\left(\Delta - \frac{1}{c^2}\frac{\partial^2}{\partial t^2}\right) G = -\delta(t)\delta(\boldsymbol{r}),$$

from products of complex exponentials with regard to time and the Cartesian directions

$$e^{\mathrm{i}\omega\, t + \mathrm{i}k_{\mathrm{x}}\, x + \mathrm{i}k_{\mathrm{y}}\, y + \mathrm{i}k_{\mathrm{z}}\, z} = e^{\mathrm{i}\boldsymbol{k}^{\mathrm{T}}\boldsymbol{r}}\, e^{\mathrm{i}\omega\, t}, \tag{A.84}$$

where the position x, y, z was gathered in a position vector $\boldsymbol{r}$, and the *wave numbers* $k_{\mathrm{x}}, k_{\mathrm{y}}, k_{\mathrm{z}}$ of the individual coordinates were gathered in a wave-number vector $\boldsymbol{k}$. From this solution, we compose the Green's function by superimposing all spatial and temporal complex exponentials in $\boldsymbol{k}$ and ω, weighted by an unknown coefficient γ:

$$G = \iint \gamma\; e^{\mathrm{i}\boldsymbol{k}^{\mathrm{T}}\boldsymbol{r}}\, e^{\mathrm{i}\omega\, t}\; \mathrm{d}\omega\, \mathrm{d}\boldsymbol{k}. \tag{A.85}$$

Because of $\Delta e^{\mathrm{i}\boldsymbol{k}^{\mathrm{T}}\boldsymbol{r}} = (-k_x^2 - k_y^2 - k_z^2)\, e^{\mathrm{i}\boldsymbol{k}^{\mathrm{T}}\boldsymbol{r}} = -k^2 e^{\mathrm{i}\boldsymbol{k}^{\mathrm{T}}\boldsymbol{r}}$ and $\frac{\partial^2}{\partial t^2} e^{\mathrm{i}\omega t} = -\omega^2 e^{\mathrm{i}\omega t}$, insertion into the inhomogeneous wave equation yields

$$-\iint \gamma\, \left[k^2 - \tfrac{\omega^2}{c^2}\right] e^{\mathrm{i}\boldsymbol{k}^{\mathrm{T}}\boldsymbol{r}}\, e^{\mathrm{i}\omega\, t}\, \mathrm{d}\omega\, \mathrm{d}\boldsymbol{k} = -\delta(t)\,\delta(\boldsymbol{r}).$$

Multiple transformations $e^{-\mathrm{i}\hat{\boldsymbol{k}}^{\mathrm{T}}\boldsymbol{r}} e^{-\mathrm{i}\hat{\omega}t}\, \mathrm{d}\boldsymbol{r}\, \mathrm{d}t \iint$ remove the integrals by orthogonality

$$-\iint \gamma\left(k^2 - \tfrac{\omega^2}{c^2}\right) \underbrace{\left[\int e^{\mathrm{i}(\boldsymbol{k}-\hat{\boldsymbol{k}})^{\mathrm{T}}\boldsymbol{r}} \mathrm{d}\boldsymbol{r}\right]}_{(2\pi)^3\delta(\boldsymbol{k}-\hat{\boldsymbol{k}})} \underbrace{\left[\int e^{\mathrm{i}(\omega-\hat{\omega})t} \mathrm{d}t\right]}_{2\pi\delta(\omega-\hat{\omega})} \mathrm{d}\omega\, \mathrm{d}\boldsymbol{k} = -\underbrace{\int \delta(t) e^{-\mathrm{i}\hat{\omega}t} \mathrm{d}t}_{1} \underbrace{\int \delta(\boldsymbol{r}) e^{-\mathrm{i}\hat{\boldsymbol{k}}^{\mathrm{T}}\boldsymbol{r}} \mathrm{d}\boldsymbol{r}}_{1},$$

and the unknown coefficient remains $\gamma = \frac{1}{(2\pi)^{3+1}} \frac{1}{k^2-\frac{\omega^2}{c^2}}$. Letting G in the frequency domain, one $\frac{1}{2\pi}$ in γ and the integral $e^{\mathrm{i}\omega t}\mathrm{d}\omega \int$ are omitted. We transform γ back via $\boldsymbol{k}$

$$G = \frac{1}{(2\pi)^3} \iint \frac{e^{\mathrm{i}\boldsymbol{k}^{\mathrm{T}}\boldsymbol{r}}}{k^2 - \frac{\omega^2}{c^2}} \,\mathrm{d}\boldsymbol{k} = \frac{1}{(2\pi)^3} \iint \frac{e^{\mathrm{i}kr\zeta}}{k^2 - \frac{\omega^2}{c^2}} \,\mathrm{d}\boldsymbol{k}.$$

By re-expressing $\boldsymbol{k}^{\mathrm{T}}\boldsymbol{r} = kr\,\boldsymbol{\theta}_{\mathrm{k}}^{\mathrm{T}}\boldsymbol{\theta}_{\mathrm{r}} = kr\cos\vartheta = kr\zeta$ we simplify the integral. *Now we already see formally that Green's function can only depend on the distance r between source and receiver location $G = G(\omega, r)$.*

The book [14, S.110-112] shows a notably compact derivation, which we will use below.

Derivation for by transforming back from the Fourier domain: For three dimensions, the transformation back from the Fourier domain is relatively easy to accomplish. Before going into details, we recognize that the substitution of $\boldsymbol{k}^{\mathrm{T}}\boldsymbol{r}$ by $kr\ \cos\vartheta$ contains the radius of the wave vector $k = \|\boldsymbol{k}\|$ and the cosine of the angle between $\boldsymbol{r}$ and $\boldsymbol{k}$. In k space, we can always define a correspondingly oriented coordinate system for any $\boldsymbol{r}$ as to simplify the integral $\iiint_{-\infty}^{\infty} \mathrm{d}\boldsymbol{k} = \int_0^\infty \int_0^{2\pi} \int_0^\pi k^2\,\mathrm{d}k\,\mathrm{d}\varphi\,\mathrm{d}\cos\vartheta = \int_0^\infty \int_0^{2\pi} \int_{-1}^{1} k^2\,\mathrm{d}k\,\mathrm{d}\varphi\,\mathrm{d}\zeta$. After re-arranging the integrals, we get

$$\begin{aligned}
G &= \frac{\int_0^{2\pi}\mathrm{d}\varphi}{(2\pi)^3} \int_0^\infty \frac{\int_{-1}^{1} e^{\mathrm{i}kr\zeta}\,\mathrm{d}\zeta}{k^2 - \frac{\omega^2}{c^2}} k^2\,\mathrm{d}k = \frac{1}{(2\pi)^2} \int_0^\infty \frac{1}{\mathrm{i}kr} \frac{e^{\mathrm{i}kr} - e^{-\mathrm{i}kr}}{k^2 - \frac{\omega^2}{c^2}} k^2\,\mathrm{d}k \\
&= \frac{1}{(2\pi)^2} \frac{1}{\mathrm{i}r} \int_0^\infty \frac{e^{\mathrm{i}kr} - e^{-\mathrm{i}kr}}{k^2 - \frac{\omega^2}{c^2}} k\,\mathrm{d}k = \frac{1}{(2\pi)^2} \frac{1}{\mathrm{i}r} \left[\int_0^\infty \frac{e^{\mathrm{i}kr}\,k}{k^2 - \frac{\omega^2}{c^2}} \,\mathrm{d}k - \int_0^\infty \frac{e^{-\mathrm{i}kr}\,k}{k^2 - \frac{\omega^2}{c^2}} \,\mathrm{d}k \right] \\
&= \frac{1}{(2\pi)^2} \frac{1}{\mathrm{i}r} \left[\int_0^\infty \frac{e^{\mathrm{i}kr}\,k}{k^2 - \frac{\omega^2}{c^2}} \,\mathrm{d}k - \int_0^{-\infty} \frac{e^{-\mathrm{i}(-k)r}\,(-k)}{(-k)^2 - \frac{\omega^2}{c^2}} \,\mathrm{d}(-k) \right] \\
&= \frac{1}{(2\pi)^2} \frac{1}{\mathrm{i}r} \left[\int_0^\infty \frac{e^{\mathrm{i}kr}\,k}{k^2 - \frac{\omega^2}{c^2}} \,\mathrm{d}k + \int_{-\infty}^{0} \frac{e^{\mathrm{i}kr}\,k}{k^2 - \frac{\omega^2}{c^2}} \,\mathrm{d}k \right] \\
&= \frac{1}{(2\pi)^2} \frac{1}{\mathrm{i}r} \int_{-\infty}^{\infty} \frac{e^{\mathrm{i}kr}}{k^2 - \frac{\omega^2}{c^2}} k\,\mathrm{d}k.
\end{aligned} \tag{A.86}$$

The denominator is expanded in partial fractions $\frac{1}{k^2-\frac{\omega^2}{c^2}} = \frac{1}{2k}\frac{1}{k-\frac{\omega}{c}} + \frac{1}{2k}\frac{1}{k+\frac{\omega}{c}}$, yielding

$$G = \frac{1}{2(2\pi)^2} \frac{1}{\mathrm{i}r} \left[\int_{-\infty}^{\infty} \frac{e^{\mathrm{i}kr}}{k - \frac{\omega}{c}} \,\mathrm{d}k + \int_{-\infty}^{\infty} \frac{e^{\mathrm{i}kr}}{k + \frac{\omega}{c}} \,\mathrm{d}k \right]. \tag{A.87}$$

To obtain causal temporal solutions, there needs to be a specific solution of the improper and singular integrals.

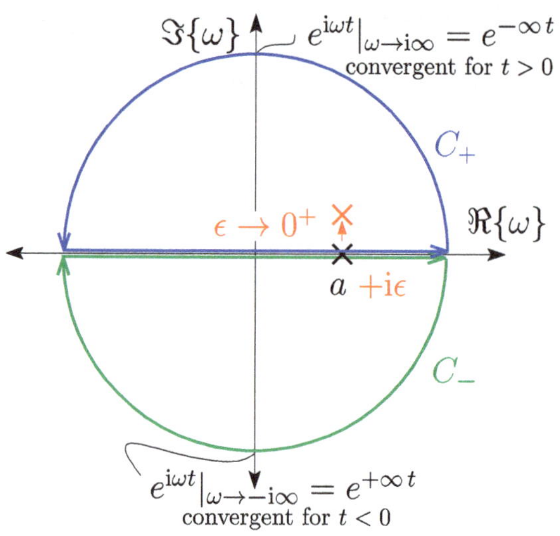

Fig. A.1 Closure of improper integration path using C_+ over the positive imaginary half plane for $t > 0$, and C_- for $t < 0$, and regularization of pole for causal result, i.e. non-zero only for C_+

Causal solutions (integral in ω***).*** Causal responses obeying the transformation

$$h(t) = \int_{-\infty}^{\infty} \frac{e^{i\omega t}}{\omega - a}\, d\omega \tag{A.88}$$

are obtained by replacing the improper integral $\int_{-\infty}^{\infty}$ by a closed integration contour, and by introducing vanishing regularization. Jordan's lemma states that improper integration $\int_{-\infty}^{\infty}$ is equivalent to a closed integration path C_+ of positive orientation involving the additional semi-circle on the upper half of the complex number plane $\int_{-\infty}^{\infty} = \oint_{C_+} = \lim_{R\to\infty}\big[\int_{-R}^{R} d\omega + \int_0^{\pi} R\, d\varphi\big]$ if the integrand of the semi-circle vanishes, i.e. $\lim_{R\to\infty} \frac{e^{(i\cos\varphi - \sin\varphi)\, R\, t}}{R\, e^{i\varphi} - a} = 0$. This is the case for positive times $t > 0$. For negative times, the integral can be closed using the lower part of the complex number plane, $\int_{-\infty}^{\infty} = \oint_{C_-} = \lim_{R\to\infty}\big[\int_{-R}^{R} d\omega + \int_0^{-\pi} R\, d\varphi\big]$, if the semi-circular integral vanishes, which is true for negative times $t < 0$ in our case, see Fig. A.1. We get

$$h(t) = \begin{cases} \oint_{C_+} \frac{e^{i\omega t}}{\omega - a} d\omega, & \text{if } t > 0, \\ \oint_{C_-} \frac{e^{i\omega t}}{\omega - a} d\omega, & \text{if } t < 0. \end{cases} \tag{A.89}$$

According to Cauchy's integral formula for analytic regular functions $f(z)$ over a single pole $\frac{1}{z-a}$, we obtain

$$\oint_{C_\pm} \frac{f(z)}{z - a}\, dz = \pm 2\pi i \begin{cases} f(a), & \text{if the path } C_\pm \text{ surrounds } a, \\ 0, & \text{if } a \text{ lies outside the path } C_\pm. \end{cases} \tag{A.90}$$

If a pole on the real axis $a \in \mathbb{R}$ is slightly shifted by a vanishing imaginary amount to $\lim_{\epsilon\to 0^+} \oint_{C_\pm} \frac{e^{i\omega t}}{\omega - i\epsilon - a}\, d\omega$ (regularization) so that it lies within the path C_+ and not in C_-, the result is perfectly causal and vanishes at negative times: $h(t) = 2\pi i \lim_{\epsilon\to 0^+} e^{i a t - \epsilon t}\, u(t)$, with the unit step function $u(t) = 1$ for $t \geq 0$ and 0 for $t < 0$, see Fig. A.1.

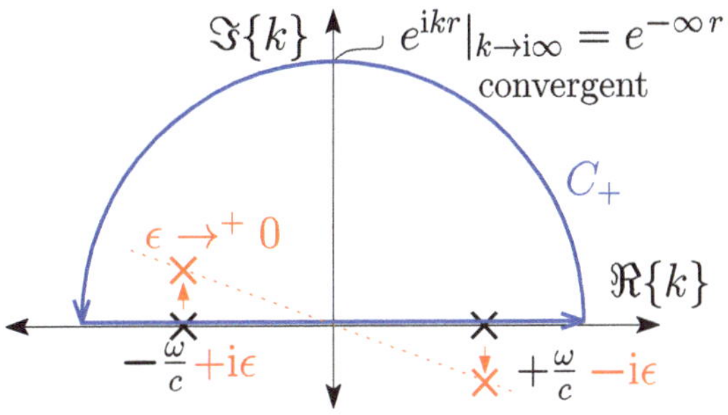

Fig. A.2 Causality-enforcing regularization $\omega = \lim_{\epsilon\to 0^+} \omega - i\epsilon$ of the poles in Green's function wrt. wave number k. For a positive radius $r \geq 0$, closure of the improper integration path requires C_+ with semicircle on the positive imaginary half plane (or C_- for $r \leq 0$)

Integral in k. Causality requires specific regularization in frequency as shown above: Replacement of ω by $\lim_{\epsilon\to 0^+} \omega - i\epsilon\, c$ guarantees causality in the partial-fraction expanded Green's function Eq. (A.87). Jordan's lemma requires to use the path C_+ to close the improper path in k for a positive radius $r \geq 0$, cf. Fig. A.2,

$$G = \frac{1}{2(2\pi)^2} \lim_{\epsilon\to 0^+} \frac{1}{ir}\left[\cancel{\int_{C_+} \frac{e^{ikr}\, dk}{k - \frac{\omega}{c} + i\epsilon}} + \int_{C_+} \frac{e^{ikr}\, dk}{k + \frac{\omega}{c} - i\epsilon}\right] = \frac{2\pi i e^{-ikr}}{2\,(2\pi)\, i\, r} = \frac{e^{-i\frac{\omega}{c}r}}{4\pi\, r}. \tag{A.91}$$

A.6.4 *Radial Solution of the Helmholtz Equation*

The radial part of the Helmholtz equation in spherical coordinates is characterized by the spherical Bessel differential equation in $x = kr$

$$y'' + 2x^{-1}\, y' + [1 - n(n+1)\, x^{-2}]y = 0. \tag{A.92}$$

Recursive construction. For $n = 0$, we know that the omnidirectional Green's function is a solution diverging from $x = 0$, and it is proportional to $y \propto \frac{e^{-ix}}{x}$. We can simplify the equation by inserting $y = x^{-1}\, u_n$, which yields with $y' = x^{-1}\, u_n' - x^{-2}\, u_n$ and $y'' = x^{-1}\, u_n'' - 2x^{-2}\, u_n' + 2x^{-3} u_n$ after multiplying with x:

$$\begin{aligned} u_n'' - 2x^{-1}u_n' + 2x^{-2}u_n \quad + 2x^{-1}u_n' - 2x^{-2}u_n \quad &+ [1 - n(n+1)x^{-2}]u_n = 0 \\ u_n'' &+ [1 - n(n+1)x^{-2}]u_n = 0. \end{aligned} \tag{A.93}$$

Moreover, we attempt to find a recursive definition for $n > 0$ using the approach

$$y_n = x^{-1}u_n, \qquad\qquad u_n = -x^a\big[x^{-a}\, u_{n-1}\big]'.$$

We evaluate the recursion for the derivatives

$$\begin{aligned}
u_n &= -u'_{n-1} + a\,x^{-1}\,u_{n-1},\\
u'_n &= -u''_{n-1} + a\,x^{-1}\,u'_{n-1} - a\,x^{-2}\,u_{n-1}, \quad \text{with} - u''_{n-1} = [1 - n(n-1)\,x^{-2}]u_{n-1}\\
&= a\,x^{-1}\,u'_{n-1} + \{1 - [n(n-1) + a]\,x^{-2}\}\,u_{n-1}\\
u''_n &= ax^{-1}u''_{n-1} - ax^{-2}u'_{n-1} + \{1 - [n(n-1) + a]x^{-2}\}u'_{n-1} + 2[n(n-1) + a]x^{-3}u_{n-1}\\
&= \{1 - [n(n-1) + 2a]x^{-2}\}u'_{n-1} + \{[2n(n-1) + 2a + an(n-1)]x^{-3} - ax^{-1}\}u_{n-1}\\
&= \{1 - [n(n-1) + 2a]\,x^{-2}\}\,u'_{n-1} + \{[n(n-1)(a+2) + 2a]\,x^{-3} - a\,x^{-1}\}\,u_{n-1}
\end{aligned}$$

The equation $u''_n + [1 - n(n+1)x^{-2}]u_n = 0$ using the above expressions becomes

$$\begin{aligned}
\{1 - [n(n-1) + 2a]\,x^{-2}\}\,u'_{n-1} + \{[n(n-1)(a+2) + 2a]\,x^{-3} - a\,x^{-1}\}\,u_{n-1}&\\
+[1 - n(n+1)x^{-2}][-u'_{n-1} + a\,x^{-1}\,u_{n-1}] = 0.&
\end{aligned}$$

Comparing coefficients for u'_{n-1} and u_{n-1} yields $a = n$

$$\begin{aligned}
u'_{n-1}: &\quad 1 - 1 - [n(n-1) + 2a - n(n+1)]x^{-2} = 2(a-n)x^{-2} = 0,\\
u_{n-1}: &\quad [-a + a]x^{-1} + [n(n-1)(a+2) + 2a - an(n+1)]x^{-3} = 0\\
&\quad an(n-1) + 2n(n-1) + 2a(1-n) - an(n-1) = 2(n-a) = 0,
\end{aligned}$$

and hereby a recurrence for y_n from $u_n = -x^n[x^{-n}u_{n-1}]'$ with $y_n = x^{-1}\,u_n$, $u_n = x\,y_n$,

$$y_n = -x^{n-1}[x^{-(n-1)}y_{n-1}]' \qquad \Rightarrow y_{n+1} = -x^n[x^{-n}y_n]'. \tag{A.94}$$

Singular and regular solution. We know from the Green's function that the omnidirectional solution should be proportional to $g_0 \propto e^{-\mathrm{i}x}$. The typical radial solution for an omnidirectional source field is chosen to be the spherical Hankel function of the second kind[2]

$$h_0^{(2)}(kr) = \frac{e^{-\mathrm{i}kr}}{-\mathrm{i}kr}, \qquad h_{n+1}^{(2)}(kr) = -(kr)^n \frac{\mathrm{d}}{\mathrm{d}(kr)}\left[\frac{1}{(kr)^n}\,h_n^{(2)}(kr)\right]. \tag{A.95}$$

However, this solution is not sufficient to solve problems without singularity at $r = 0$. We know that the function $\frac{\sin(kr)}{kr}$ is finite at kr, and so are all real parts of the spherical Hankel functions of the second kind, the spherical Bessel functions

$$j_0(kr) = \frac{\sin(kr)}{kr}, \qquad j_{n+1}(kr) = -(kr)^n \frac{\mathrm{d}}{\mathrm{d}(kr)}\left[\frac{1}{(kr)^n}\,j_n(kr)\right]. \tag{A.96}$$

[2]Note that some scholars use the Fourier expansion $e^{\mathrm{i}\omega t}$ with opposite sign $e^{-\mathrm{i}\omega t}$ and require to use the complex conjugate in every expression containing imaginary constants $h_n^{(1)} = h_n^{(2)*}$.

The solutions are linearly independent. One check after some calculation that their Wronski determinant is non-zero [15, Eq. 10.50.1]

$$\begin{vmatrix} j_n(kr) & h_n^{(2)}(kr) \\ j_n'(kr) & h_n'^{(2)}(kr) \end{vmatrix} = j_n(kr)h_n'^{(2)}(kr) - j_n'(kr)h_n^{(2)}(kr) = -\frac{\mathrm{i}}{(kr)^2}. \tag{A.97}$$

Below, the Frobenius method is shown as alternative way to get these functions.

Alternative way: Frobenius method. Given a second-order differential equation with singular coefficients, it can be solved by a generalized infinite power series:

$$y'' + \left(\sum_{l=0}^{\infty} a_l\, x^l\right) x^{-1}\, y' + \left(\sum_{l=0}^{\infty} b_l\, x^l\right) x^{-2}\, y = 0, \quad \text{solution: } y = \sum_{k=0}^{\infty} c_k\, x^{k+\gamma}. \tag{A.98}$$

Insertion of the solution yields

$$\sum_{k=0}^{\infty} (k+\gamma-1)\,(k+\gamma)c_k\, x^{k+\gamma-2} + \sum_{k'=0}^{\infty}\sum_{l=0}^{\infty}\left[(k'+\gamma)\,a_l + b_l\right]c_{k'}\, x^{k'+l+\gamma-2} = 0,$$

an index shift $k' + l = k$, and $l = 0 \dots k$ allows to pull out the common factor $x^{k+\gamma-2}$

$$\sum_{k=0}^{\infty}\left\{(k+\gamma-1)\,(k+\gamma)c_k + \sum_{l=0}^{k}\left[(k-l+\gamma)\,a_l + b_l\right]c_{k-l}\right\} x^{k+\gamma-2} = 0$$

$$\sum_{k=0}^{\infty}\left\{\left[(k+\gamma+a_0-1)\,(k+\gamma)+b_0\right]c_k + \sum_{l=1}^{k}\left[(k-l+\gamma)\,a_l + b_l\right]c_{k-l}\right\} x^{k+\gamma-2} = 0.$$

The coefficient of every exponent of x in the above equation must be zero:

indical equation for $k = 0$: $$\left[(\gamma+a_0-1)\gamma + b_0\right]c_0 = 0, \tag{A.99}$$

indical equation for $k = 1$: $$\left[(\gamma+a_0)(\gamma+1)+b_0\right]c_1 + \left[a_1\,\gamma + b_1\right]c_0 = 0, \tag{A.100}$$

recurrence for $k > 1$: $$-\frac{\sum_{l=1}^{k}\left[(k-l+\gamma)\,a_l + b_l\right]c_{k-l}}{(k+\gamma+a_0-1)\,(k+\gamma)+b_0} = c_k. \tag{A.101}$$

Depending on the specific values found for γ, the recurrence, etc. the Frobenius method suggests how to find or construct an independent pair of solutions.

Spherical Bessel differential equation. In $y'' + 2x^{-1}y' + [-n(n+1) + x^2]x^{-2}y = 0$, all a_l and b_l are zero except $a_0 = 2$, $b_0 = -n(n+1)$, and $b_2 = 1$. Indical equations and recurrence become

$$\left[\gamma(\gamma+1)-n(n+1)\right]\,c_0=0, \tag{A.102}$$

$$\left[(\gamma+1)(\gamma+2)-n(n+1)\right]\,c_1=0, \tag{A.103}$$

$$(k+\gamma+1)(k+\gamma)\,c_k=-c_{k-2}. \tag{A.104}$$

We see that the recurrence is again a two-step recurrence, so that one can choose between an even solution using $c_0 \neq 0, c_1 = 0$ yielding $\gamma = n$ or $\gamma = -(n+1)$, [or an odd solution that won't be used, with $c_0 = 0, c_1 \neq 0$ yielding $\gamma + 1 = n$ or $\gamma + 1 = -(n+1)$].

Spherical Bessel functions. The choice $\gamma = n$ yields a solution converging everywhere: Powers of x are all positive, the recurrences $c_k = -\frac{c_{k-2}}{(n+k+1)(n+k)} = \frac{(-1)^k c_0}{(n+k+1)!}$ yield a convergence radius $R = \lim_{k\to\infty}\left|\frac{c_{k-2}}{c_k}\right| = \lim_{k\to\infty}(n+k+1)(n+k) = \infty$. With a starting value $c_0 = \frac{2^n\,n!}{(2n+1)!}$, solutions are called spherical Bessel functions [5, Chap. 3.4]

$$j_n=(2x)^n\sum_{k=0}^{\infty}\frac{(-1)^k(n+k)!}{k![2(n+k)+1]!}x^{2k}. \tag{A.105}$$

which are a physical set of regular solutions with n-fold zero at 0. The spherical Bessel function for $n = 0$ is

$$j_0(x)=\frac{\sin x}{x}. \tag{A.106}$$

With the above recursive definition iterated [5, Eq. 3.4.15] one could define

$$j_{n+1}=-x^n\frac{\mathrm{d}}{\mathrm{d}x}\left(\frac{1}{x^n}j_n\right), \qquad j_n=(-x)^n\left(\frac{1}{x}\frac{\mathrm{d}}{\mathrm{d}x}\right)^n j_0. \tag{A.107}$$

Spherical Neumann functions. For $\gamma = -(n+1)$ and $c_0 \neq 0, c_1 = 0$, the recurrences are $c_k = -\frac{c_{k-2}}{(n-k+1)(n-k)} = \frac{(-1)^k c_0}{(n-k+1)!}$ and yield the spherical Neumann functions with an $(n+1)$-fold pole at 0. They also obey the recursive definition from above,

$$y_0=-\frac{\cos x}{x}, \qquad y_n=(-x)^n\left(\frac{1}{x}\frac{\mathrm{d}}{\mathrm{d}x}\right)^n y_0. \tag{A.108}$$

Spherical Hankel functions. The spherical Neumann and Bessel functions based on either sin or cos are clearly linearly independent. The spherical Bessel functions are useful to representing fields convergent everywhere. Physical source fields (Green's function) diverge at the source location and exhibit a specific way phase radiates, with $G \propto \frac{e^{-\mathrm{i}x}}{x}$. The spherical Bessel and Neumann functions are asymptotically similar to [15, Eq. 10.52.3]

$$\lim_{x\to\infty} j_n = (-1)^n \frac{\sin(x)}{x}, \qquad \lim_{x\to\infty} y_n = (-1)^{n+1} \frac{\cos(x)}{x}, \tag{A.109}$$

therefore only their combination to spherical Hankel functions of the second kind

$$h_n^{(2)} = j_n - \mathrm{i}\, y_n \tag{A.110}$$

yields useful physical set of singular solutions. They inherit their $(n+1)$-fold pole from the spherical Neumann functions at 0. Their limiting form for large arguments are

$$\begin{aligned}
\lim_{r\to\infty} h_n^{(2)}(x) &= -x^{n-1} \lim_{r\to\infty} \frac{\mathrm{d}}{\mathrm{d}x}\left(\frac{1}{x^{n-1}} h_{n-1}^{(2)}\right) \\
&= -x^{n-1} \lim_{r\to\infty}\left(-\frac{n-1}{x^n} h_{n-1}^{(2)} + \frac{1}{x^{n-1}} \frac{\mathrm{d}}{\mathrm{d}x} h_{n-1}^{(2)}\right) = -\lim_{r\to\infty} \frac{\mathrm{d}}{\mathrm{d}x} h_{n-1}^{(2)} \\
&= (-1)^n \lim_{r\to\infty} \frac{\mathrm{d}^n}{\mathrm{d}x^n} h_0^{(2)} = \mathrm{i}^n\, h_0^{(2)}(x).
\end{aligned} \tag{A.111}$$

With Eqs. (A.110), (A.106) and (A.108), the zeroth-order spherical Hankel function is $h_0^{(2)}(x) = \frac{e^{-\mathrm{i}x}}{-\mathrm{i}x}$.

Alternative implementation by cylindrical functions. We can transform the spherical Bessel differential equation by inserting $y = x^\alpha\, u$ and obtain after division by x^α

$$\begin{aligned}
x^\alpha u'' + 2\,\alpha x^{\alpha-1} u' + \alpha(\alpha-1)u \quad + 2x^{\alpha-1}u' + 2\alpha x^{\alpha-2}u \quad + [1 - n(n+1)x^{-2}]u = 0 \\
u'' + 2\frac{\alpha+1}{x}u' + \left[1 + \frac{\alpha(\alpha+1) - n(n+1)}{x^2}\right] u = 0.
\end{aligned}$$

For $\alpha = -\frac{1}{2}$, the equation for u becomes the Bessel differential equation with $\alpha(\alpha+1) - n(n+1) = -(n^2 + n + \frac{1}{4}) = -(n + \frac{1}{2})^2$

$$u'' + \frac{1}{x}u' + \left[1 - \frac{(n+\frac{1}{2})^2}{x^2}\right] u = 0. \tag{A.112}$$

Consequently, the spherical Bessel functions and spherical Hankel functions of the second kind can be implemented using the Bessel and Hankel functions that can be found in any standard maths programming library. The specific relations are:

$$j_n(x) = \sqrt{\frac{\pi}{2}\frac{1}{x}}\, J_{n+\frac{1}{2}}(x), \qquad h_n^{(2)}(x) = \sqrt{\frac{\pi}{2}\frac{1}{x}}\, H_{n+\frac{1}{2}}^{(2)}(x). \tag{A.113}$$

A.6.5 Green's Function in Spherical Solutions, Angular Distributions, Plane Waves

We can write the inhomogeneous Helmholtz equation $(\Delta + k^2)G = -\delta$ to be excited by a source at the direction $\boldsymbol{\theta}_0$ at the radius r_0. We decompose the excitation into a Delta function in radius and direction $-r_0^{-2}\delta(r - r_0)\delta(\boldsymbol{\theta}_0^{\mathrm{T}}\boldsymbol{\theta} - 1)$. The directional part needs not be restricted to the spherical Dirac delta function, so we can take a distribution of sources at r_0, weighted by the panning function $g(\boldsymbol{\theta})$,

$$\left(\Delta + k^2\right) p = -r_0^{-2}\delta(r - r_0)\, g(\boldsymbol{\theta}). \tag{A.114}$$

From the spherical basis solutions, we know that at a radius other than r_0, p can be expanded into spherical harmonics

$$p = \sum_{n=0}^{\infty}\sum_{m=-n}^{n} \psi_{nm}\; Y_n^m(\boldsymbol{\theta}). \tag{A.115}$$

Acting on the decomposition of p, the directional part of the Laplacian will yield the eigenvalue $\Delta_{\varphi,\zeta} Y_n^m = -n(n+1)\, r^{-2}\, Y_n^m$ of the spherical harmonics, and its radial part $r^2\Delta_{\mathrm{r}} = \frac{\partial^2}{\partial r^2} + 2r^{-1}\frac{\partial}{\partial r}$, as around Eq. (6.11), hence

$$(\Delta + k^2)\sum_{n=0}^{\infty}\sum_{m=-n}^{n} \psi_{nm}\; Y_n^m(\boldsymbol{\theta}) =$$

$$\sum_{n=0}^{\infty}\sum_{m=-n}^{n}\left[\frac{\partial^2}{\partial r^2} + \frac{2\partial}{r\partial r} + k^2 - \frac{n(n+1)}{r^2}\right]\psi_{nm}\; Y_n^m(\boldsymbol{\theta}) = -r_0^{-2}\delta(r - r_0)\, g(\boldsymbol{\theta}).$$

Obviously, ψ_{nm} must depend on k and r, so we may pull the factor k into the differentials $\frac{\mathrm{d}}{\mathrm{d}r} = k\frac{\mathrm{d}}{\mathrm{d}kr}$ to get the differential operator $k^2\big[\frac{\mathrm{d}^2}{\mathrm{d}(kr)^2} + \frac{2}{kr}\frac{\mathrm{d}}{\mathrm{d}(kr)} + 1 - \frac{n(n+1)}{(kr)^2}\big]$ and observe kr as its variable on the left, and we replace kr by x for brevity. Applying the factor k^{-2} and the spherical harmonics transform $\int_{\mathbb{S}^2} Y_{n'}^{m'}(\boldsymbol{\theta})\,\mathrm{d}\boldsymbol{\theta}$ on the equation removes the double sum on the left (orthogonality) and decomposes the panning function $g(\boldsymbol{\theta})$ on the right into γ_{nm}

$$\left[\frac{\mathrm{d}^2}{\mathrm{d}x^2} + \frac{2}{x}\frac{\mathrm{d}}{\mathrm{d}x} + 1 - \frac{n(n+1)}{x}\right]\psi_{nm} = -(kr_0)^{-2}\delta(r - r_0)\,\gamma_{nm}.$$

We collect the x-independent term γ_{nm} as factors of the solution $\psi_{nm} = y\;\gamma_{nm}$ and get

$$y'' + \frac{2}{x}y' + \left[1 - \frac{n(n+1)}{x^2}\right] y = -x_0^{-2}\delta(r - r_0), \tag{A.116}$$

the inhomogeneous spherical Bessel differential equation. As described, e.g., in [16, 17], the inhomogeneous differential equation can be solved by the Lagrangian *variation of the parameters* for equations of the type $y'' + py' + qy = r$, knowing its independent homogeneous solutions $y_1 = h_n^{(2)}(x)$ and $y_2 = j_n(x)$.

It uses a solution $y = uy_1 + vy_2$ with variable parameters u and v, which upon first and second-order differentiation becomes

$$\begin{aligned} y = uy_1 + vy_2, \qquad & y' = uy_1' + u'y_1 \quad + vy_2' + v'y_2 \\ & y'' = uy_1'' + 2u'y_1' + u''y_1 \quad + vy_2'' + 2v'y_2' + v''y_2. \end{aligned}$$

Inserted into the equation $y'' + py' + qy = r$, this yields

$$\begin{aligned} & u\,\overbrace{(y_1'' + py_1' + qy_1)}^{=0} + v\,\overbrace{(y_2'' + py_2' + qy_2)}^{=0} + u''y_1 + 2u'y_1' + v''y_2 + 2v'y_2' \\ & \qquad + p(u'y_1 + v'y_2) = \\ & (u'y_1 + v'y_2)' + u'y_1' + v'y_2' + p\,(u'y_1 + v'y_2) = u'y_1' + v'y_2' + (\tfrac{\mathrm{d}}{\mathrm{d}x} + p)(u'y_1 + v'y_2) = r. \end{aligned}$$

Now two functions u and v are to be determined from only one equation, so we may pose an additional constraint. The above equation would simplify if the term $(u'y_1 + v'y_2)$ vanished. By this and the simplified equation, we get two conditions

$$\begin{aligned} \mathrm{I}: \qquad & u'y_1 + v'y_2 = 0 \\ \mathrm{II}: \qquad & u'y_1' + v'y_2' = r \end{aligned}$$

and obtain by elimination with either $\mathrm{A} = \mathrm{I}\,y_1' - \mathrm{II}\,y_1$ or $\mathrm{B} = \mathrm{I}\,y_2' - \mathrm{II}\,y_2$

$$\begin{aligned} \mathrm{A}: \qquad & v'\,\underbrace{(y_1'y_2 - y_1y_2')}_{-W} = -r\,y_1 \\ \mathrm{B}: \qquad & u'\,\underbrace{(y_1'y_2 - y_1y_2')}_{-W} = -r\,y_2. \end{aligned}$$

So that the solution $y = uy_1 + vy_2$ uses $u = \int \frac{r\,y_2}{W}\,\mathrm{d}x$ and $v = \int \frac{r\,y_1}{W}\,\mathrm{d}x$. In our case, we have $y_1 = h_n^{(2)}(x)$, $y_2 = j_n(x)$, $r = -x_0^{-2}\delta(r - r_0)$, and the Wronskian $W = (\mathrm{i}x^2)^{-1}$ from Eq. (A.97), hence with integration constants enforcing the physical solutions:

$$y = -h_n^{(2)}(x)\int_0^x \mathrm{i}x^2\,j_n(x)\,x_0^{-2}\delta(r - r_0)\,\mathrm{d}x - j_n(x)\int_x^\infty \mathrm{i}x^2\,h_n^{(2)}(x)\,x_0^{-2}\delta(r - r_0)\,\mathrm{d}x.$$

To convert $\delta(r - r_0)$ into $\delta(x - x_0)$ with $x = kr$, we use $\int\delta(x)\,\mathrm{d}x = \int\delta(r)\,\mathrm{d}r = 1$ with the integration constant replaced, $\frac{\mathrm{d}x}{\mathrm{d}r} = k$, hence $\mathrm{d}x = k\,\mathrm{d}r$ and obviously by $\int\delta(x)\,k\,\mathrm{d}r = \int\delta(r)\,\mathrm{d}r$ we find $\delta(r) = k\,\delta(x)$,

$$y = -h_n^{(2)}(x)\int_0^x \mathrm{i}x^2 j_n(x)\,k\,x_0^{-2}\delta(x-x_0)\,\mathrm{d}x - j_n(x)\int_x^\infty \mathrm{i}x^2 h_n^{(2)}(x)\,k\,x_0^{-2}\delta(x-x_0)\,\mathrm{d}x$$
$$= -\mathrm{i}\,k\begin{cases} h_n^{(2)}(x)\,j_n(x_0), & \text{for } x \geq x_0,\\ j_n(x)\;\;h_n^{(2)}(x_0) & \text{for } x \leq x_0.\end{cases}$$

The solution becomes after re-substituting $x = kr$ and expanding $\psi_{nm} = y\,\gamma_{nm}$ over the spherical harmonics $p = \sum_{n=0}^{\infty}\sum_{m=-n}^{n}\psi_{nm}Y_n^m(\boldsymbol{\theta})$:

$$p = -\mathrm{i}\,k\sum_{n=0}^{\infty}\sum_{m=-n}^{n}\gamma_{nm}\,Y_n^m(\boldsymbol{\theta})\begin{cases} h_n^{(2)}(kr)\,j_n(kr_0), & \text{for } r \geq r_0,\\ j_n(kr)\;\;h_n^{(2)}(kr_0) & \text{for } r \leq r_0.\end{cases} \tag{A.117}$$

Green's function. For the Green's function at the direction $\boldsymbol{\theta}_0$, the angular panning function is expanded as $\phi_{nm} = Y_n^m(\boldsymbol{\theta}_0)$, and we get the formulation of the Green's function in terms of spherical basis functions:

$$G = -\mathrm{i}\,k\sum_{n=0}^{\infty}\sum_{m=-n}^{n}Y_n^m(\boldsymbol{\theta}_0)\,Y_n^m(\boldsymbol{\theta})\begin{cases} h_n^{(2)}(kr)\,j_n(kr_0) & \text{for } r \geq r_0,\\ j_n(kr)\;\;h_n^{(2)}(kr_0) & \text{for } r \leq r_0.\end{cases} \tag{A.118}$$

Plane waves/far field approximation. Equation (6.7) in Sect. 6.3.1 formulates plane waves $p = e^{\mathrm{i}k\,\boldsymbol{\theta}_0^{\mathrm{T}}\boldsymbol{r}}$ as far-field limit $p = 4\pi\lim_{r_0\to\infty}\frac{r_0}{e^{-\mathrm{i}kr_0}}\,G = \lim_{r_0\to\infty}\frac{1}{-\mathrm{i}k\,h_0^{(2)}(kr_0)}\,G$. Using Eq. (A.117), a distribution of plane waves driven by the gains $g(\boldsymbol{\theta}) = \sum_n\sum_m\gamma_{nm}Y_n^m$ consequently yields with $\lim_{r_0\to\infty}h_n^{(2)}(kr_0) = \mathrm{i}^n\,h_0^{(2)}(kr_0)$,

$$p = 4\pi\sum_{n=0}^{\infty}\sum_{m=-n}^{n}j_n(kr)\left[\lim_{r_0\to\infty}\frac{h_n^{(2)}(kr_0)}{h_0^{(2)}(kr_0)}\right]Y_n^m(\boldsymbol{\theta})\,\gamma_{nm}$$
$$= 4\pi\sum_{n=0}^{\infty}\sum_{m=-n}^{n}\mathrm{i}^n\,j_n(kr)\,Y_n^m(\boldsymbol{\theta})\,\gamma_{nm}. \tag{A.119}$$

or for a single plane-wave direction $\gamma_{nm} = Y_n^m(\boldsymbol{\theta}_0)$

$$p = 4\pi\sum_{n=0}^{\infty}\sum_{m=-n}^{n}\mathrm{i}^n\,j_n(kr)\,Y_n^m(\boldsymbol{\theta})\,Y_n^m(\boldsymbol{\theta}_0). \tag{A.120}$$

A.7 Sine and Tangent Law

The sine and tangent law [18] observes the sound pressure of plane waves at to locations $x = 0,\ y = \pm d$ at ear distance in order to simulate the ear signals. A plane wave from the left half of the room from the angle $\varphi > 0$ first arrives at the left ear

$p_{\mathrm{left}} = e^{\mathrm{i}kd\,\sin\varphi}$ and later on the right one $p_{\mathrm{right}} = e^{-\mathrm{i}kd\,\sin\varphi}$. The phase difference is $\Phi_\varphi = 2\,kd\,\sin\varphi$.

A superimposed pair of plane waves from the directions $\pm\alpha$ arrives at the left ear as $p_{\mathrm{left}} = g_1\,e^{\mathrm{i}kd\,\sin\alpha} + g_2\,e^{-\mathrm{i}kd\,\sin\alpha}$, right as $p_{\mathrm{right}} = g_1\,e^{-\mathrm{i}kd\,\sin\alpha} + g_2\,e^{\mathrm{i}kd\,\sin\alpha} = p_{\mathrm{left}}^*$. The phase difference $\Phi_{\pm\alpha} = 2\angle p_{\mathrm{left}} = 2\,\arctan\frac{(g_1-g_2)\,\sin(kd\,\sin\alpha)}{(g_1+g_2)\cos(kd\,\sin\alpha)}$ can be linearized for long wave lengths $kd \to 0$ to $\Phi_{\pm\alpha} \approx 2\,\arctan\left(kd\,\frac{g_1-g_2}{g_1+g_2}\,\sin\alpha\right) \approx 2\frac{g_1-g_2}{g_1+g_2}\,kd\,\sin\alpha$.

Comparing the phase difference of the single plane wave with the one of the superimposed pair, $2\,kd\,\sin\varphi = 2\,kd\,\frac{g_1-g_2}{g_1+g_2}\,\sin\alpha$, one arrives at the sine law

$$\sin\varphi = \frac{g_1 - g_2}{g_1 + g_2}\,\sin\alpha.$$

If we claim our hearing to possess the ability to not only estimate the interaural phase difference Φ but also its derivative with regard to head rotation $\frac{\partial\Phi}{\partial\delta}$, we arrive at a value pair of binaural features $(\Phi_\varphi, \frac{\partial\Phi_\varphi}{\partial\delta}) = 2\,kd\,(\sin\varphi, \cos\varphi)$ than should match the one of the stereophonic plane-wave pair. For stereo, the phase difference derived with regard to head rotation is $2\,kd\,\frac{\partial}{\partial\delta}\Phi_{\pm\alpha} \approx 2\,kd\,\frac{g_1\,\frac{\partial}{\partial\delta}\,\sin(\alpha+\delta)|_{\delta=0} + g_2\,\frac{\partial}{\partial\delta}\,\sin(-\alpha+\delta)|_{\delta=0}}{g_1+g_2} = 2\,kd\,\frac{g_1+g_2}{g_1+g_2}\,\cos\alpha = 2\,kd\,\cos\alpha$, and yields $(\Phi_{\pm\alpha}, \frac{\partial\Phi_{\pm\alpha}}{\partial\delta}) = 2\,kd\,(\frac{g_1-g_2}{g_1+g_2}\,\sin\alpha,\,\cos\alpha)$. In polar coordinates, the radius of both value pairs differs. While the plane wave yields a value pair at the radius $2\,kd$ in the binaural feature space, the stereophonic waves is of the radius $2\,kd$ only at $\pm\alpha$, at which one of the two gains must vanish, and amplitude panning can be used to connect these two points $2kd\,(\pm\sin\alpha,\,\cos\alpha)$ by a straight line. The plane wave with the most similar feature pair must lie on the same polar angle. We may equate the tangents of both points $\frac{\Phi_\varphi}{\frac{\partial\Phi_\varphi}{\partial\delta}} = \frac{\Phi_{\pm\alpha}}{\frac{\partial\Phi_{\pm\alpha}}{\partial\delta}}$ and obtain the tangent law:

$$\tan\varphi = \frac{g_1 - g_2}{g_1 + g_2}\,\tan\alpha.$$

If instead of the angle of a plane wave with the closest features to those of a given amplitude difference is searched, but the closest features of an amplitude difference matching those of a given plane wave, then the sine law is the best match, even in the two-dimensional feature space.

References

1. S. Zeki, J.P. Romaya, D.M.T. Benincasa, M.F. Atiyah, The experience of mathematical beauty and its neural correlates. Front. Hum. Neurosci. **8**, e00068 (2014)
2. Mathematical signs and symbols for use in physical sciences and technology (1978)
3. Iso 80000-2, quantities and units? part 2: mathematical signs and symbols to be used in the natural sciences and technology (2009)

4. G. Dickins, Sound field representation, reconstruction and perception. Master's thesis, Australian National University, Canberra (2003)
5. W. Cassing, H. van Hees, Mathematische Methoden für Physiker, Göthe-Universität Frankfurt (2014), https://th.physik.uni-frankfurt.de/~hees/publ/maphy.pdf
6. A. Sommerfeld, *Vorlesungen über Theoretische Physik (Band 6) Partielle Differentialgleichungen in der Physik*, 6th edn. (Harri-Deutsch, 1978) (reprint 1992)
7. F. Zotter, M. Frank, All-round ambisonic panning and decoding. J. Audio Eng. Soc. (2012)
8. J.R. Driscoll, D.M.J. Healy, Computing fourier transforms and convolutions on the 2-sphere. Adv. Appl. Math. **15**, 202–250 (1994)
9. W. Contributors, Ideal gas, https://en.wikipedia.org/wiki/Ideal_gas. Accessed Dec 2017
10. G. Franz, *Physik III: Thermodynamik* (Hochschule, München, 2014), http://www.fb06.fh-muenchen.de/fbalt/queries/download.php?id=6696
11. W. Contributors, Enthalpy, https://en.wikipedia.org/wiki/Enthalpy. Accessed Dec 2017
12. W. Contributors, First law of thermodynamics, https://en.wikipedia.org/wiki/First_law_of_thermodynamics. Accessed Dec 2017
13. W. Contributors, Adiabatic process, https://en.wikipedia.org/wiki/Adiabatic_process. Accessed Dec 2017
14. D.G. Duffy, *Green's Functions with Applications* (CRC Press, Boca Raton, 2001)
15. F.W.J. Olver, R.F. Boisvert, C.W. Clark (eds.), *NIST Handbook of Mathematical Functions* (Cambridge University Press, Cambridge, 2000), http://dlmf.nist.gov. Accessed June 2012
16. E. Kreyszig, *Advanced Engineering Mathematics*, 8th edn. (Wiley, New York, 1999)
17. U.H. Gerlach, *Linear Mathematics in Infinite Dimensions: Signals, Boundary Value Problems, and Special Functions*, Beta edn. (2007), http://www.math.ohio-state.edu/~gerlach/math/BVtypset/
18. H. Clark, G. Dutton, P. Vanerlyn, The 'stereosonic' recording and reproduction system, reprinted in the J. Audio Eng. Soc., from Proceedings of the Institution of Electrical Engineers, 1957, vol. 6(2) (1958), pp. 102–117

www.ingramcontent.com/pod-product-compliance
Lightning Source LLC
LaVergne TN
LVHW082226200926
848484LV00008B/2310